Computer Control
of Machines
and Processes

Computer Control of Machines and Processes

John G. Bollinger

Neil A. Duffie

University of Wisconsin

 ADDISON-WESLEY PUBLISHING COMPANY

Reading, Massachusetts • Menlo Park, California • New York
Don Mills, Ontario • Wokingham, England • Amsterdam • Bonn
Sydney • Singapore • Tokyo • Madrid • San Juan

This book is in the Addison-Wesley Series in Mechanical Engineering

The programs and applications presented in this book have been included for their instructional value. They have been tested with care, but are not guaranteed for any particular purpose. The publisher does not offer any warranties or representations, nor does it accept any liabilities with respect to the programs or applications.

T J
2 1 3
B5952
1988

Library of Congress Cataloging-in-Publication Data

Bollinger, John G.
    Computer control of machines and processes.

    Includes index.
    1. Automatic control—Data processing.   I. Duffie,
Neil A.   I. Title.
J213.B5952   1988   629.8'95   87-14019
ISBN 0-201-10645-0

To our families

John and Neil

Preface

We are living in an analog world surrounded by a universe of digital technology. For many years control engineers integrated analog devices, machines, and processes wherein the variables that were input, the manipulations and control actions computed, and the variables that were output were continuously varying signals in time. Thus, analog devices were controlled by analog controllers. There was also the need to be concerned with discrete events and the logical intelligence for machines and processes, commonly referred to as sequence control. These two different aspects of automatic machine and process control were traditionally handled by two separate bodies of knowledge and reflected vastly different technologies and levels of theoretical treatment. On one hand, feedback control theory formed a significant body of knowledge on which much research has been and continues to be done. It was not uncommon for a control engineer to work on only this aspect of a control problem. On the other hand, sequence control formed another body of knowledge that was originally implemented with little theory (originally in the form of relay or pneumatic logic) and was generally implemented by the practitioner of logic systems.

Technological developments in computers and associated input/output hardware have changed much of the way we look at automatic control problems. Digital computers are now used to control machines and processes in feedback control and logic control modes. In fact, a computer can perform both types of control concurrently, and is as adept at logical computations as it is at arithmetic. The design of appropriate software to provide effective feedback control and logic control actions, and the design of computer input/output devices to interface machines and processes to computers to meet complete control requirements, therefore, are key issues to be considered along with appropriate analysis and design methodologies.

The goal of this text is to make the process of creation of effective controller designs as comprehensive, yet as straightforward, as possible. This has been achieved throughout the text by starting with fundamental principles in all of the areas that are important in controlling machines and processes with digital computers, and then using those fundamental principles to quickly develop theories, methods, and techniques that are used by computer control system designers.

TIME-DOMAIN APPROACH

From inside a digital computer, the world looks like a stream of discrete numbers and logic values. They are interpreted as data in any way that the computer is programmed to interpret them. Algebraic and logical computations are performed on the data, and the computer gives back to the world a similar set of discrete numbers and logic values. Mathematically, the computer is dealing in time series, or sequences of numbers. Using the digital computer stimulates a natural desire to view the dynamic behavior of machines and processes as a computer would view them. A discrete system model responds to this need by providing a defining relationship for the behavior between input and output time series. Because all information is in the time domain, it is logical to begin by working in the time domain.

There are conceptual advantages in working in the time domain, and for incorporating discussions of logic design and computer hardware and software, in an introduction to computer control of machines and processes. Some of these advantages include:

1. Only algebra is needed as a mathematical foundation.

2. The entire topic can be approached without any prior knowledge of classical control theory.

3. A large part of what is called control of machines and processes is, in fact, sequential logic control. Logic is designed in the time domain, and the treatment of this subject in an integrated-textbook format provides the student with a very effective background to approach a wide range of computer control problems.

4. A significant portion of the solution of a total, control engineering problem involves the generation of command information such as that needed for machine motion control. Time-series modeling sets the stage for developing

computer-generated command series to satisfy a wide range of functional-system requirements.

5. Computer hardware and software function in the time domain. A time-domain approach reduces the level of abstraction between theory and implementation.

There are a number of well known frequency-domain techniques for feedback control system analysis and design that the more advanced student should be aware of. These are discussed in the closing chapters of the text. We have found that it is not necessary to discuss these topics at the beginning of introductory courses on computer control systems for machines and processes, but have included them in courses where the students have already had a first course in feedback control theory.

CHAPTER CONTENTS

Each chapter in the text starts with basic principles and develops the analysis, design, and implementation methods required to solve problems in important areas of computer control.

1. Introduction to Computer Control   Chapter 1 briefly describes the history of computer control and the explosion in the application of electronic technology since the 1960s.

2. Elements of Discrete Modeling   Chapter 2 begins with a discussion of basic process types, and develops the idea of representing process dynamics using difference equations. The use of difference equations to represent the process leads to the topics of discrete controllers and the difference equations that represent them. Methods of algebraically combining and solving the difference equations that represent the components of closed-loop systems are introduced, and this leads to the concept of the system transfer function—the ability to find system output as a function of system input, and the ability to analyze and predict the stability of a system.

3. System Response   Two fundamental concepts are introduced in Chapter 3 —the representation of a time-varying signal sampled at specific intervals of time as a sample sequence or time series, and the use of a time-shift operator to calculate the response of a system to an input sample sequence. The method of partial fraction expansion is introduced, and a final value theorem is developed that allows steady-state values of dynamic system variables to be determined.

4. Discrete Controller Design   Chapter 4 focuses on the design of control algorithms for desired closed-loop response to specified command and disturbance inputs. Critical issues in the selection of a sample period are also discussed. Finally, the design of feedforward, cascade, and noninteracting control algorithms is described.

5. Control Computers   Chapter 5 deals with control computer hardware and software. Topics include binary logic, basic computer hardware, registers, the concepts of instructions and data, computer input/output (I/O), and the program-driven and interrupt-driven options for control algorithm implementation at the assembly language level. A set of high-level language procedures are defined for analog I/O, logic I/O, and interrupt handling. These are then used to further illustrate important controller implementation options.

6. Computer Interfacing   Chapter 6 describes how interfaces for devices such as push-buttons, limit switches, lamps, and numeric displays can be constructed. Interfaces for logic-level I/O are described, as are digital-to-analog conversion and analog-to-digital conversion. Computer bus architectures are discussed along with issues such as address decoding, device selection, and interrupt interfacing.

7. Sensors for Computer Control   Chapter 7 describes a spectrum of sensors that are often found in computer control systems for machines and processes.

8. Command Generation for Machine and Process Control   Command generation often represents a major portion of the "intelligence" of a system, and several techniques for command generation for machine and process control are described in Chapter 8.

9. Sequential Control Using Programmable Logic Controllers   Implementation of logic control and solution of logic equations on a control computer are discussed in Chapter 9. The implications of serial solution of logic on a digital computer are then discussed, together with the relationship between computer solution of logic and relay-logic systems defined using ladder diagrams. A number of design methods for logic control are described, including the use of flowcharts, switching tables, and state diagrams.

10. Process Modeling   Chapter 10 reviews a number of approaches to process modeling. These include physical modeling, wherein mathematical definitions of the dynamic properties of all the components that make up a process are combined to create a complete process model. Two alternative approaches are also described: the step-response and stochastic methods of process model identification. It is shown that the transient and steady-state characteristics of the response of the process to a step input can be used to identify and develop a straightforward discrete model for the process. A stochastic method is then presented in which linear least-squares estimation is used to find the values of coefficients in a discrete process model that "best fit" observed process inputs and outputs.

11. Analysis Using Transform Methods   The Laplace and z transformations are defined in Chapter 11 and used in the analysis of continuous and discrete control systems and their components. The concepts of ideal samplers, hold elements, and representation of a sample sequence by a train of impulses are introduced. The frequency response of a closed-loop system and its elements is discussed, and it is shown that the delay implicit in a hold element (such as a digital-to-analog converter) has an adverse effect on the stability of the system under computer control.

12. Design Using Transform Methods   A number of methods for designing discrete controllers for machines and processes are described in Chapter 12. The design techniques are developed for both continuous and discrete systems, illustrating their similarities and differences. Root-locus and frequency response methods are discussed first. Pole-zero cancellation and integral control compensation techniques are then described, as are techniques for anticipating and compensating for the detrimental effects of hold elements in a system. PID controller approximation is described, as are the bilinear transformation, hold equivalence, and pole-zero mapping techniques for continuous controller approximation. Direct design of controllers to obtain specified closed-loop transfer functions is also reviewed.

Appendix A   Appendix A gives a brief review of state-variable methods for control system analysis and design. Both continuous and discrete systems are discussed.

Appendix B   Appendix B provides a review of binary arithmetic.

TECHNOLOGY-BASED INFORMATION

Throughout the text, technology-based information has been avoided wherever posssible. The approach taken has been to describe computer architectures, interfaces, and programming languages at a generic level. The simplified computer architecture that is used in Chapters 5 and 6 allows fundamental concepts to be illustrated, and allows system-level issues to be discussed, without describing complex computer architectures in commercial use. There are a great number of books, many of them inexpensive paperbacks, that can be used to supplement this text if additional information on a computer system's specific hardware and software is required.

EXAMPLES AND PROBLEMS

A variety of processes are used in examples and problems in the text to illustrate the analysis, design, and implementation techniques that are introduced. Many of the problems reiterate basic concepts illustrated in examples. Others challenge the reader to extend basic concepts to more complex situations. Still others require the integration of concepts presented in one or more chapters, leading toward broader concepts at the system level. Many of the examples and problems are manufacturing related, and most refer to practical application of concepts that are introduced.

PROGRAMS

It is important to view the computer as a tool in the control system development process as well as the hardware basis for discrete controller implementation. Programs, therefore, are included at many locations in the text to illustrate the use of the computer for analysis, design, simulation, and modeling as well as for control. The programs given are usually greatly simplified to improve the illustration of important concepts. We have chosen Pascal as the language in which to present these software concepts. The reader will find little fundamental difference between the simple programs presented and programs written in other structured languages. It is expected that the reader will use the programs as prototypes for development of more elaborate computer software. An understanding of the basic techniques illustrated will promote use and familiarity with future and existing computer-aided design and analysis software for computer control systems.

COURSES BASED ON THE TEXT

This text, and the courses that it supports at the University of Wisconsin—Madison, have been in a continuous state of development, evolution, and refinement since the mid-1960s. The students who have enrolled in our courses have included nearly equal numbers of graduate and undergraduate students, nearly equal numbers of electrical and mechanical engineering students, and students from all other engineering disciplines including aerospace, agricultural, chemical, civil, industrial, and nuclear. In addition, we have presented the material in the text in our industry outreach programs and in many short courses given for industry both in the United States and abroad.

The spectrum of topics that are important in computer control is exceedingly broad, and instructors using this text in formal courses will want to tailor the material they present to the background of their students and the time available. A previous course in control theory is not required in the use of this text, and Chapters 1 through 9 can be covered in one semester at this introductory level. Instructors at institutions using the quarter system may wish to cover Chapters 1 through 4 together with Chapters 11 and 12 in one quarter at the introductory level. Chapters 5 through 10 can then be covered in a second quarter.

Instructors of courses in which students have already had an introductory course in control theory may wish to cover the material in Chapters 11 and 12 before covering Chapters 5 through 10, de-emphasizing Chapters 2 through 4 if desired. Similarly, if students have already had experience with computer hardware and programming at the assembly language level, an instructor may wish to de-emphasize Chapters 5 and 6, and spend more time on Chapters 8 through 12. We suggest that instructors use inexpensive hardware and software documentation to supplement the text if a specific computer system is to be used by the student.

Throughout the text, the succession of topics and ideas presented in each area follows a logical progression based on the premise that the student has little prior knowledge in the area. This has proved advantageous in our experience because of the great differences in backgrounds of students who enroll in computer control courses. We have found that even with a very heterogeneous student population with wide disparities in backgrounds and experiences, all of the students can leave courses in which this text is used with the confidence that they have been exposed to most of the major issues in the analysis, design, and implementation of computer control systems for machines and processes.

ACKNOWLEDGEMENTS

We are sincerely thankful for the contributions of many people during the years in which this text has been in development. Professors David D. Bedworth, A. H. Haddad, David E. Hardt, Stephen P. Krosner, Robert D. Lorenz, and Kim A. Stelson have reviewed the manuscript, and we have tried to incorporate their valuable ideas and suggestions wherever possible. Special thanks go to Professors Howard L. Harrison and Thomas J. Higgins for their contributions, criticisms, and encouragement. Many students at the University of Wisconsin—Madison made significant contributions. These include Bruce R. Beadle, Dongik Joo, Richard A. Lund, J. Kelly Lee, Aun-Noew Poo, Mahesh K. Seth, Rafael Rangel-Sostmann, Jose Rodriguez-Ortiz, and the students of ME 547 who made so many helpful suggestions for improvements in the manuscript.

Madison, Wisconsin   J. G. B. and N. A. D.

Contents

Chapter 3

SYSTEM RESPONSE   69

Chapter 4

DISCRETE CONTROLLER DESIGN   111

Contents

Chapter 5

CONTROL COMPUTERS   **173**

Chapter 7

SENSORS FOR COMPUTER CONTROL 270

Chapter 8

COMMAND GENERATION IN MACHINE AND PROCESS CONTROL   314

Chapter 9

SEQUENTIAL CONTROL USING PROGRAMMABLE LOGIC CONTROLLERS   369

Chapter 12

DESIGN USING TRANSFORM METHODS   **508**

Appendix A

ANALYSIS AND DESIGN USING STATE VARIABLES   551

Appendix B

BINARY NUMBERS AND ARITHMETIC   579

Chapter 1

Introduction to Computer Control

1.1

INTRODUCTION

Everywhere we look today we see evidence of the impact of computers and other electronic devices on our lives. In our homes, our automobiles, and our offices and factories, these devices are changing the way we work, entertain ourselves, and communicate with others.

The impact of this rapidly developing electronics and computer technology is also accelerating the trend toward automation and the use of computer control in the manufacturing industry. A landmark of early manufacturing automation was the A. O. Smith Corporation's fully automated automobile frame production facility built in 1921. Over 550 operations in the facility were synchronized via line shafts. There was, of course, no computer. Electronic sensor and actuator technologies had not yet progressed beyond their infancy as components of automatic systems, and control functions were performed by mechanical mechanisms and pneumatics.

Although electronic measurement, control, and actuation technology continued to develop and improve during succeeding decades, it had reached only a marginal level of "credibility" in the manufacturing industry by 1950. Most electronic devices at that time were basically analog in nature, functioning with electronic inputs or outputs proportional to desired physical quantities.

In the mid-1950s digital devices began to appear, including digital voltmeters and N/C (numerically controlled) machine tools. The latter represented a significant step forward, because they not only enabled machining operations to be preprogrammed rather than performed manually, but also allowed for rapid reprogramming when it was necessitated by changes in parts to be manufactured. This flexibility (previously obtainable only in manual operations) demonstrated the future potential of automation in the manufacturing industry. The electronic logic in these systems, although "hard-wired" rather than computer-based, proved that "intelligence" could be designed into production processes, resulting in greater sophistication, higher rates, and greater precision than could be obtained manually.

Since 1960 there has been an explosion in the application of electronic technology. Electronic transducers are found in virtually all manufacturing industries, including the metalworking, plastics processing, chemical processing, paper, and textiles industries. The advent of the digital computer has provided the core about which complex data-acquisition, monitoring, communication, and control systems can be constructed. One key element that emerged in the eighties was the development of "smart sensors"—that is, sensors with microcomputer-based calibration, computation, and decision-making power built into the unit. This trend has greatly enhanced the use of many physical measurement techniques, particularly those that are inherently nonlinear.

The service industry has also become automated. The banking industry was one of the first to undergo computer automation of accounting, bookkeeping, and data-entry functions. These innovations were followed by the trend toward electronic money. In another domain, computers access a worldwide commercial air travel data-base to assist airlines and travel agents in making flight reservations. New aircraft have more than 100 on-board computers and can be flown and landed automatically.

Where will this lead us? Clearly, automation of the service industry will continue, spurred by the desire to provide improved services at lower cost. In the factory of the future, computers will run machines using complex algorithms that achieve a level of machine intelligence necessary to operate, monitor, and diagnose the condition of both the manufacturing process and the product. This will require sophisticated electronic devices such as tool-failure sensors, operating parameter sensors (torque, speed, flow, and so on) and such inspection devices as tactile and optical sensors.

Much of the technology for the factory of the future already exists, and electronic devices play leading roles in it. Digital computers, sensors, and actuators are distributed throughout complex systems. Sensors of all types provide the computer with the information it needs to perform monitoring and control tasks. Developments in the use of laser and microelectronic sensors make possible the rapid measurement of physical properties that were previously difficult to measure. Actuator technology also continues to improve through developments in servo-drives, torque and force motors, and piezoelectric actuators for high-precision positioning systems.

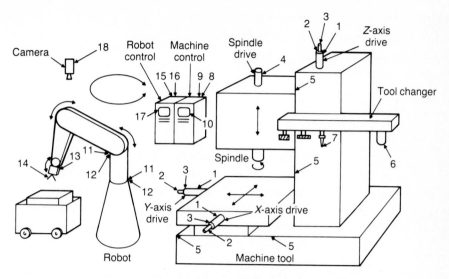

Figure 1.1 Automated machining cell with loading/unloading robot.

As an example of the numerous electronic devices that may be found in an advanced manufacturing system, consider the machining cell shown in Fig. 1.1. The cell includes a machine tool and a robot for loading and unloading parts. The various electronic devices typically found in such a cell, and the purposes they serve, are as follows:

Machine Tool

1. Axis Drive Servo-Motor—machine motion
2. Tachometer—motor velocity sensor
3. Resolver—machine position sensor
4. Spindle Drive Motor—tool rotation
5. Limit Switch—overtravel protection
6. Stepping Motor—tool changer positioning
7. Tactile Probe—measurement of workpiece
8. Servo-Amplifiers—machine drives
9. Control Computer—machine control
10. Display—machine status

Robot

11. Servo-Motor—arm motion
12. Optical Encoder—arm position sensor

13. Control Valve—pneumatic gripper
14. Tactile Sensor—gripper force
15. Servo-Amplifiers—robot arms
16. Control Computer—robot control
17. Display—robot status
18. Camera—part identification, guidance

As you can see, these devices include control computers, position sensors, velocity sensors, limit switches, and actuators such as servo-motors and stepping motors. Also shown are a vision system for part identification and guidance of the robot and a tactile sensor on the machine tool for measurement of workpiece dimensions and automatic tool adjustment. In this regard, we should note that control computers are used to implement system communication, sequencing, and decision-making functions as well as feedback control functions.

Numerous ideas about the application of digital computers to the manipulation of elements in automated manufacturing systems have been advanced, published, discussed, and marketed. It is interesting that many early computer systems, designed as either total-automation systems or process control systems, failed to gain immediate respect or failed to be commercially successful for a number of important reasons. First and foremost, many early systems were designed, sold, and installed before essential computer programing, control theory, and maintenance capabilities were developed. Second, it was not uncommon to see large computer installations originally dedicated only to process monitoring and data collection rather than incorporating control. An anxious management, expecting to be rewarded by financial return on the computer investment, became discouraged by the inability of the computer staff to show prompt and tangible progress. Third, computer reliability, including the massive interface systems required for on-line monitoring and control, had not yet developed to the point where they generated firm confidence. Fourth, not a great deal was known about the overall dynamics of systems on which control was to be implemented, and there were relatively few persons knowledgeable in the art and science of discrete systems modeling and control theory.

The analytical treatment of computer-based feedback control systems is a culmination of a historical development that began with consideration of continuous analog control concepts. The mathematical foundations of differential equations and complex variable theory, including the Laplace transform, afforded designers of continuous linear control systems a means of model formulation, equation manipulation, analysis, design, and controller synthesis.

As interest in discrete systems grew during the late 1950s, the z-transform and certain associated concepts such as the ideal sampler came into use. As a result, it became generally accepted that one should be well versed in continuous-system control theory and in digital control theory before one proceeded to the design of discrete control systems. This situation can be most discouraging to both practicing engineers and students who wish to accomplish feedback control tasks with a digital computer. Fortunately, it need not be so.

The authors propose that we reflect on the basic nature of the computer——namely, its handling of discrete information—and begin with the basic mathematics that describes its nature. This text is based on the idea that a time-domain approach utilizing difference equations, sequences, and series provides an efficient and productive introduction to the implementation of computer control. The backward shift operator of classical finite-difference equation theory is used here as the basic manipulation tool. It provides a direct, efficient approach to learning how to design computer feedback control systems for practical applications and allows rapid utilization of computers in the solution of a broad spectrum of practical problems.

The material that follows first develops an approach to computer control design and synthesis using the backward shift operator. The concepts of process modeling, system stability, controller algorithm synthesis, and analytical manipulation of control dynamical equations are discussed. This presentation is divided into chapters on the elements of discrete modeling, system response, and discrete controller design. Important topics in the implementation of computer control are then presented in chapters on control computers, computer interacting, sensors, command generation, sequential control using programmable logic controllers and process modeling. The final chapters and Appendix A present approaches to the analysis and design of discrete systems and controllers via transform and state-variable techniques.

1.2
CONCEPTS OF COMPUTER CONTROL

The application of a digital computer for control may take many forms and may entail a wide range of ideas and principles. The simplest approach is to use the computer as a large and efficient data collector and calculator. This concept is illustrated schematically in Fig. 1.2, wherein the system shown is described as a computer monitor. The terms *process, plant,* and *machine* may be used interchangeably, depending on that which is to be controlled. In control literature, the terms *plant* and *process* are commonly employed. The plant or process itself may involve a great number of internal, individual, or interrelated feedback control loops, or it may simply be a specific physical process. For example, a machine tool or robot has a number of axes, each of which can have control loops for position, velocity, current, and so on. These axes are combined with other machine

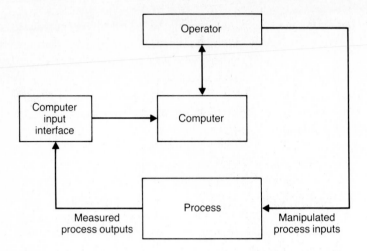

Figure 1.2 Schematic diagram of a computer monitoring system.

components to perform processing functions that are to be monitored and manipulated.

In a computer monitoring system, human operators manipulate individual process controllers directly. The computer simply observes the process via an interface that is capable of detecting physical signals such as contact closures and analog voltages, converting analog signals to digital words, and converting digital words of various formats into meaningful binary information for computer processsing. Operators make decisions and take actions on the basis of analyses that the computer can be programmed to prepare. Many early computer installations were undertaken with this approach in the hope that the logging and processing of large amounts of data would lead to a better understanding of the overall behavior of the process and that, ultimately, a more sophisticated level of computer control could be developed.

Figure 1.3 illustrates the concept of **open-loop computer control**. In this scheme the computer is used as an on-line processor of programs and data to generate specific commands for the manipulation of machine and process actuators. Open-loop principles are also employed in the use of computers to manipulate electric or electro-hydraulic stepping motors for machine tool control.

In order to automate a process or plant completely, it is necessary to develop computer systems that are capable of comparing a desired set of results with actual results and of taking corrective action. This is the essence of **closed-loop computer control** or **feedback control.** Figure 1.4 depicts a computer control system wherein the computer is placed in a control loop by collecting operating data from sensors in the process, accepting desired reference states (the set point) from an operator, and closing the control loop by establishing manipulation

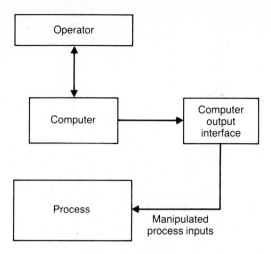

Figure 1.3 Schematic diagram of computer open-loop control.

commands that cause specific changes in the process outputs. Within the computer, two distinct operations must usually be performed: The difference or error between the operating point and the set point or error is calculated from the equation

$$\text{Error} = \text{Set Point} - \text{Actual Output}$$

and a corrective action is determined by solving an equation or set of equations that have been established as characterizing a desirable way to control the system. The law governing determination of this manipulation can be referred to as the **control algorithm**. These operations are illustrated in block diagram form in Fig. 1.5.

Figure 1.4 Schematic diagram of computer closed-loop control.

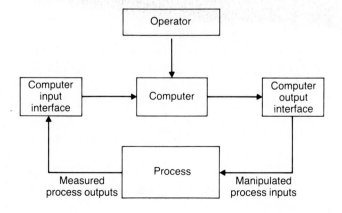

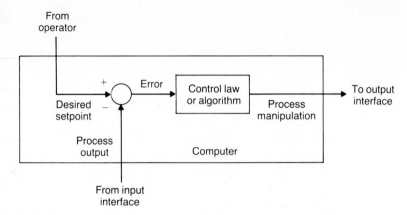

Figure 1.5 Block diagram of computer functions for closed-loop control.

Computer closed-loop feedback control is similar in concept to closed-loop feedback control of continuous analog systems. The difference lies primarily in the fact that the digital computer operates with specific sets of numbers at discrete points in time. The computer is programmed to sample values of the process output variables at certain instants in time. Calculation then proceeds in such a way as to establish the error and the desired manipulation corresponding to that error, using an appropriate control law or algorithm. The manipulation is then applied to the process. The process outputs are again sampled after a predetermined interval called the **sample interval**, and the procedure is repeated. Clearly, the speed at which the computer can perform many operations relative to the speed at which the process can respond is a significant factor in how well the process can be kept under control. Techniques for designing control algorithms and establishing sampling rates are discussed in considerable detail in subsequent chapters.

Digital control techniques have been utilized extensively in machine tools, chemical process plants, rolling mills, paper machines, and many other processes. The paper machine serves as a good example. Two of the primary output variables in the Fourdrinier papermaking process are basis weight (the weight of a certain area of paper) and moisture content. These two parameters can be related to the amount of fiber stock, which is controlled by the position of a stock valve, and to the amount of moisture being taken out in the dryer section, which may be considered in a simple analysis as proportional to the setting of the dryer steam valve. The computer can sample the output transducers (namely, a weight gauge and moisture indicator), compute the necessary valve actuations, and directly manipulate the stock and steam valves.

Another scheme that has been widely used is **feedforward control**. Processses are often subjected to disturbance inputs that affect the output of the process but can be neither predicted in advance nor controlled. Where a process is subjected to external disturbances that can be readily measured and monitored, it is possible to compute process manipulations and implement them directly without waiting for information to return via a feedback path. For example, feedforward control has proved to be a powerful technique in the digestion of wood chips in a pressure vessel under controlled conditions of temperature, pressure, time, and liquor composition. (Because of long process times at various stages in the digester, feedback control is incapable of producing alterations in process inputs in time to prevent significant loss of product when disturbances such as changes in the properties of incoming wood chips occur.) The computer can monitor and compute the optimal "cooking conditions" for material entering the digester and can alter control functions as these constituents proceed through the process.

Quality and productivity are important elements in any modern manufacturing system. Consider the machining of a cylindrical object. One aspect that may be important to quality is diameter. The diameter of the cylinder may need to be within a certain tolerance in order to fit into a hole in another part. If cylinders are produced with a diameter outside of this tolerance, the condition must be detected and the parts scraped. Material and processsing time are wasted and productivity suffers. An even more serious situation arises when it is not detected that out-of-tolerance parts are being manufactured, and an attempt is made to produce assemblies with parts that do not fit. Productivity losses are magnified whenever errors are allowed to persist while further value is being added.

Statistical quality control based on checking samples of parts was developed because it was often too expensive to check *every* machined part. Today, quality can be built into manufacturing processes by using appropriate sensors, computer decision making, and actuators. For the cylinder machining process, this can be accomplished using opto-electronic gauging and robotic manipulators. Not only is it possible to provide 100 % inspection, but the information gleaned can also be used in the control of variables such as tool position. And by studying the successive corrections required to sustain part accuracy, engineers can make further inferences about tool wear and breakage. Remember, as you study this text, that information collected and not used to its fullest extent is information partially wasted.

SUMMARY

This general introduction outlines only a few of the many important concepts in computer control of machines and processes. In the material to follow, these and other significant concepts are developed in detail, with emphasis on design techniques and examples. The current "state of the art" has evolved over several decades of improvement in computer speed and reliability, reduction in computer size and cost, and increased availability of interface hardware. Modern computers can accomplish a great deal in real-time control, as the concepts and techniques developed in subsequent chapters will illustrate.

First we must develop a mathematical basis for analyzing systems as viewed by a computer. Then we can investigate techniques for synthesizing control algorithms. Chapter 2 provides the concepts necessary to create mathematical models of real elements and systems that are compatible with the discrete nature of the computer and provide the analyst and designer with a view of the process to be controlled as "seen" by the computer. The more information that can economically be gathered about a process or a product, and the more we understand about how that information reflects performance, quality, and productivity, the better our opportunity to achieve improvement through computer control. This improvement can be the difference between success and failure of a system. In industry, achieving improvement over the competition is a key strategic objective.

BIBLIOGRAPHY

Astrom, K. J., and B. Wittenmark. *Computer Controlled Systems: Theory and Design.* Englewood Cliffs, NJ: Prentice-Hall, 1984.

Beeby, W., and P. Collier. *New Directions through CAD/CAM.* Dearborn, MI: Society of Manufacturing Engineers, 1986.

Besant, C. B. *Computer Aided Design and Manufacture, 2/e.* New York, NY: Wiley, 1983.

Bollinger, J. G. "Computer Control of Machine Tools," *Annals of the CIRP,* Vol. 21, No. 2, 1972.

Cadzow, J. A., and H. R. Martens. *Discrete Time and Computer Control Systems.* Englewood Cliffs, NJ: Prentice-Hall, 1970.

Costa, A., and M. Garetti. "Design of a Control System for a Flexible Manufacturing Cell," *Journal of Manufacturing Systems,* Vol. 4, No. 1, 1985, pp. 65–84.

Desrochers, A. A., "A Comparison of Adaptive Control Strategies for Hot Steel Rolling Mills," *Journal of Manufacturing Systems,* Vol. 1, No. 2, November 1982, pp. 183–194.

Dorf, R. C. *Modern Control Systems, 4/e.* Reading, MA: Addison-Wesley, 1986.

Dornfeld, D. A., D. M. Auslander, and P. M. Sagues. "Microprocessor-Controlled Manufacturing Processes," *Mechanical Engineering,* Vol. 102, No. 13, December 1980, pp. 34–41.

Duffie, N. "An Approach to the Design of Distributed Machinery Control Systems," *IEEE Transactions on Industry Applications*, Vol. IA–18, No. 4, July/August 1982, pp. 435–442.

ElMaraghy, H. "Automated Tool Management in Flexible Manufacturing," *Journal of Manufacturing Systems*, Vol. 4, No. 1, 1985, pp. 1–13.

Fotsch, R. "Machine Tool Justification Policies: Their Effect on Productivity and Profitability," *Journal of Manufacturing Systems*, Vol. 3, No. 2, 1984, pp. 169–195.

Groover, M. P. *Automation, Production Systems and Computer-Integrated Manufacturing.* Englewood Cliffs, NJ: Prentice-Hall, 1987.

Harrison, H. L., and J. G. Bollinger. *Introduction to Automatic Controls.* New York, NY: Harper & Row, 1969.

Hutchinson, G. K., and J. R. Holland. "The Economic Value of Flexible Automation," *Journal of Manufacturing Systems*, Vol. 1, No. 2, 1982, pp. 215–228.

International Trends in Manufacturing Technology. Bedford, UK: IFS (Publications) Ltd, 1983–87 (series).

Jones, A., and C. McLean. "A Proposed Hierarchical Control Model for Automated Manufacturing Systems," *Journal of Manufacturing Systems*, Vol. 5, No. 1, 1986, pp. 15–25.

Kalpakjian, S. *Manufacturing Processes for Engineering Materials.* Reading, MA: Addison-Wesley, 1984.

Koren, Yoram. *Computer Control of Manufacturing Systems.* New York, NY: McGraw-Hill, 1983.

Maimon, O. "Real-Time Operational Control of Flexible Manufacturing Systems," *Journal of Manufacturing Systems*, Vol. 6, No. 2, 1987, pp. 125–136.

Ránky, P. G. *Computer-Integrated Manufacturing.* Englewood Cliffs, NJ: Prentice-Hall, 1986.

Rembold, U., C. Blume, and R. Dillmann. *Computer-Integrated Manufacturing Technology and Systems.* New York, NY: Marcel Dekker, 1985.

Riley, F. J. *Assembly Automation.* New York, NY: Industrial Press Inc., 1983.

Roger, A. "The Microprocessor Invades the Production Line," *IEEE Spectrum*, January 1979, pp. 53–59.

Spur, G., G. Seliger, and B. Viehweger. "Cell Concepts for Flexible Automated Manufacturing," *Journal of Manufacturing Systems*, Vol. 5, No. 3, 1986, pp. 171–179.

Weck, M. "Machine Diagnostics in Automated Production," *Journal of Manufacturing Systems*, Vol. 2, No. 2, 1983, pp. 101–106.

Chapter 2

Elements of
Discrete Modeling

INTRODUCTION

The purpose of this chapter is to explore some of the basic concepts and techniques applicable to the analysis and design of digital computer control systems. As a starting point, let us consider the way in which a computer can be utilized in controlling a process with many variables to be controlled. The nature of the computer is such that it can work with only one controlled variable at a time. Because of this, having sampled a particular variable (obtained its current value), the computer usually must initiate an appropriate action or correction quickly and then move on to the next controlled variable. Thus, although a given variable is continuous with time, the computer has knowledge of its value only at discrete points in time.

Much of control theory is concerned with continuous systems characterized by analog signals, and much of it uses differential equations as a basis for modeling. However, the nature of the digital computer makes it convenient to think in terms of discrete systems for which difference equations provide the appropriate mathematical basis. Thus, this chapter introduces the concepts of **discrete modeling**. Examples of how difference equations can be derived for specific first-order, second-order, and delay processes are presented first, followed by a discussion of block diagrams for discrete systems and of how a table of transfer functions can be used to obtain the transfer function of a given system. The concept of proportional, integral, and derivative control actions performed by a computer is introduced, and we show how closed-loop system models are obtained. Finally, a discussion of the solution of difference equations leads to the analysis of stability for discrete systems.

12

2.2

DISCRETE PROCESS MODELS

Only processes with a single manipulated variable (input) and a single controlled variable (output) will be considered here, but the results are applicable to processes with many controlled variables, provided that there are no interactions between variables. The interaction problem is discussed in Chapter 4, and the more generalized state-variable modeling approach is developed in Appendix A.

The immediate objective is to develop a suitable mathematical relationship between some output variable and a corresponding input variable on the basis that the dynamics of the process are known. The essential ideas are illustrated in Fig. 2.1. A process with a single input and a single output is shown schematically in Fig. 2.1(a). For a large class of computer control applications, the input can be considered a series of steps, because it is the nature of computer control hardware to establish a level of input and to have this level held fixed until one sample interval later, at which time a new level can be produced. Normally, the process output is sampled via computer control hardware that nearly instantaneously "captures" the value and makes it available for subsequent control calculations.

Figure 2.1(b) shows a motor/amplifier process for which the input is a voltage $v(t)$ and the output is the motor shaft position $\theta(t)$. A hypothetical input sequence is plotted against time in Fig. 2.1(c). T is the time period between samples of the output; it is assumed that the time required to sample the process output and produce a new process input is small compared to the sample period T. The process input results in an output, as illustrated in Fig. 2.1(d).

The mathematical problem can be stated as follows: Given the defining equation for the process, the value of the input at some arbitrary time $t = 0$, and the value of the output at time $t = 0$, together with any necessary previous values, determine the value of the output one sample period later—that is, at time $t = T$. The relationship can also be thought of in terms of normalized time n, where n is an integer index and $n = t/T$. We will now illustrate concepts of discrete modeling by considering several basic models that can be used either to define certain simple processes or to approximate the behavior of more complicated processes.

Example 2.1 An integration process   To provide an example of what is commonly known as an **integration** in control terminology, assume that the process illustrated in Fig. 2.1(b) can be defined approximately by the equation

$$\frac{d\theta(t)}{dt} = Kv(t)$$

where $v(t)$ is the input, $\theta(t)$ is the output, and K is a constant that depends on motor/amplifier characteristics. Assume that $\theta(0) = \theta_0$ at time $t = 0$ and that a

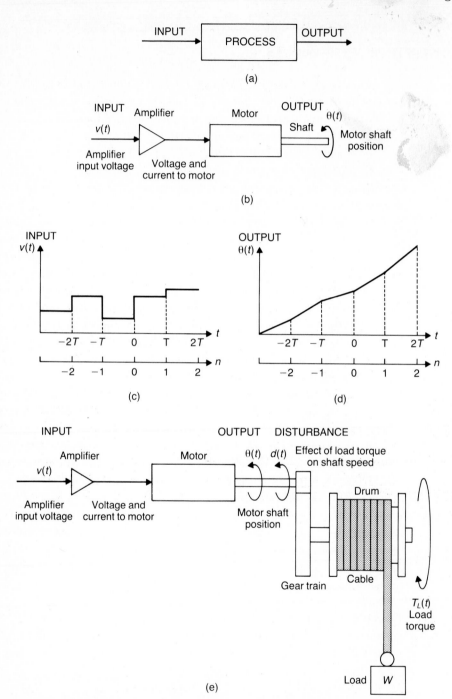

Figure 2.1 (a) Single-input, single-output process; (b) motor/amplifier process; (c) example of process input; (d) corresponding process output; (e) and hoist system with load modeled as a disturbance.

constant input $v(t) = v_o$ for $0 \le t < T$ is introduced. In this case,

$$\theta(T) = \int_{-\infty}^{o} Kv(t)\, dt + \int_{0}^{T} Kv_o\, dt$$

Integrating and using the initial condition yields

$$\theta(T) = \theta_o + Kv_o T$$

Generalizing by the formal replacement of v_0 and θ_0 with v_{n-1} and θ_{n-1}, respectively, and by the replacement of $\theta(T)$ with θ_n, produces the first-order difference equation

$$\theta_n - \theta_{n-1} = KTv_{n-1}$$

or

$$\theta_n = \theta_{n-1} + KTv_{n-1}$$

which is a discrete model for the system. This equation defines the process output at the beginning of the current sample period on the basis of given values of the process input and process output at the beginning of the previous sample period.

Processes may have inputs that are not computer-manipulated. These inputs are often referred to as **disturbances**. They affect the process output and often cannot be sensed conveniently, much less controlled. It can be useful to include them in a model so that response to disturbances can be predicted. For example, consider the application of the motor/amplifier in the hoist mechanism shown in Fig. 2.1(e). The load W generates a torque $T_L(t)$ that is transmitted through the gear train to the motor. Because it cannot be predicted how large a load will be carried, or when, a convenient approach is to treat load as a disturbance, $d(t)$. If the load disturbance directly affects the speed of the motor and is constant between samples, the process model becomes

$$\frac{d\theta(t)}{dt} = Kv(t) + d(t)$$

and the discrete process model becomes

$$\theta_n - \theta_{n-1} = KTv_{n-1} + Td_{n-1} \quad \square$$

Example 2.2 A first-order process The liquid-level system illustrated in Fig. 2.2 can be modeled as a first-order process. If the relationship between flow $q_i(t)$ into the tank and valve position $\theta(t)$ is

$$q_i(t) = K_v \theta(t)$$

where K_v is the flow control valve constant and the relationship between flow $q_o(t)$ out of the tank and height of the liquid $h(t)$ is

$$q_o(t) = K_o h(t)$$

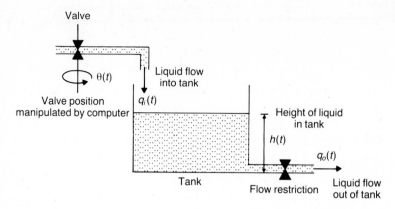

Figure 2.2 Liquid-level system with inlet valve and flow restriction in outlet.

when K_o is the outlet flow constant, then the process can be defined by the equation

$$K_v \theta(t) - K_o h(t) = A \frac{dh(t)}{dt}$$

where A is the surface area of the tank. This equation can be rewritten in the form

$$\tau \frac{dh(t)}{dt} + h(t) = K\theta(t)$$

where $\tau = A/K_o$ is the process time constant and $K = K_v/K_o$ is the process gain. With $h(0) = h_0$ and a constant input $\theta(t) = \theta_0$ for $0 \le t < T$, a solution is obtained as

$$h(t) = Ce^{-t/\tau} + K\theta_0$$

which for $t = T$ is

$$h(T) = Ce^{-T/\tau} + K\theta_0$$

The constant C is determined from the initial condition as

$$C = h_0 - K\theta_0$$

Thus

$$h(T) = h_0 e^{-T/\tau} + K\theta_0 (1 - e^{-T/\tau})$$

The corresponding first-order difference equation is

$$h_n - e^{-T/\tau} h_{n-1} = K(1 - e^{-T/\tau}) \theta_{n-1}$$

It is instructive at this point to illustrate that the foregoing difference equation yields the correct response values at the sampling instants. Assume that

$\tau = 1$ minute, $K = 0.2$ meters/degree, $\theta(t) = 10°$ for $t \geq 0$, and the initial condition is $h(0) = 0$ meters. The corresponding differential equation is

$$\frac{dh(t)}{dt} + h(t) = 2$$

Its solution is

$$h(t) = 2(1 - e^{-t})$$

and is plotted in Fig. 2.3. From this solution, Program 2.1 yields the following values for $h(t)$ in meters, obtained at 1-minute increments in time.

$$h(0) = 0$$

$$h(1) = 1.264$$

$$h(2) = 1.729$$

$$h(3) = 1.900$$

$$h(4) = 1.963$$

If the sample period is chosen as $T = 1$ minute, the constants in the difference equation are

$$e^{-T/\tau} = e^{-1} = 0.368$$

$$K(1 - e^{-T/\tau}) = 0.2(0.632) = 0.1264 \text{ meters/degree}$$

Thus the difference equation for the process is

$$h_n = 0.368h_{n-1} + 0.1264\theta_{n-1}$$

Figure 2.3 Response in height of liquid in tank for constant valve position of $10°$.

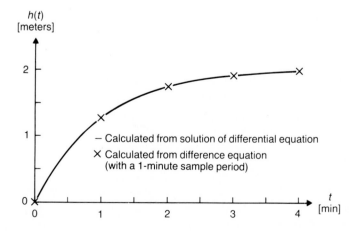

$h(t)$
[meters]

2

1

— Calculated from solution of differential equation
X Calculated from difference equation
  (with a 1-minute sample period)

0

0   1   2   3   4

t
[min]

```
program Continuous_tank_model;
{Height of liquid in a tank is calculated for a constant input valve position.}
   const
      dt = 0.1;                                {0.1-minute time increment}
      tau = 1.0;                               {1-minute process time constant}
      k = 0.2;                                 {0.2-meters/degree process gain}
   var
      h, theta, t : real;                      {height [m], position [deg], time [min]}
   begin
      t := 0.0;                                {initial values}
      h := 0.0;
      theta := 10.0;                           {10-degree valve position}
      repeat
         t := t + dt;                          {increment time}
         h := K * (1.0 - exp(-t / tau)) * theta;   {solution of differential equation}
         writeln(t, h)
      until t >= 10.0                          {stop after 10 minutes}
   end.
```

Program 2.1 Continuous tank model with constant input for Example 2.2.

With the initial condition $h_o = 0$ meters, and with $\theta_n = 10°$ for $n = 0, 1, 2, 3, \ldots$, values at several sample times can be calculated, by using the difference equation. Some of these values are

$$h_o = 0$$

$$h_1 = 0 + 1.264 = 1.264$$

$$h_2 = 0.368(1.264) + 1.264 = 1.729$$

$$h_3 = 0.368(1.729) + 1.264 = 1.900$$

$$h_4 = 0.368(1.900) + 1.264 = 1.963$$

This technique can be referred to as the **recursion method** for calculating the process output from the input. These values, which can be obtained via Program 2.2, are plotted in Fig. 2.3. Note that they correspond exactly to those evaluated previously from the solution of the differential equation. However, unlike the solution of the differential equation, the recursion method does not yield information about the response of the process at times other than the sample times $t = 0, 1, 2, 3, \ldots$ minutes. □

Example 2.3 Varying process input   It must not be inferred that the process input is required to be the same constant value from sample period to sample period. In fact, the opposite is usually the case. The necessary assumption is that the input is constant during any sample period. To illustrate this point, the

```
program Discrete_tank_model;
{Height of liquid in a tank is calculated for a constant input valve position.}
  const
    T = 1.0;                                {1-minute sample period}
    tau = 1.0;                              {1-minute process time constant}
    K = 0.2;                                {0.2 meters/degree process gain}
  var
    n : integer;
    h0, h1, theta : real;                   {height h [m], position theta [deg]}
    a1, b1 : real;
  begin
    h0 := 0.0;                              {initial height}
    theta := 10.0;                          {valve position is 10 degrees}
    a1 := exp(-T / tau);                    {coefficients for difference equation}
    b1 := K * (1.0 - exp(-T / tau));
    for n := 1 to 10 do                     {10 1-minute sample periods}
      begin
        h1 := h0;                           {update previous height}
        h0 := a1 * h1 + b1 * theta;         {next height from difference equation}
        writeln(n, h0)
      end
end.
```

Program 2.2 Discrete tank model with constant input for Example 2.2.

response of the system in Example 2.2 can be determined when the input values are as follows:

$$\theta_0 = 10°$$

$$\theta_1 = 5°$$

$$\theta_2 = 2°$$

$$\theta_3 = 0°$$

$$\theta_n = 0° \quad (\text{for } n \geq 3)$$

Using the recursion method with $h_0 = 0$ meters,

$$h_0 = 0$$

$$h_1 = 0.368(0) + 0.1264(10) = 1.264$$

$$h_2 = 0.368(1.264) + 0.1264(5) = 1.097$$

$$h_3 = 0.368(1.097) + 0.1264(2) = 0.656$$

$$h_4 = 0.368(0.656) + 0.1264(0) = 0.242$$

$$h_5 = 0.368(0.242) + 0 = 0.089$$

$$h_6 = 0.368(0.089) + 0 = 0.033$$

$$\vdots$$

```
program Discrete_tank_model;
{Height of liquid in a tank is calculated with varying input valve position.}
  const
    T = 1.0;                              {1-minute sample period}
    tau = 1.0;                            {1-minute process time constant}
    K = 0.2;                              {0.2-meter/degree process gain}
  var
    n : integer;
    theta : array[0..10] of real;        {valve position sequence [deg]}
    h : array[0..10] of real;            {liquid height sequence [m]}
    a1, b1 : real;
begin
  theta[0] := 10.0;                      {create valve position sequence}
  theta[1] := 5.0;
  theta[2] := 2.0;
  for n := 3 to 10 do
    theta[n] := 0.0;
  a1 := exp(-T / tau);                   {coefficients for difference equation}
  b1 := K * (1.0 - exp(-T / tau));
  h[0] := 0.0;                           {initial height}
  writeln(0, theta[0], h[0]);
  for n := 1 to 10 do                    {10 1-minute sample periods}
    begin
      h[n] := a1 * h[n - 1] + b1 * theta[n - 1]; {next h from difference equation}
      writeln(n, theta[n], h[n]);
    end
end.
```

Program 2.3 Discrete tank model with varying input for Example 2.3.

Program 2.3 illustrates one way in which these values can be calculated. The process input and output are plotted versus time in Fig. 2.4. □

Example 2.4 A double-integration process   A simplified approach to modeling the motion of the spacecraft illustrated in Fig. 2.5 is to use the equation

$$M \frac{d^2 y(t)}{dt^2} = F(t)$$

where $y(t)$ represents spacecraft motion, M is the mass of the spacecraft, and $F(t)$ is the reaction force generated by the spacecraft's engine. This form of second-order process model is often referred to as **double integration**.

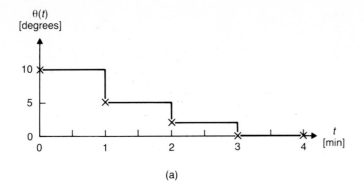

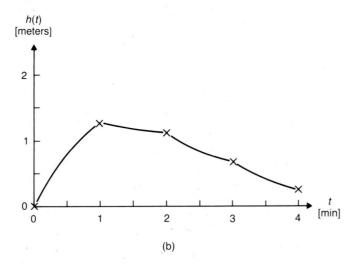

Figure 2.4 (a) Valve position input for Example 2.3 (constant between samples); (b) response in height of liquid in tank to varying valve position.

If at time $t = 0$ a constant input $F(t) = F_o$ is introduced, then, proceeding with the solution by successive integrations,

$$\frac{d^2y(t)}{dt^2} = KF_o$$

$$\frac{dy(t)}{dt} = KF_o t + \dot{y}_o$$

$$y(t) = \frac{KF_o}{2} t^2 + \dot{y}_o t + y_o$$

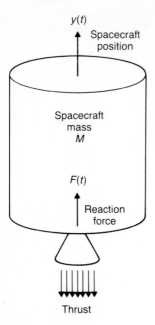

$y(t)$

Spacecraft
position

Spacecraft
mass
M

$F(t)$

Reaction
force

Thrust

Figure 2.5 Spacecraft with
mass M and thrust force
$F(t)$.

where $K = 1/M$ and y_0 and $\dot{y}_0$ are initial conditions. At $t = T$,

$$\dot{y}(T) = KF_0 T + \dot{y}_0$$

$$y(T) = \frac{KF_0}{2} T^2 + \dot{y}_0 T + y_0$$

The corresponding difference equations are

$$\dot{y}_n - \dot{y}_{n-1} = KTF_{n-1}$$

$$y_n - y_{n-1} = T\dot{y}_{n-1} + \frac{KT^2}{2} F_{n-1}$$

Unfortunately, the equation for y is not a useful form for computer application because of the $\dot{y}$ term. The derivative $\dot{y}$ is not, in general, available for direct measurement, and modification of the equation is necessary.

First, the equation for y is solved for $\dot{y}_{n-1}$, producing

$$\dot{y}_{n-1} = \frac{1}{T} y_n - \frac{1}{T} y_{n-1} - \frac{KT}{2} F_{n-1}$$

Shifting this equation one interval forward in time yields

$$\dot{y}_n = \frac{1}{T} y_{n+1} - \frac{1}{T} y_n - \frac{KT}{2} F_n$$

Substituting these equations into the original equation for $\dot{y}$ yields

$$\frac{1}{T}y_{n+1} - \frac{1}{T}y_n - \frac{KT}{2}F_n - \frac{1}{T}y_n + \frac{1}{T}y_{n-1} + \frac{KT}{2}F_{n-1} = KTF_{n-1}$$

Simplifying and shifting back one interval in time yields

$$y_n - 2y_{n-1} + y_{n-2} = \frac{KT^2}{2}F_{n-1} + \frac{KT^2}{2}F_{n-2} \quad \square$$

Example 2.5 A dead-time process   Inherent in many processes is the phenomenon of fluid or material moving from one point to another without
undergoing change. This phenomenon is called a **transportation lag** or **dead
time**. In Fig. 2.6, the input temperature T_{in} and output temperature T_{on} of a fluid
flowing through a pipe are shown plotted against n. For this specific case, it can
be seen in Fig. 2.6(b) that the output is identical to the input except that the
output lags the input by two sample periods. The appropriate characterizing
equation is therefore

$$T_{o_n} = T_{in-2}$$

For the general case of the system shown in Fig. 2.6(a), the dead time D is
found from

$$D = L\frac{A}{q}$$

Figure 2.6 (a) Pipe through which fluid of varying temperature flows at a constant
rate; (b) input and output temperature sequences showing dead time of two sample
periods.

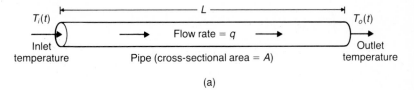

(a)

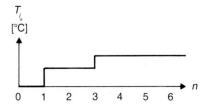 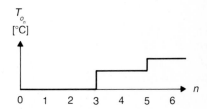

(b)

where L is the length of the pipe, A is its cross-sectional area, and q is the volumetric flow rate of fluid through the pipe. The number of sample periods this delay represents is found from

$$d = \frac{D}{T}$$

and

$$T_{O_n} = T_{i_{n-d}}$$

Here d must be an integer, and if D/T is not an integer, d should usually be rounded to the next highest integer. This conservative approximation of dead time is desirable when one is designing control systems, because dead time adversely affects the stability of a closed-loop system. Underestimating the amount of dead time can lead to unreasonable expectations for the performance of a closed-loop system. □

Other Process Types

Difference equations characterizing other basic processes can be obtained by following procedures similar to those we have just illustrated. This section has been concerned primarily with first-order and second-order processes, and, although there are many processes of this nature, it is not to be inferred that the typical process is so simple. However, many complex processes can be satisfactorily modeled either with lower-order models or with combinations of lower-order and dead-time models. One very useful model is the first-order dead-time model, which is discussed in Chapter 10. Higher-order discrete process models can be developed using the techniques described in Chapter 11 and Appendix A.

Note that the development of models in this section involved no approximations where values at sampling instants were concerned. The discrete models obtained yield exact values at the sample times. Keep in mind, however, that the discrete models are based on the assumption that inputs are constant between successive sample times.

2.3

THE DISCRETE CONTROLLER

Although processes usually are continuous in time, the control computer exists in a discrete world because it has knowledge of process outputs only at the discrete points in time when sample values are obtained. It was for this reason that the previous section was devoted to obtaining discrete models for continuous systems, so that both the process and the computer can be treated on a similar basis.

In general, the control computer performs the following tasks:

1. Obtains a sample value of the process output c_n.
2. Calculates the error e_n from the relationship

$$e_n = r_n - c_n \tag{2.1}$$

where r_n is the reference (or desired) value stored in the computer.

3. Computes the proper value for the manipulated process input m_n.
4. Outputs m_n to the appropriate control element.
5. Continues with the next controlled variable.

The time required to achieve a new level of the manipulated variable typically is short compared with the time between samples. It is therefore reasonable to assume that the input to the process is a sequence of constant values that change instantaneously at the beginning of each sample period.

A control algorithm must be provided so that the computer can calculate values for the manipulated variable. Because the current error value e_n and stored values of previous errors are available, a useful form for control algorithms is

$$m_n - m_{n-1} = K_0 e_n + K_1 e_{n-1} + K_2 e_{n-2} + \cdots \tag{2.2}$$

For the benefit of readers who are familiar with the continuous control actions that can be obtained with analog controllers, it will be instructive to develop the discrete approximations of these control actions. The point is to show that the discrete approximations are of the form of Eq. (2.2).

Proportional Control

With **proportional control** action, the correction made is proportional to the error, or

$$m(t) = K_p e(t) \tag{2.3}$$

where K_p is the controller proportionality constant or gain. The discrete form of Eq. (2.3) is

$$m_n = K_p e_n \tag{2.4}$$

Shifting this equation backward one sample period yields

$$m_{n-1} = K_p e_{n-1} \tag{2.5}$$

Subtracting Eq. (2.5) from Eq. (2.4) yields the incremental form, which is

$$\Delta m_n = m_n - m_{n-1} = K_p(e_n - e_{n-1}) \tag{2.6}$$

Note that Eq. (2.6) is of the same form as Eq. (2.2).

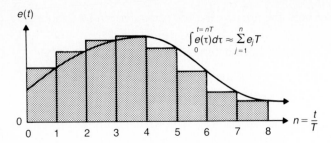

Figure 2.7 Discrete integration using backward rectangular rule.

Integral Control

With **integral control** action, the correction made is proportional to the time integral of the error, or

$$m(t) = K_i \int_0^t e(\tau)\, d\tau \tag{2.7}$$

where K_i is the integral controller gain. As shown in Fig. 2.7, the area under the continuous error curve implied by the integration can be approximated by the sum of rectangular areas, each being computed as the product of a discrete error value e_j and the sample interval T. Thus

$$m_n = K_i \sum_{j=1}^{n} Te_j \tag{2.8}$$

This can be rewritten as

$$m_n = K_i \sum_{j=1}^{n-1} Te_j + K_i Te_n \tag{2.9}$$

Also note that

$$m_{n-1} = K_i \sum_{j=1}^{n-1} Te_j \tag{2.10}$$

Subtracting Eq. (2.10) from Eq. (2.9) yields

$$\Delta m_n = m_n - m_{n-1} = K_i Te_n \tag{2.11}$$

which is the incremental discrete approximation of integral control action.[1]

---

[1] Note that the integral of the error is approximated here using the backward rectangular rule, which uses the current value of the error, rather than with the previous value, as would be the case if the result of Example 2.1 were used.

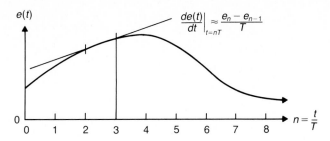

Figure 2.8 Discrete differentiation using backward difference rule.

Derivative Control

Derivative control action is defined by

$$m(t) = K_d \frac{de(t)}{dt} \tag{2.12}$$

As illustrated in Fig. 2.8, backward difference approximation can be used to obtain the associated discrete equation

$$m_n = K_d \left(\frac{e_n - e_{n-1}}{T} \right) \tag{2.13}$$

and then

$$\Delta m_n = m_n - m_{n-1} = \frac{K_d}{T} (e_n - 2e_{n-1} + e_{n-2}) \tag{2.14}$$

PID Control

A widely used control action is **proportional plus integral plus derivative (PID) control**. PID control action combines the effects of each control action to obtain a composite control manipulation according to the equation

$$\Delta m_n = (\Delta m_n)_p + (\Delta m_n)_i + (\Delta m_n)_d \tag{2.15}$$

Combining Eqs. (2.6), (2.11), and (2.14) yields

$$\Delta m_n = m_n - m_{n-1} = K_0 e_n + K_1 e_{n-1} + K_2 e_{n-2} \tag{2.16}$$

where

$$K_0 = K_p + K_i T + \frac{K_d}{T} \tag{2.17}$$

$$K_1 = -K_p - \frac{2K_d}{T} \tag{2.18}$$

$$K_2 = \frac{K_d}{T} \tag{2.19}$$

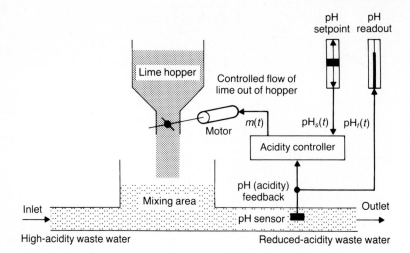

Figure 2.9 System for controlling acidity of waste water from an industrial process.

A discussion of control actions has been included to show that their various corrective characteristics can be obtained with control algorithms of the form of Eq. (2.2). It should not be inferred that the design of discrete controllers is based solely on a control-action point of view. The subject of discrete controller design is covered more thoroughly in Chapters 4 and 12.

Example 2.6 Process with PID control   A system for controlling the level of acidity in the waste-water effluent from an industrial process is illustrated in Fig. 2.9. Suppose that an existing continuous (analog-circuit–based) controller uses a command value set by the system operator and feedback from a pH (acidity) sensor in implementing the following control laws:

$$c(t) = 5.0 - pH_f(t)$$

$$r(t) = 5.0 - pH_s(t)$$

$$e(t) = r(t) - c(t)$$

$$m(t) = 400e(t) + 8\int e(t)\, dt + 804\, de(t)/dt$$

The first two equations convert the pH scale to a scale that increases in numerical value as acidity increases (the pH scale decreases as acidity increases). The last equation represents proportional plus integral plus derivative (PID) control.

If it is desired to replace the continuous controller with a computer-based controller, Eq. (2.16) could be used as an approximation of the continuous PID control equation. If a sample period of 0.1 second were chosen (a sample of the feedback pH_f would be taken and the control manipulation calculated every 0.1

second), the resulting discrete PID control equation would be

$$m_n - m_{n-1} = [400 + 8(0.1) + 804/0.1]e_n - [400 + 2(804)/0.1]e_{n-1} \\ + (804/0.1)e_{n-2}$$

or

$$m_n = m_{n-1} + 8440.8e_n - 16480.0e_{n-1} + 8040.0e_{n-2}$$

If a sample period of 0.5 second had been chosen rather than 0.1 second, the result would have been

$$m_n = m_{n-1} + 2012.0e_n - 3616.0e_{n-1} + 1608.0e_{n-2}$$

Two things should be noted regarding these results. First, the coefficients of the discrete control equation are functions of the sample period. Hence the discrete control equation must be changed whenever the sample period is changed. Second, a relatively high number of significant figures may be required in the parameters of the discrete control equation. In the case of the 0.1-second sample period, rounding the coefficient of the e_n term from 8440.8 up to 8441 would be equivalent to increasing the integral gain approximated by the controller from 8 to 10 (a factor of 25%). The differences between the magnitudes of the coefficients in a discrete control equation are often just as important as their relative magnitudes. □

2.4
TRANSFER FUNCTIONS AND BLOCK DIAGRAMS

It has been shown that difference equations provide a suitable mathematical basis for defining the behavior of both processes and control computers. It now will be shown that through the use of the backward-shift operator B it is possible to obtain discrete transfer functions, develop block diagrams, and use algebraic techniques for manipulation purposes in obtaining transfer functions and differ- ence equations describing closed-loop systems.

Discrete Transfer Functions

The **transfer function** of a system (or element of a system) can be defined as the ratio of its output to its input, with all initial conditions assumed to be zero. Transfer functions can be obtained from difference equations by using the backward-shift operator B, which is defined such that

$$By_n \equiv y_{n-1} \tag{2.20}$$

$$B^2 y_n \equiv y_{n-2} \tag{2.21}$$

$$B^j y_n \equiv y_{n-j} \tag{2.22}$$

Note that B is an operator, not a variable, and that numerical values should not be assigned to it. Equation (2.22) can be used to obtain process transfer functions by substitution into the process difference equation and solution for the process output c_n as a function of the input process m_n in the form

$$G_p(B) = \frac{c_n}{m_n} \tag{2.23}$$

$G_p(B)$ is the resulting process transfer function. Similarly, the controller transfer function can be written as

$$G_c(B) = \frac{m_n}{e_n} \tag{2.24}$$

where e_n is the error in the process output being controlled and m_n is the manipulation output to the process.

Example 2.7 Processor and controller transfer functions   As an illustrative example, assume that an integrating process that is to be controlled is characterized by the equation

$$\frac{dc(t)}{dt} = Km(t)$$

where c is the process output variable to be controlled and m is the manipulated process input variable. This equation is identical in form to the equation describing the process in Example 2.1, and the corresponding difference equation is therefore

$$c_n - c_{n-1} = KTm_{n-1}$$

Introduction of the B operator via Eq. (2.20) yields

$$c_n - Bc_n = KTBm_n$$

or

$$(1 - B)c_n = KTBm_n$$

Forming the ratio of output to input results in the discrete process transfer function

$$G_p(B) = \frac{c_n}{m_n} = \frac{KTB}{1 - B}$$

Assume that the control algorithm has the form of Eq. (2.16) with $K_d = 0$:

$$m_n - m_{n-1} = (K_p + K_i T)e_n - K_p e_{n-1}$$

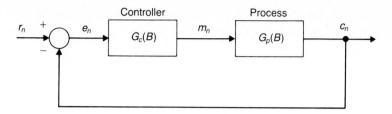

Figure 2.10 Block diagram for closed-loop process control.

Using the B operator, we obtain the discrete controller transfer function as follows:

$$m_n - Bm_n = (K_p + K_i T)e_n - K_p Be_n$$

$$(1 - B)m_n = [(K_p + K_i T) - K_p B]e_n$$

$$G_c(B) = \frac{m_n}{e_n} = \frac{(K_p + K_i T) - K_p B}{1 - B} \quad \square$$

Block Diagrams

A closed-loop system **block diagram** is shown in Fig. 2.10. Equation (2.1) is represented by the summing point, whereas Eqs. (2.23) and (2.24) are represented by the blocks labeled process and controller, respectively. The use of the discrete process transfer function in the diagram is valid, because the computer produces a sequence of constant values that serve as the input to the process.

The Closed-Loop Transfer Function

The block diagram of Fig. 2.10 can be used to obtain the **closed-loop transfer function** c_n/r_n and, from it, the system difference equation. The closed-loop transfer function is

$$G(B) = \frac{c_n}{r_n} \tag{2.25}$$

and is obtained from the elements of the block diagram. Because $e_n = r_n - c_n$ at the sample times,

$$c_n = G_c(B)G_p(B)(r_n - c_n) \tag{2.26}$$

The closed-loop transfer function is therefore

$$G(B) = \frac{c_n}{r_n} = \frac{G_c(B)G_p(B)}{1 + G_c(B)G_p(B)} \tag{2.27}$$

Example 2.8 Obtaining closed-loop transfer functions   In the closed-loop block diagram shown in Fig. 2.11(a),

$$G_c(B) = \frac{1.5 - B}{1 - B}$$

$$G_p(B) = \frac{B}{1 - B}$$

In this case,

$$G_c(B)G_p(B) = \frac{(1.5 - B)B}{(1 - B)^2}$$

The resulting closed-loop transfer function from Eq. (2.27) is

$$G(B) = \frac{c_n}{r_n} = \frac{(1.5 - B)B}{1 - 0.5B}$$

as shown in Fig. 2.11(b).

This can be rewritten as

$$(1 - 0.5B)c_n = (1.5 - B)Br_n$$

or

$$c_n - 0.5c_{n-1} = 1.5r_{n-1} - r_{n-2}$$

which is the system difference equation. □

Figure 2.11 (a) Closed-loop block diagram for Example 2.8; (b) closed-loop transfer function for Example 2.8.

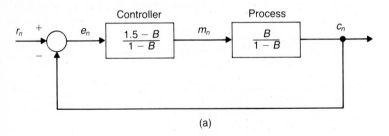

(a)

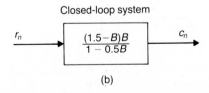

(b)

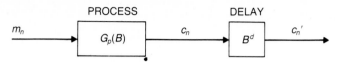

Figure 2.12 Process with delay.

Example 2.9 Process with dead time   Dead time is easily included in the process transfer function. Consider a process that is followed by a delay of D time units before measurement of a controlled variable is possible. The effect of the delay is represented through the use of the relationship

$$c_n' = c_{n-d}$$

or

$$c_n' = B^d c_n$$

where $d = D/T$ (rounded up to the next largest integer), c_n is the process output ahead of the delay, and c_n' is the controlled variable available for measurement as indicated in Fig. 2.12. The entire process transfer function, including dead time, then, is

$$G_p'(B) = \frac{c_n'}{m_n} = B^d G_p(B) \quad \square$$

Table of Transfer Functions

Table 2.1 lists appropriate discrete transfer functions for dead-time, first-order, and second-order processes, thus eliminating the need to derive a transfer function each time it is required. A cautionary note is in order concerning the correct use of the discrete transfer functions given in Table 2.1. A fundamental assumption that has been made is that the input to the process being modeled is constant between samples. This leads to a potential difficulty in attempting to combine transfer functions of process elements. The problem is best illustrated through the use of an example.

Example 2.10 Invalid combination of transfer functions   Suppose that a process is clearly composed of two integrations in series and can be defined by the equations

$$\frac{dy(t)}{dt} = K_1 m(t)$$

and

$$\frac{dc(t)}{dt} = K_2 y(t)$$

Table 2.1

Discrete Transfer Functions for Dead-Time, First-Order, and Second-Order Processes

DEFINING EQUATION	TRANSFER FUNCTION, y_n/x_n SAMPLE TIME = T
1. $y(t) = Kx(t - D)$	KB^d where $d = D/T$ (an integer)
2. $\dfrac{dy(t)}{dt} = Kx(t)$	$\dfrac{KTB}{1 - B}$
3. $\tau \dfrac{dy(t)}{dt} + y(t) = Kx(t)$	$\dfrac{K(1 - e^{-T/\tau})B}{1 - e^{-T/\tau}B}$
4. $\dfrac{d^2y(t)}{dt^2} = Kx(t)$	$\dfrac{\dfrac{KT^2}{2}B(1 + B)}{(1 - B)^2}$
5. $\tau \dfrac{d^2y(t)}{dt^2} + \dfrac{dy(t)}{dt} = Kx(t)$	$\dfrac{K(b_1 B + b_2 B^2)}{1 - a_1 B - a_2 B^2}$

where $a_1 = 1 + e^{-T/\tau}$

$$a_2 = -e^{-T/\tau}$$

$$b_1 = [T - \tau(1 - e^{-T/\tau})]$$

$$b_2 = -[Te^{-T/\tau} - \tau(1 - e^{-T/\tau})]$$

6. $\dfrac{1}{\omega_n^2}\dfrac{d^2y(t)}{dt^2} + \dfrac{2\zeta}{\omega_n}\dfrac{dy(t)}{dt} + y(t) = Kx(t)$	$\dfrac{K(b_1 B + b_2 B^2)}{1 - a_1 B - a_2 B^2}$

$$\tau_1\tau_2 \dfrac{d^2y}{dt^2} + (\tau_1 + \tau_2)\dfrac{dy(t)}{dt} + y(t) = Kx(t)$$

a. Overdamped: $\zeta > 1$

$$a_1 = e^{-T/\tau_1} + e^{-T/\tau_2}$$

$$a_2 = -e^{-(1/\tau_1 + 1/\tau_2)T}$$

$$b_1 = 1 + \frac{\dfrac{1}{\tau_2}e^{-T/\tau_1} - \dfrac{1}{\tau_1}e^{-T/\tau_2}}{\dfrac{1}{\tau_1} - \dfrac{1}{\tau_2}}$$

$$b_2 = e^{-(1/\tau_1 + 1/\tau_2)T} + \frac{\dfrac{1}{\tau_2}e^{-T/\tau_2} - \dfrac{1}{\tau_1}e^{-T/\tau_1}}{\dfrac{1}{\tau_1} - \dfrac{1}{\tau_2}}$$

b. Critically Damped: $\zeta = 1$

$$a_1 = 2e^{-\omega_n T}$$

$$a_2 = -e^{-2\omega_n T}$$

$$b_1 = 1 - e^{-\omega_n T} - \omega_n Te^{-\omega_n T}$$

$$b_2 = e^{-\omega_n T}(e^{-\omega_n T} + \omega_n T - 1)$$

c. Underdamped: $\zeta < 1$

$$\omega_d = \omega_n\sqrt{1 - \zeta^2}$$

$$a_1 = 2e^{-\zeta\omega_n T}\cos \omega_d T$$

$$a_2 = -e^{-2\zeta\omega_n T}$$

$$b_1 = 1 - \frac{\zeta\omega_n}{\omega_d}e^{-\zeta\omega_n T}\sin \omega_d T - e^{-\zeta\omega_n T}\cos \omega_d T$$

$$b_2 = e^{-\zeta\omega_n T}\left(e^{-\zeta\omega_n T} + \frac{\zeta\omega_n}{\omega_d}\sin \omega_d T - \cos \omega_d T\right)$$

in which $m(t)$ is the manipulated variable, $c(t)$ is the controlled variable, and $y(t)$ is an intermediate variable. It is tempting to use entry 2 in Table 2.1 to obtain

$$\frac{y_n}{m_n} = \frac{K_1 TB}{1 - B}$$

and, finally,

$$\frac{c_n}{m_n} = \frac{y_n}{m_n} \cdot \frac{c_n}{y_n} = \frac{K_1 K_2 T^2 B^2}{(1 - B)^2}$$

But this result is erroneous because, although the manipulated variable $m(t)$ meets the requirement of remaining constant throughout each interval between samples, the intermediate variable $y(t)$ does not.

The correct procedure is to combine the defining equations, which yields

$$\frac{d^2 c(t)}{dt^2} = K_2 \frac{dy(t)}{dt} = K_1 K_2 m(t)$$

and to use entry 4 in Table 2.1 to obtain

$$\frac{c_n}{m_n} = \frac{\dfrac{K_1 K_2 T^2}{2} B(1 + B)}{(1 - B)^2}$$

Although the erroneous transfer function is similar to the correct result, it is obviously not the same. In general, continuous processes in series must be combined into one continuous model before the discrete transfer function or difference equation is obtained. □

Example 2.11 Control of a DC motor/amplifier system Figure 2.13(b) shows the physical elements in the rotary-table positioning system illustrated in Fig. 2.13(a). A DC motor and amplifier are used to drive the rotary table, and a potentiometer is used to provide a position feedback signal that can be sampled by a control computer. This discrete control system computes the error between the position of the rotary table and the commanded table position and outputs manipulation voltages to the amplifier that are proportional to the error in position. A tachometer is incorporated in the motor/amplifier system.

The variables and constants that define the elements of the system are as follows:

Variables

$v(t)$ = amplifier input voltage (volts)

$e(t)$ = amplifier output voltage (volts)

$i(t)$ = motor current (amps)

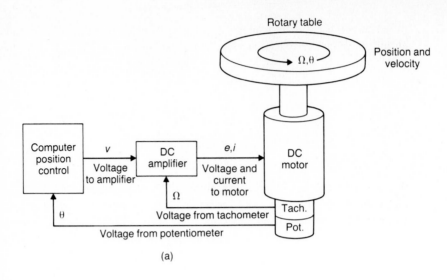

(a)

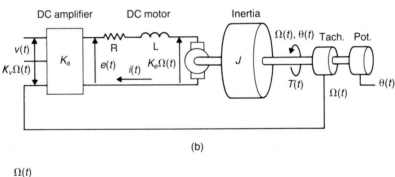

(b)

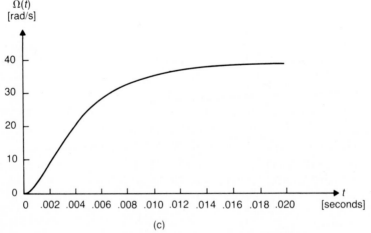

(c)

Figure 2.13 (a) Computer-controlled rotary table driven by DC motor/amplifier;
(b) motor/amplifier model for rotary-table positioning system; (c) response of
table velocity to a 1-volt step in amplifier input $v(t)$;

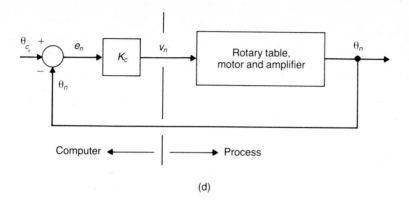

(d)

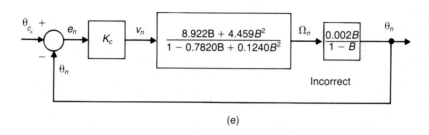

(e)

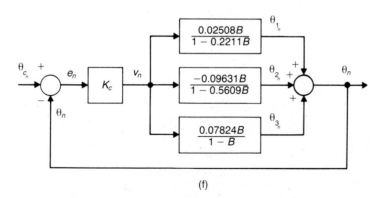

(f)

Figure 2.13 (d) proportional computer control for rotary-table position; (e) *incorrect* closed-loop block diagram for sample period $T = 0.002$ seconds [$\Omega(t)$ is *not constant* between samples]; (f) block diagram resulting from the decomposition of a third-order process model into the sum of three first-order components [$v(t)$ is *constant* between samples];

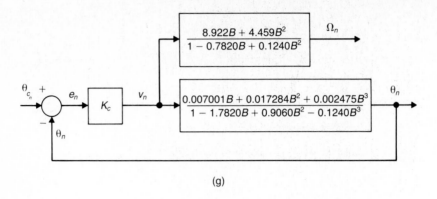

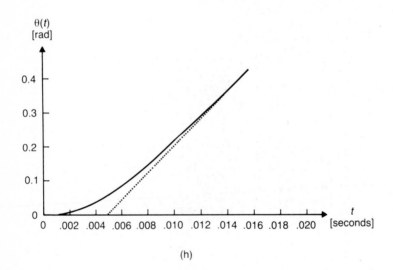

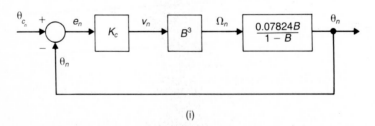

Figure 2.13 (g) block diagram *correctly* showing position and velocity transfer functions; (h) response of table position to a step in amplifier input $v(t)$ of 1 volt; (i) simplified closed-loop block diagram approximating the velocity dynamics of the system by a delay of three sample periods.

$T(t)$ = motor torque (N-m)

$\Omega(t)$ = motor and table velocity (rad/s)

$\theta(t)$ = motor and table position (rad)

Parameters

K_a = amplifier gain (119 volts/volt)

R = armature resistance (1.67 ohms)

L = armature inductance (1.6×10^{-3} henrys)

K_e = back-emf constant (4.32×10^{-2} volts-s/rad)

K_t = torque constant (4.2×10^{-2} N-m/amp)

J = motor and table inertia (36.6×10^{-5} kg-m²)

K_v = tachometer gain (2.52×10^{-2} volts-s/rad)

The equations that define the interaction between elements of the system are

$$e(t) = Ri(t) + L\frac{di(t)}{dt} + K_e\Omega(t)$$

$$T(t) = K_t i(t)$$

$$J\frac{d\Omega(t)}{dt} = T(t)$$

$$\Omega(t) = \frac{d\theta(t)}{dt}$$

$$e(t) = K_a[v(t) - K_v\Omega(t)]$$

Combining these equations results in the following differential equation, which relates the table velocity to the amplifier input voltage.

$$\frac{LJ}{(K_e + K_aK_v)K_t}\frac{d^2\Omega(t)}{dt^2} + \frac{RJ}{(K_e + K_aK_v)K_t}\frac{d\Omega(t)}{dt} + \Omega(t) = \frac{K_a}{(K_e + K_aK_v)}v(t)$$

Substituting the values of the system parameters into this equation yields

$$4.582 \times 10^{-6}\frac{d^2\Omega}{dt^2} + 4.782 \times 10^{-3}\frac{d\Omega}{dt} + \Omega(t) = 39.12v(t)$$

A plot of the response of this system to a step of 1 volt in $v(t)$ is shown in Fig. 2.13(c). The system is overdamped with a damping ratio of 1.1. Entry 6a in

Table 2.1 can be used to obtain the corresponding discrete transfer function, which for sample period $T = 0.002$ second is

$$\frac{\Omega_n}{v_n} = \frac{8.922B + 4.459B^2}{1 - 0.7820B + 0.1240B^2}$$

If it is desired to implement proportional computer control of table position with feedback from the potentiometer, as illustrated in Fig. 2.13(d), the discrete process transfer function relating table position to amplifier input voltage is required. Unfortunately, this transfer function *cannot* be found by using entry 2 Table 2.1 to establish the relationship between table velocity and table position, as indicated in the block diagram in Fig. 2.13(e). This approach is incorrect because table velocity *is not constant between samples.* Instead, we must find the discrete process transfer function by using the differential equation relating table position to amplifier input voltage. This equation is

$$\frac{LJ}{(K_e + K_a K_v)K_t}\frac{d^3\theta(t)}{dt^3} + \frac{RJ}{(K_e + K_a K_v)K_t}\frac{d^2\theta(t)}{dt^2} + \frac{d\theta(t)}{dt} = \frac{K_a}{(K_e + K_a K_v)}v(t)$$

There are a number of ways to solve this problem, two of which will be described here. Table 2.1 contains transfer functions for first-order and second-order systems. One approach, therefore, is to rewrite the foregoing third-order differential equation as a sum of lower-order differential equations.[2] The result of expanding it into the sum of three first-order differential equations and substituting the values of the system parameters is

$$0.001325\frac{d\theta_1(t)}{dt} + \theta_1(t) = 0.03219v(t)$$

$$0.003459\frac{d\theta_2(t)}{dt} + \theta_2(t) = -0.2193v(t)$$

$$\frac{d\theta_3(t)}{dt} = 39.12v(t)$$

with

$$\theta(t) = \theta_1(t) + \theta_2(t) + \theta_3(t)$$

The right-hand side of each of these differential equations is the amplifier input voltage that is generated by the computer and hence is constant between

---

[2] This corresponds to partial fraction expansion, which is discussed in Section 3.3 and in more detail in Section 11.4. It is possible to verify the validity of the resulting set of equations by recombining them by using either the Laplace transformation described in Section 11.2 or the differential operator defined by

$$D^j y(t) \equiv d^j y(t)/dt^j$$

samples. Using Table 2.1, we can find corresponding discrete transfer functions for each of these differential equations:

$$\frac{\theta_{1_n}}{v_n} = \frac{0.02508B}{1 - 0.2211B}$$

$$\frac{\theta_{2_n}}{v_n} = \frac{-0.09631B}{1 - 0.5609B}$$

$$\frac{\theta_{3_n}}{v_n} = \frac{0.07824B}{1 - B}$$

Also,

$$\theta_n = \theta_{1_n} + \theta_{2_n} + \theta_{3_n}$$

The resulting block diagram is shown in Fig. 2.13(f).

The transfer functions that we have just found can be combined (they are in parallel rather than in series) to obtain the process transfer function, which is

$$\frac{\theta_n}{v_n} = \frac{0.007001B + 0.017284B^2 + 0.002475B^3}{1 - 1.7820B + 0.9060B^2 - 0.1240B^3}$$

This result is shown in the block diagram in Fig. 2.13(g), which also contains the discrete velocity transfer function. This block diagram correctly shows the discrete position and velocity models of the system, unlike the incorrect block diagram in Fig. 2.13(e). The closed-loop transfer function for this system with proportional control gain K_c is

$$\frac{\theta_n}{\theta_{c_n}} = \frac{K_c(0.007001B + 0.01784B^2 + 0.002475B^3)}{1 - (1.7820 - 0.007001K_c)B + 8(0.9060 + 0.017284K_c)B^2 - (0.1240 - 0.002475K_c)B^3}$$

Process models tend to become unwieldy as their order increases, and it is often desirable to obtain a simplified, lower-order model that approximates the process dynamics.[3] Figure 2.13(h) shows the response of table position to a step in amplifier input voltage. This response can be approximated by the asymptote shown in the figure, and the process can therefore be approximated by the equation

$$\frac{d\theta(t)}{dt} \approx 39.12v(t - 0.005)$$

which represents the motor and amplifier by a delay that is followed by the integration of velocity to position. The delay of 0.005 second corresponds to

[3] This topic is addressed in Section 10.3.

delay $d = 2.5$ sample periods with the sample period of 0.002 second. A conservative model is obtained by rounding the delay up to $d = 3$. Using entries 1 and 2 in Table 2.1 reveals the discrete transfer function corresponding to the foregoing equation to be

$$\frac{\theta_n}{v_n} \approx \frac{0.07824B^4}{1 - B}$$

The resulting block diagram is shown in Fig. 2.13(i). (This diagram is valid because both the input and the output of the delay are constant between samples.) The closed-loop transfer function for this approximate system model is

$$\frac{\theta_n}{\theta_{c_n}} \approx \frac{K_c(0.07824)B^4}{1 - B + K_c(0.07824)B^4}$$

The corresponding approximate closed-loop difference equation is

$$\theta_n \approx \theta_{n-1} - K_c(0.07824)\theta_{n-4} + K_c(0.07824)\theta_{c_{n-4}} \quad \square$$

2.5
THE SOLUTION OF DIFFERENCE EQUATIONS

Just as there is a general approach for solving linear differential equations, there is a general approach for solving linear difference equations with constant coefficients. Such equations have the form

$$y_n - a_1 y_{n-1} - a_2 y_{n-1} - \cdots - a_{i-1}y_{n-i+1} - a_i y_{n-i} = f(n) \quad (2.28)$$

in which the a's are the constant coefficients, y_n is the response, and $f(n)$ is the forcing function. The order of the difference equation is i.

A homogeneous difference equation is one in which the forcing function $f(n)$ is zero. Consider first the solutions for first-order, second-order, and higher-order homogeneous equations.

First-Order Equations

The general form of the **first-order homogeneous difference equation** is

$$y_n - a_1 y_{n-1} = 0 \tag{2.29}$$

With the introduction of the B operator defined by Eq. (2.22), Eq. (2.29) can be written as

$$y_n - a_1 B y_n = (1 - a_1 B)y_n = 0 \tag{2.30}$$

Because y_n is the sought-after solution and is not, in general, equal to zero,

$$1 - a_1 B = 0 \tag{2.31}$$

Equation (2.31) is the characteristic equation expressed in terms of the operator B. The root r of this algebraic equation is

$$r = \frac{1}{a_1} \tag{2.32}$$

The general solution for homogeneous first-order difference equations is of the form

$$y_n = Cr^{-n} \tag{2.33}$$

or

$$y_n = C\left(\frac{1}{a_1}\right)^{-n} \tag{2.34}$$

where C is a constant that can be determined from a known initial condition. This solution can be checked by substitution into Eq. (2.29). First, from Eq. (2.34),

$$y_{n-1} = C\left(\frac{1}{a_1}\right)^{-(n-1)} \tag{2.35}$$

Substitution of Eqs. (2.34) and (2.35) into Eq. (2.29) yields

$$C\left(\frac{1}{a_1}\right)^{-n} - a_1 C\left(\frac{1}{a_1}\right)^{-n}\left(\frac{1}{a_1}\right) = 0 \tag{2.36}$$

which verifies the solution obtained.

Example 2.12 Solution of a first-order equation   Solve the first-order homogeneous difference equation

$$y_n - 2y_{n-1} = 0$$

Using the B operator, we find the characteristic equation to be

$$1 - 2B = 0$$

and that the single root is

$$r = \tfrac{1}{2}$$

From Eq. (2.34), the solution is

$$y_n = C\left(\frac{1}{2}\right)^{-n} = C2^n$$

where the constant C is determined from initial conditions. Note that regardless of the initial conditions, this equation represents an unstable condition because y_n increases without bound for $n = 1, 2, 3, \ldots.$ □

Second-Order Equations

The general form of the **second-order homogeneous difference equation** is

$$y_n - a_1 y_{n-1} - a_2 y_{n-2} = 0 \tag{2.37}$$

Again introducing the B operator,

$$y_n - a_1 B y_n - a_2 B^2 y_n = 0 \tag{2.38}$$

for which the characteristic equation is

$$1 - a_1 B - a_2 B^2 = 0 \tag{2.39}$$

The two roots of Eq. (2.39) are obtained from

$$r_{1,2} = \frac{a_1 \pm \sqrt{a_1^2 + 4a_2}}{-2a_2} \tag{2.40}$$

There are three possible cases for the values of these roots, depending on the value of the quantity under the radical: The two roots can be real and unequal, real and equal, or complex conjugates.

    Case 1   Real and unequal roots, $a_1^2 + 4a_2 > 0$ If the quantity under the radical in Eq. (2.40) is positive, the roots of the characteristic equation are real and unequal. In this case, the solution has the form

$$y_n = C_1 r_1^{-n} + C_2 r_2^{-n} \tag{2.41}$$

where roots r_1 and r_2 are obtained from Eq. (2.40). C_1 and C_2 are constants that are determined from initial conditions.

Example 2.13 Solution with unequal real roots   As an example, consider the homogeneous equation

$$y_n - y_{n-1} - 6y_{n-2} = 0$$

Here $a_1 = 1$ and $a_2 = 6$. Upon introduction of the B operator, the characteristic equation is found to be

$$1 - B - 6B^2 = 0$$

The roots are

$$r_1 = -\tfrac{1}{2}$$

and

$$r_2 = \tfrac{1}{3}$$

These roots are real and unequal and, from Eq. (2.41), the solution is therefore

$$y_n = C_1\left(-\frac{1}{2}\right)^{-n} + C_2\left(\frac{1}{3}\right)^{-n}$$

This solution can be verified by substitution into the original equation. □

Case 2 *Real and equal roots,* $a_1^2 + 4a_2 = 0$ For the case in which the quantity under the radical in Eq. (2.40) is zero, the two roots are real and equal, and the solution has the form

$$y_n = C_1 r^{-n} + C_2 nr^{-n} \tag{2.42}$$

where

$$r = \frac{-a_1}{2a_2} \tag{2.43}$$

Example 2.14 Solution with equal real roots Consider the difference equation

$$y_n - 2y_{n-1} + y_{n-2} = 0$$

for which the characteristic equation is

$$1 - 2B + B^2 = 0$$

From Eq. (2.40), the roots of the characteristic equation are determined to be

$$r_1 = r_2 = 1$$

Equation (2.42) is the general solution for the case of real and equal roots. Thus the solution is

$$y_n = C_1(1)^{-n} + C_2 n(1)^{-n} = C_1 + C_2 n \quad □$$

Case 3 *Complex conjugate roots,* $a_1^2 + 4a_2 < 0$ When the quantity under the radical in Eq. (2.40) is negative, the roots of the characteristic equation are a complex conjugate pair. It is useful in this case to rewrite Eq. (2.40) as

$$r_{1,2} = \alpha \pm j\beta \tag{2.44}$$

where $j = \sqrt{-1}$ and

$$\alpha = \frac{-a_1}{2a_2} \tag{2.45}$$

$$\beta = \frac{\sqrt{-(a_1^2 + 4a_2)}}{2a_2} \tag{2.46}$$

The solution in this case is

$$y_n = A_1(\alpha + j\beta)^{-n} + A_2(\alpha - j\beta)^{-n} \tag{2.47}$$

where A_1 and A_2 are constants and are complex conjugates.
    This solution can be expressed more conveniently as

$$y_n = |r|^{-n}(C_1 \sin \theta n + C_2 \cos \theta n) \tag{2.48}$$

where C_1 and C_2 are real constants and

$$|r| = (\alpha^2 + \beta^2)^{1/2} \tag{2.49}$$

$$\theta = \tan^{-1}(\beta/\alpha) \tag{2.50}$$

Alternatively, Eq. (2.48) can be modified to obtain

$$y_n = C|r|^{-n} \sin(\theta n + \phi) \tag{2.51}$$

where

$$C = (C_1^{\,2} + C_2^{\,2})^{1/2} \tag{2.52}$$

$$\phi = \tan^{-1}\!\left(\frac{C_2}{C_1}\right) \tag{2.53}$$

Example 2.15 Solution with complex roots   Solve the difference equation

$$y_n - 2y_{n-1} + 2y_{n-2} = 0$$

The characteristic equation in this case is

$$1 - 2B + 2B^2 = 0$$

and the roots are

$$r_{1,2} = \tfrac{1}{2} \pm j\tfrac{1}{2}$$

With complex roots, the general solution from Eq. (2.51) is

$$y_n = C|r|^{-n} \sin(\theta n + \phi)$$

Here,

$$|r| = \frac{1}{\sqrt{2}}$$

and

$$\theta = \frac{\pi}{4}$$

Thus the solution of the difference equation is

$$y_n = C\left(\frac{1}{\sqrt{2}}\right)^{-n} \sin\left(\frac{\pi}{4}n + \phi\right)$$

This result can be verified by substitution into the original equation. □

Higher-Order Equations

The solutions of many **higher-order homogeneous difference equations** can be obtained as combinations of those appropriate for the first-order and second-order cases. This is illustrated in the following example.

Example 2.16 Third-order solution Consider the homogeneous difference equation

$$y_n - 8y_{n-1} + 21y_{n-2} - 18y_{n-3} = 0$$

The characteristic equation for this difference equation is

$$1 - 8B + 21B^2 - 18B^3 = 0$$

which has the following roots:

$$r_1 = \tfrac{1}{2}$$
$$r_2 = \tfrac{1}{3}$$
$$r_3 = \tfrac{1}{3}$$

The solution is therefore

$$y_n = C_1\left(\frac{1}{2}\right)^{-n} + C_2\left(\frac{1}{3}\right)^{-n} + C_3 n\left(\frac{1}{3}\right)^{-n}$$

or

$$y_n = C_1 2^n + C_2 3^n + C_3 n 3^n \quad \square$$

Consideration of Initial Conditions

Initial conditions are used to evaluate the constants in the solution resulting from an initial state of the system. In general, as many initial conditions must be provided as there are constants to be evaluated. As is illustrated in the following examples, substitution of these values into the difference equation makes possible solution for the constants by algebraic methods.

Example 2.17 First-order equation with initial conditions Determine the value of the constant C for the solution to $y_n - 2y_{n-1} = 0$, given the initial condition $y_0 = 1$. As determined in Example 2.12,

$$y_n = C\left(\frac{1}{2}\right)^{-n}$$

At $n = 0$,

$$y_0 = C$$

We apply the initial condition $y_0 = 1$. The required solution is therefore

$$y_n = \left(\frac{1}{2}\right)^{-n} = 2^n$$

or

$$y_0 = 1, \qquad y_1 = 2, \qquad y_2 = 4, \text{ etc. } \square$$

Example 2.18 Second-order equation with initial conditions   As another example, consider the equation

$$y_n - y_{n-1} - 6y_{n-2} = 0$$

with the initial conditions $y_0 = 0$ and $y_{-1} = -1$. The homogeneous equation was solved earlier (in Example 2.13) and noted to have the solution

$$y_n = C_1\left(-\frac{1}{2}\right)^{-n} + C_2\left(\frac{1}{3}\right)^{-n}$$

At $n = 0$ and $n = -1$,

$$y_0 = C_1 + C_2$$

$$y_{-1} = -\tfrac{1}{2}C_1 + \tfrac{1}{3}C_2$$

Substituting the initial conditions for y_0 and y_{-1} yields

$$C_1 + C_2 = 0$$

$$-\tfrac{1}{2}C_1 + \tfrac{1}{3}C_2 = -1$$

Constants C_1 and C_2 are then determined to be

$$C_1 = \tfrac{6}{5}$$

$$C_2 = -\tfrac{6}{5}$$

and the required solution is

$$y_n = \frac{6}{5}\left(-\frac{1}{2}\right)^{-n} - \frac{6}{5}\left(\frac{1}{3}\right)^{-n}$$

or

$$y_n = \tfrac{6}{5}(-2)^n - \tfrac{6}{5}(3)^n \quad \square$$

Nonhomogeneous Equations—Particular Solution

The general solution of a nonhomogeneous equation is obtained as the sum of the general solution of the corresponding homogeneous equation and a particular solution that is closely related to the forcing function. This sum can be expressed as

$$y_n = (y_n)_h + (y_n)_p \tag{2.54}$$

where $(y_n)_h$ is the solution of the homogeneous equation, and $(y_n)_p$ is the particular solution. Solutions for homogeneous equations have already been discussed. We can now turn our attention to the determination of particular solutions.

A particular solution $(y_n)_p$ is obtained from an assumed solution on the basis of the form of the forcing function. The assumed solution is then substituted into the nonhomogeneous difference equation for the purpose of evaluating certain constants. Table 2.2 provides appropriate assumed solutions for several common forcing functions. Note that for each entry in the table, the assumed solution is similar in form to the forcing function. The method of obtaining the particular solution for a nonhomogeneous difference equation will be illustrated by the examples that follow.

Example 2.19 Constant forcing function   Obtain the particular solution for the equation

$$y_n - y_{n-1} - 6y_{n-2} = 12$$

The forcing function is

$$f(n) = 12$$

From Table 2.2, we know that the assumed solution for a constant forcing function is

$$(y_n)_p = K_0$$

Table 2.2

Assumed Solutions for Obtaining the Particular Solution
y_n (K, K_0, K_1, K_2, and A are constants)

FORCING FUNCTION $f(n)$	ASSUMED SOLUTION $(y_n)_p$
K	K_0
Kn	$K_1 n + K_0$
Kn^2	$K_2 n^2 + K_1 n + K_0$
$K\sin(An)$	$K_1 \sin(An) + K_2 \cos(An)$
KA^n	$K_0 A^n$

Also,

$$(y_{n-1})_p = K_0$$

and

$$(y_{n-2})_p = K_0$$

Substituting into the difference equation yields

$$K_0 - K_0 - 6K_0 = 12$$

Therefore

$$K_0 = -2$$

The particular solution is then

$$(y_n)_p = -2$$

which is a solution that satisfies the original nonhomogeneous difference equation. □

Example 2.20 Nonconstant forcing function   Obtain $(y_n)_p$ for the difference equation

$$y_n - 2y_{n-1} = n$$

Here the forcing function is $f(n) = n$ and, from the second entry in Table 2.2,

$$(y_n)_p = K_1 n + K_0$$

Also,

$$(y_{n-1})_p = K_1(n-1) + K_0$$

Substitution into the difference equation yields

$$K_1 n + K_0 - 2K_1 n + 2K_1 - 2K_0 = n$$

and equating the coefficients of like terms yields

$$K_1 n - 2K_1 n = n$$
$$K_0 + 2K_1 - 2K_0 = 0$$

Thus

$$K_1 = -1$$
$$K_0 = -2$$

and

$$(y_n)_p = -(n+2) □$$

Nonhomogeneous Equations—Complete Solution

We use Eq. (2.54) to find the complete solution to a difference equation by adding the particular solution to the homogeneous solution. Then we evaluate any constants in the resulting complete solution by using initial conditions. These constants cannot, in general, be evaluated using only the homogeneous solution.

Two examples are now offered to illustrate the procedure for obtaining complete solutions to nonhomogeneous difference equations.

Example 2.21 Complete solution of first-order equation   Consider the equation

$$y_n - 2y_{n-1} = 1$$

with the initial condition $y_0 = 0$. The difference equation is first-order, and the homogeneous equation is $y_n - 2y_{n-1} = 0$. The homogeneous solution of this equation is

$$(y_n)_h = C2^n$$

The forcing function is a constant and, from the first entry in Table 2.2, we obtain the particular solution

$$(y_n)_p = -1$$

The complete solution is the sum of the homogeneous and particular solutions:

$$y_n = (y_n)_h + (y_n)_p = C2^n - 1$$

With the initial condition $y_0 = 0$,

$$y_0 = C2^0 - 1 = 0$$

$$C - 1 = 0$$

$$C = 1$$

Thus the complete solution is

$$y_n = 2^n - 1$$

Note that the value of the constant C is evaluated *after* the homogeneous and particular solutions are combined in order to include all of the information in the complete solution.

The foregoing solution can be checked by substituting the solution into the difference equation. The result is

$$(2^n - 1) - 2(2^{n-1} - 1) = 1 \quad \square$$

Example 2.22 Difficulty in form of particular solution   The following example illustrates a difficulty that arises in assuming the form of the particular solution. Consider the equation

$$y_n - y_{n-1} = 1$$

The homogeneous solution is

$$(y_n)_h = C$$

The assumed particular solution $(y_n)_p = K_0$ cannot be used, because this choice has the same form as the homogeneous solution. The assumed particular solution is therefore multiplied by n to eliminate the duplication. Thus

$$(y_n)_p = K_0 n$$

and

$$(y_{n-1})_p = K_0(n-1)$$

Upon substitution into the difference equation, we find that

$$K_0 n - K_0 n + K_0 = 1$$

Therefore

$$K_0 = 1$$

and

$$(y_n)_p = n$$

The complete solution, then, is

$$y_n = (y_n)_h + (y_n)_p = C + n$$

If the initial condition is $y_0 = 0$, then

$$y_0 = C + 0 = 0$$

and

$$C = 0$$

The required solution is therefore

$$y_n = n \quad \square$$

From these examples, the reader should conclude that once a difference equation has been established for a system, a solution can be obtained for the response of the system to a defined forcing function starting from prescribed initial conditions. Noting the parallelism between the solution of difference equations and the solution of differential equations should enhance our understanding of the dynamics of discrete systems. System stability, a closely related topic, is discussed in the next section. The key to grasping the basic concept of system stability lies in understanding the solution of difference equations and the nature of the roots of the solution.

2.6

STABILITY ANALYSIS

Stability analysis is based on the roots of the characteristic equation. With real roots, the form of solutions for single, nonrepeated roots is

$$y_n = C_1 r_1^{-n} + C_2 r_2^{-n} + \cdots \tag{2.55}$$

For complex roots, the form is

$$y_n = C|r|^{-n} \sin(\theta n + \phi) + \cdots \tag{2.56}$$

In both cases, if $|r| < 1$ for any root r, then y_n increases without bound. The criterion for **absolute stability** therefore is that $|r| > 1$ so that r^{-n} will converge to zero.

Note that repeated roots pose no difficulty, provided that $|r| > 1$, because r^{-n} will approach zero. There are cases wherein there may be one or more roots for which $|r| = 1$. The presence of one root $r = 1$ leads to a constant term in the solution given; this results in **limited stability**. The presence of more than one root $r = 1$ results in **instability**. For example, if $r_1 = r_2 = 1$, then from Eq. (2.42),

$$y_n = C_1 + C_2 n + \cdots \tag{2.57}$$

and y_n increases without bound.

The basic mathematical theory of linear difference equations with constant coefficients makes it possible to relate stability to the roots of the characteristic equation associated with the homogeneous form of the difference equation. In general, the characteristic equation corresponding to Eq. (2.28) has the form

$$1 - a_1 B - a_2 B^2 - \cdots - a_{k-1} B^{k-1} - a_k B^k = 0 \tag{2.58}$$

A necessary and sufficient condition for absolute stability of a kth-order system is

$$|r_i| > 1 \qquad \text{for all } i = 1, 2, \ldots, k \tag{2.59}$$

where $r_1, r_2, \ldots, r_k$ are the roots of Eq. (2.58). Systems that do not satisfy Eq. (2.59) are unstable.

As pointed out earlier, systems with one root of magnitude equal to 1 can be said to exhibit limited stability. These systems do not satisfy the stability criterion and hence are unstable. However, for some inputs these systems may produce bounded outputs; hence the term **limited stability**.

Another and perhaps more convenient way of obtaining the characteristic equation is to equate the denominator of the closed-loop transfer function to zero. From Eq. (2.27), the characteristic equation can be found from

$$1 + G_c(B)G_p(B) = 0 \tag{2.60}$$

If $N(B)$ represents the polynomial in the numerator of $G_c(B)G_p(B)$, and $D(B)$ represents the polynomial in the denominator, then

$$G_c(B)G_p(B) = \frac{N(B)}{D(B)} \tag{2.61}$$

and the characteristic equation is

$$N(B) + D(B) = 0 \tag{2.62}$$

Example 2.23 Stability of a first-order system The closed-loop transfer function found in Example 2.8 was

$$\frac{c_n}{r_n} = \frac{(1.5 - B)B}{1 - 0.5B}$$

The characteristic equation can be found directly from the denominator of this transfer function as

$$1 - 0.5B = 0$$

This equation has only one root

$$r_1 = 2$$

The system satisfies Eq. (2.59) and therefore is stable. $\square$

Example 2.24 Stability of rotary-table system The computer-controlled rotary-table system shown in Fig. 2.13(a) was analyzed in Example 2.11, resulting in the closed-loop transfer function

$$\frac{\theta_n}{\theta_{c_n}} = \frac{K_c(0.007001B + 0.017284B^2 + 0.002475B^3)}{1 - (1.7820 - 0.007001K_c)B + (0.9060 + 0.01728K_c)B^2 - (0.1240 - 0.002475K_c)B^3}$$

where K_c is the gain of the proportional position controller. The characteristic equation for this system is therefore

$$1 - (1.7820 - 0.007001K_c)B + (0.9060 + 0.017284K_c)B^2 - (0.1240 - 0.002475K_c)B^3 = 0$$

The stability of this system depends on what value is chosen for K_c. The three cases that follow represent choices of relatively low, relatively high, and excessively high values of K_c. Not only do the roots of the characteristic equation change as K_c is increased (eventually resulting in instability), but the response of the closed-loop system also changes significantly—from overdamped, to oscillatory, to unstable—as K_c is increased.

Case 1 $K_c = 0.8$ For $K_c = 0.8$ the characteristic equation is

$$1 - 1.776B + 0.9198B^2 - 0.1220B^3 = 0$$

which has the roots

$$r_1 = 1.08$$

$$r_2 = 1.54$$

$$r_3 = 4.92$$

The magnitude of each of these roots is greater than 1, so the system is stable. It can be seen in Fig. 2.14(a) that for this relatively low gain, the system does not oscillate (all of the roots are real) when subjected to a step change in position command input.

Case 2 $K_c = 3$ For $K_c = 3$ the characteristic equation is

$$1 - 1.7610B + 0.9578B^2 - 0.1166B^3 = 0$$

which has the roots

$$r_1 = 1.16 + 0.34j$$

$$r_2 = 1.16 - 0.34j$$

$$r_3 = 5.90$$

The magnitude of roots r_1 and r_2 is

$$|r_{1,2}| = \sqrt{1.16^2 + 0.34^2} = 1.21$$

Because the magnitude of each of the roots is greater than 1, the system is stable. The gain is relatively high in this case, and the effect can be seen in Fig. 2.14(b) where it is apparent that the response of the system to a step input is more rapid than that for the lower gain shown in Fig. 2.14(a). The response is slightly oscillatory (two of the roots are complex), but it settles rapidly to its final value.

Case 3 $K_c = 14.0$ For $K_c = 14.0$ the characteristic equation is

$$1 - 1.684B + 1.148B^2 - 0.08934B^3 = 0$$

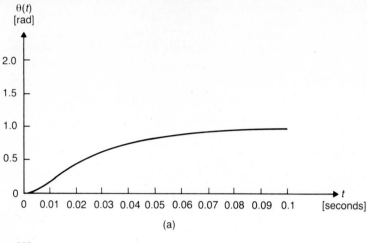

(a)

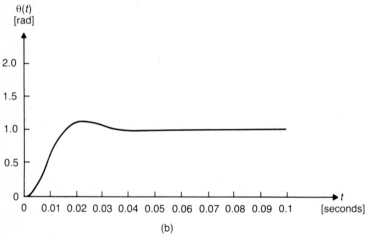

(b)

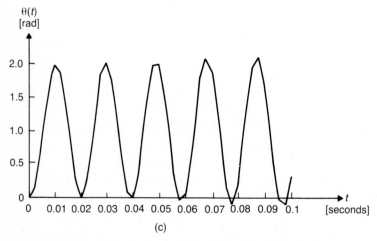

(c)

Figure 2.14 Response of system in Figure 2.13(a) to step position command with $T = 0.002$ second and (a) $K_c = 0.8$; (b) $K_c = 3.0$; and (c) $K_c = 14$.

The roots of this equation are

$$r_1 = 0.79 + 0.60j$$

$$r_2 = 0.79 - 0.60j$$

$$r_3 = 11.26$$

In this case the magnitude of roots r_1 and r_2 is

$$|r_{1,2}| = \sqrt{0.79^2 + 0.60^2} = 0.997$$

which is less than 1. Therefore the system is unstable. This fact can be verified in Fig. 2.14(c); the step response for this gain does not reach a final constant value. Instead, it oscillates rapidly with a magnitude that continues to grow with time. □

SUMMARY

This chapter has introduced some of the basic aspects of discrete modeling and response computation that are useful for the analysis and design of digital computer control systems. Of primary importance is the table of transfer functions, Table 2.1, which enables us to determine the discrete transfer function for a process without going through a lengthy derivation procedure. We can derive discrete equations describing processes with higher-order models that are not in Table 2.1 by using partial fraction expansion, a technique that is discussed in Chapters 3 and 11.

The solution of difference equations was presented in order to demonstrate how one can compute the response of a system and to develop an understanding of stability. These concepts are extended in Chapter 3, which treats the response of discrete systems to various inputs. Using the recursion method involves successively calculating process outputs, given input and output values at current and previous sample periods. Entire systems can be simulated by implementing difference equations in computer software and calculating system response using the recursion method.

The subject of computer control algorithms, discussed briefly in this chapter, is treated in more detail in Chapter 4, which is concerned with discrete controller design. Implementation of computer control algorithms is covered in subsequent chapters: Software is discussed in Chapter 5, interfacing in Chapter 6, and sensors for feedback data acquisition in Chapter 7.

The time-domain approach to discrete process and closed-loop control system modeling enables us to develop a representation of a system that is based on time series—fundamentally the same series of data that would be sampled and generated by a control computer in the system. An alternative approach to representing and designing discrete control systems is the z-transform approach presented in Chapters 11 and 12. The need to introduce such mathematical concepts as ideal samples, impulse trains, and zero-order holds can detract from the basic simplicity of the actions of closed-loop computer control systems. Accordingly, they are not introduced until these later chapters.

BIBLIOGRAPHY

Astrom, K. J., and B. Wittenmark. *Computer Controlled Systems: Theory and Design.* Englewood Cliffs, NJ: Prentice-Hall, 1984.

Cadzow, J. A., and H. R. Martens. *Discrete Time and Computer Control Systems.* Englewood Cliffs, NJ: Prentice-Hall, 1970.

Close, C. M., and D. K. Frederick. *Modeling and Analysis of Dynamic Systems.* Boston, MA: Houghton-Mifflin, 1978.

Doebelin, E. O. *Systems Modeling and Response.* New York, NY: Wiley, 1980.

Dorf, R. C. *Modern Control Systems, 4/e.* Reading, MA: Addison-Wesley, 1986.

Franklin, G. F., and J. D. Powell. *Digital Control of Dynamic Systems.* Reading, MA: Addison-Wesley, 1981.

Grimaldi, R. *Discrete and Combinatorical Mathematics.* Reading, MA: Addison-Wesley, 1985.

Harrison, H. L., and J. G. Bollinger. *Introduction to Automatic Controls.* New York, NY: Harper & Row, 1969.

Kuo, B. C. *Automatic Control Systems, 5/e.* Englewood Cliffs, NJ: Prentice-Hall, 1987.

PROBLEMS

2.1 A vertically oriented, cylindrical water tank has an area of 10 m². When the valve at its inlet is opened one revolution, it allows 0.1 m³ of water to flow into the tank per second. Find a discrete model that relates height of water in the tank (in meters) to inlet valve position (in revolutions). Identify any significant limitations in the model.

2.2 The positioning table shown in Fig. 2.15 can be modeled using the following equations:

$$0.02 \frac{d^2\theta(t)}{dt^2} + \frac{d\theta(t)}{dt} = 10v(t)$$

$$x(t) = 5\theta(t)$$

Find the difference equation relating output $x(t)$ to input $v(t)$ for a sample period of $T = 0.01$ s.

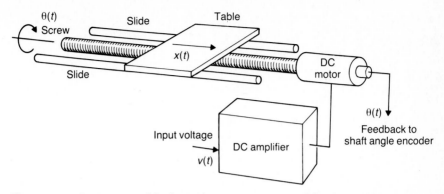

Figure 2.15 Positioning table driven by a DC motor/amplifier for Problem 2.2.

2.3   The water tank illustrated in Fig. 2.16 has a manually operated outlet chute and a computer-controlled inlet valve. The diameter of the tank is 20 ft. The valve allows water to flow into the tank at a rate of 4 ft³/s when it is fully opened. Flow out of the tank, $q_o(t)$, is to be modeled as a disturbance and is nearly constant over any given sample period. For a sample period of 0.5 s, find the difference equation describing height of water in the tank $h(t)$ as a function of percent valve opening $p(t)$ and disturbance $q_o(t)$.

2.4   Following the general procedure used in Section 2.2, obtain the discrete model for a process defined by the equation

$$\tau \frac{d^2 y(t)}{dt^2} + \frac{dy(t)}{dt} = K x(t)$$

Check the result with that given in entry 5 of Table 2.1.

Figure 2.16 Water tank with computer-controlled inlet valve for Problem 2.3.

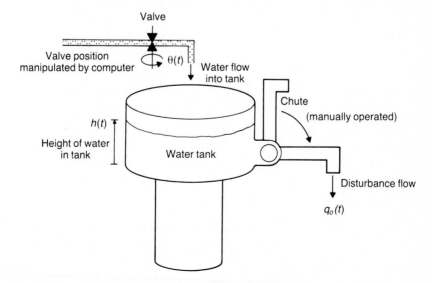

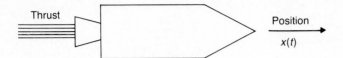

Figure 2.17 Spacecraft for Problem 2.6.

2.5 A double-integration process is initially at rest with $y(0) = \dot{y}(0) = 0$. A step input of amplitude 2 is applied. Thus

$$\frac{d^2 y(t)}{dt^2} = 2$$

Solve this equation and evaluate $y(t)$ for values of $t = 0, 1, 2, \ldots$. Obtain the discrete model with the assumption that $T = 1$. Use the recursion method to find values at $n = 0, 1, 2, \ldots$, and compare the results with values from the continuous case to show that the discrete model yields exact values at the sample instants.

2.6 The spacecraft shown in Fig. 2.17 has a mass of 1000 kg and an engine that generates a thrust of 50,000 N. Develop a discrete model for spacecraft position (in kilometers) for a sample period of 0.001 s. The input should be a variable $m(t)$ that will be restricted in practice to either 1 (when the engine is on) or 0 (when the engine is off.)

2.7 The product being delivered in an oil pipeline is periodically changed. (Products are fuel oil, regular gasoline, premium gasoline, diesel #1, diesel #2, and so on.) The pipeline is 950 km long, the flow rate is 2.8 m³/s, the area of the pipe is 0.5 m², and the sample period is 0.1 s. Develop a difference equation relating the product inserted at the pipeline input, $p_i(t)$, to the product arriving at the pipeline output, $p_o(t)$. Assume plug flow.

2.8 The conveyor shown in Fig. 2.18 is used to move manufactured parts between two machines. A sensor located at the midpoint of the conveyor checks the dimensional accuracy of each part produced. Parts are placed on the conveyor immediately after production at 10-s intervals, and they take 120 s to travel the length of the conveyor. Obtain a discrete model describing the delay between the production of a part and the sensing of its dimensional accuracy.

Figure 2.18 Conveyor with part-dimension sensor at midpoint of travel for Problem 2.8.

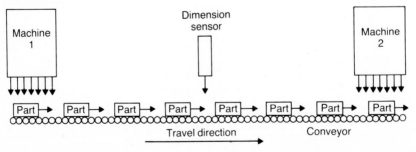

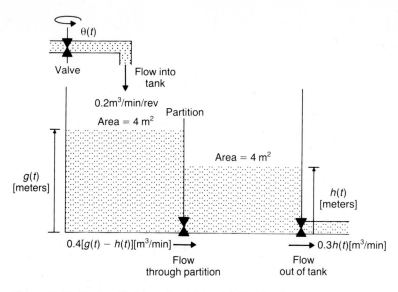

Figure 2.19 Tank with internal partition for Problem 2.9.

2.9 Consider the tank shown in Fig. 2.19.
   a. Obtain a differential equation that models the process with input $\theta(t)$ and output $h(t)$.
   b. Obtain a difference equation that models the process.
   c. Use the difference equation to calculate and plot the response to a step input of one valve revolution for a sample period of 0.5 min. Assume zero initial conditions.
   d. Plot the solution of the differential equation for the same input.
   e. Compare the results.

2.10 Calculate and plot the response of process output $h(t)$ shown in Fig. 2.19 to the input shown in Fig. 2.20. The sample period is 0.5 min. Assume zero initial conditions.

Figure 2.20 Input to system in Figure 2.19 for Problem 2.10.

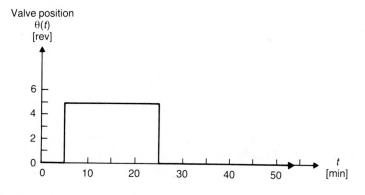

2.11 In Example 2.2, several values of h_n are obtained using the recursion method from

$$h_n = 0.368h_{n-1} + 0.1264\theta_{n-1}$$

on the basis that $h_0 = 0$ and $\theta_n = 10°$ for $n \geq 0$. Solve this equation by the techniques presented in Section 2.5. Compare numerical results with those obtained in Example 2.2.

2.12 A process is described by the following set of equations:

$$c(t) = ky(t)$$

$$\tau \frac{dy(t)}{dt} + y(t) = x(t)$$

$$m(t) = \frac{dx(t)}{dt}$$

where $m(t)$ is the manipulated variable and $c(t)$ is the response. Draw a block diagram for the process with m_n as the input and c_n as the output.

2.13 Draw a block diagram for proportional closed-loop control of the system in Problem 2.12 with r_n as the input and proportional controller gain k_c. Formulate the discrete closed-loop transfer function c_n/r_n.

2.14 A computer control algorithm is given as

$$m_n = m_{n-1} + 1.5e_n - e_{n-1}$$

with sample time T. Determine a continuous control equation for which this is a discrete model. What type of control actions does this algorithm represent?

2.15 A PI (proportional plus integral) motor velocity controller is described by the equation

$$\frac{dm(t)}{dt} = K\left[\tau \frac{de(t)}{dt} + e(t)\right]$$

where $m(t)$ is the output of the motor/amplifier and $e(t)$ is the velocity error. Here τ can be chosen to have the same value as the longest time constant in the differential equation for motor velocity, effectively canceling motor/amplifier dynamics with controller dynamics. Assuming that the appropriate motor/amplifier time constant is 0.016 s and that a discrete controller is to be implemented using a sample period of 0.0004 s, give a discrete control equation that approximates the foregoing PI control equation. K is to remain a constant.

2.16 A chemical process has a continuous reaction rate controller that can be described as being proportional plus integral plus derivative with $K_p = 20$, $k_i = 100$, and $k_d = 5$. The process is to be retrofitted with a computer controller with a sample period of 0.25 s. Develop a control equation to calculate the process manipulation $m(t)$ as a function of the difference between the desired reaction rate $r_d(t)$ and the actual rate sensed $r_a(t)$.

2.17 Draw a block diagram for the tank system shown in Fig. 2.21 with a proportional discrete controller of the form

$$q_n = 0.2e_n$$

The sample period is 0.05 min.

2.18 The block diagram shown in Fig. 2.22(a) has a discrete transfer function $H(B)$ in the feedback loop. What is the equation for the closed-loop transfer function corresponding to Eq. (2.27) for this case? Show that the block diagram in Fig. 2.22(a) can be redrawn as shown in Fig. 2.22(b).

2.19 Give the closed-loop transfer function and closed-loop difference equation for each of the block diagrams in Fig. 2.23.

2.20 Find the discrete transfer function for each of the systems shown in Fig. 2.24.

Figure 2.21 Tank with internal filter for Problem 2.17.

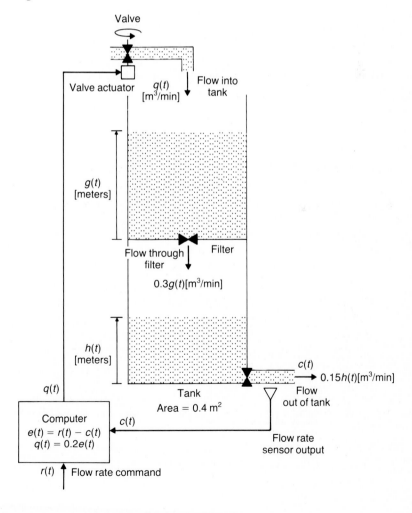

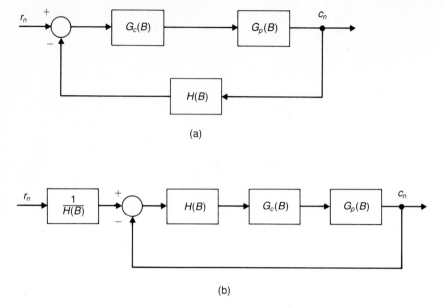

(a)

(b)

Figure 2.22 (a) Block diagram with non-unity feedback for Problem 2.18; (b) reorganized block diagram for Problem 2.18.

2.21 Determine the discrete transfer function c_n/m_n for the process composed of three physical elements that is shown in Fig. 2.25. The sample period is $T = 2$ s, and the manipulated variable is held constant during sample periods.

2.22 A second-order system has a natural frequency of 400 radians/s, a damping ratio of 0.8, and a gain of 50. It is to be controlled using a PID control strategy with an integral gain of 10, a proportional gain of 0.04, and a derivative gain of 6.25×10^{-5}. The sample period is to be 0.002 s. Draw a block diagram for the system and obtain the closed-loop transfer function.

2.23 It takes 0.2 s to process the output of the vision sensor used to determine the relative position of the robot gripper shown in Fig. 2.26 with respect to a hot ingot to be picked up and placed in a quenching tank. The sample period of the discrete proportional controllers to be used to position the arm is 0.02 s, and the transfer functions of the arm-positioning drives can be approximated as integrations with gains K_x and K_y. Draw block diagrams for the x and y axes of the system and determine the corresponding closed-loop transfer functions. What is the steady-state x and y axis error when the proportional controller gains are P_x and P_y; respectively, and the ingots move on the conveyors at a constant velocity V in the negative x direction.

2.24 Show that the decomposition in Example 2.11 of the third-order differential equation for $\theta(t)$ into a sum of three first-order equations $\theta_1(t)$, $\theta_2(t)$, and $\theta_3(t)$ is valid.

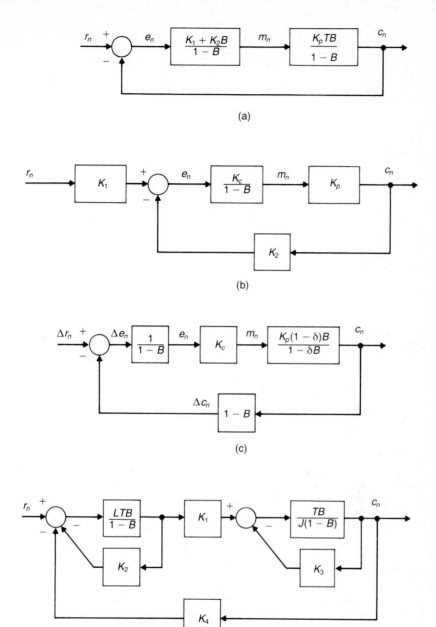

Figure 2.23 Block diagrams for Problem 2.19.

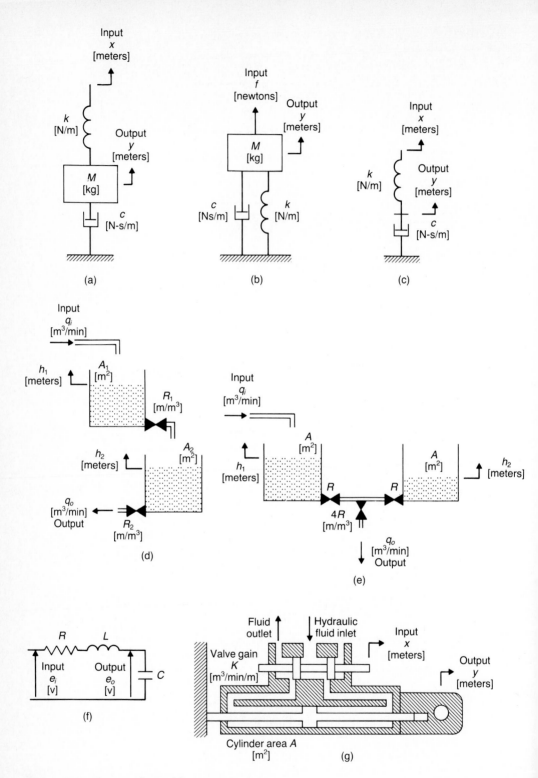

Figure 2.24 Systems for Problem 2.20.

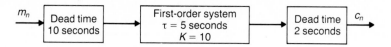

Figure 2.25 Process for Problem 2.21.

2.25 Suppose a load torque disturbance $T_L(t)$ is applied to the rotary table shown in Fig. 2.13(a). In this case, the equation for motor velocity in Example 2.11 becomes

$$J\frac{d\Omega(t)}{dt} = T(t) + T_L(t)$$

If it is assumed that the position command input $\theta_c(t)$ is zero, the response of the closed-loop system can be studied for disturbance inputs.
a. Obtain a complete velocity transfer function and approximate position transfer function for the motor/amplifier (without the position controller).
b. Obtain an approximate closed-loop transfer function for the system with proportional position control.
c. Plot the response to a unit step in a load torque for each of the controller gains investigated in Example 2.24, and compare the results to Figures 2.14(a), (b) and (c).

2.26 Repeat Problem 2.25 without approximating the position transfer function.

2.27 Check the solution of Example 2.12 by substituting the solution into the given difference equation.

Figure 2.26 Robot for manipulating hot ingots in Problem 2.23.

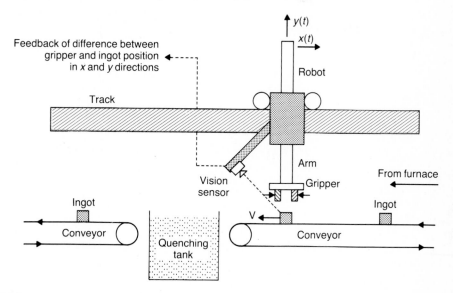

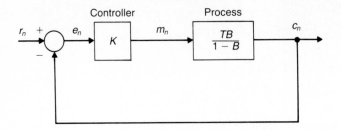

Figure 2.27 Block diagram for Problem 2.30.

2.28 Solve the following homogeneous difference equations.

a. $y_n - 0.75y_{n-1} = 0$

b. $y_n + 1.25y_{n-1} = 0$

c. $y_n - 1.65y_{n-1} + 0.7y_{n-2} = 0$

d. $y_n - 3y_{n-1} + 2.35y_{n-2} = 0$

e. $y_n - 2.44y_{n-1} + 0.88y_{n-2} = 0$

f. $y_n - 2y_{n-1} + 2y_{n-2} - y_{n-3} = 0$

2.29 Determine the stability of the following difference equations.

a. $y_n - 1.6y_{n-1} + 1.13y_{n-2} = 0$

b. $y_n + 0.5y_{n-1} = 2$

c. $y_{n+3} - 2y_{n+2} + 1.5y_{n+1} - 0.5y_n = 0$

d. $y_n = y_{n-1} + 1.05x_n - x_{n-1}$

e. $y_n - 2y_{n-1} + y_{n-2} = 10x_{n-1}$

2.30 For the system shown in the block diagram in Fig. 2.27, show that instability results for $KT > 2$. (Note that in discrete systems, both sample time and gain influence stability.)

2.31 To illustrate the adverse effect of dead time on system stability, suppose that the process described in Problem 2.30 is changed such that it has a dead time equal to one sample time $(D = T)$. The process transfer function then is

$$\frac{c_n}{m_n} = \frac{(TB)B}{1 - B}$$

Show that instability now results for $KT > 1$.

2.32 In Problem 2.30, proportional control is used with an integration process. Show that absolute stability cannot be attained with integral control [Eq. (2.11)].

2.33 It is desired to use a proportional control algorithm to control a process defined as follows:

$$0.1\frac{dc(t)}{dt} + c(t) = 2m(t)$$

Assume that $T = 0.1$.

a. Draw a suitable block diagram.

b. Determine the maximum controller gain possible for absolute stability.

Chapter 3
System Response

INTRODUCTION

The ability to predict the response of a dynamic system to various inputs is very important when one is designing and analyzing control systems. How rapidly a system responds to an input and the nature of the response (Is it oscillatory? Are errors significant?) are features that can be used as criteria in evaluating the performance of a controlled system and as objectives to be achieved in the design of a controller and in the selection of its gains. For discrete systems, the modeling techniques presented in Chapter 2 allow the dynamics of a system to be characterized in terms of difference equations and transfer functions that relate system or process input and output for any input sequence. What is needed is a means of (1) representing an input in a discrete form compatible with the discrete modeling techniques already developed and (2) allowing the system output resulting from that input to be derived. For example, given the transfer function of a building temperature control system, it should be possible to calculate sequences of heating/cooling variables and temperatures inside the building in response to a given sequence of outside air temperatures.

Two fundamental ideas are presented in this chapter. First, any time-varying signal sampled at specific intervals of time can be represented by a sequence of values called a **sample sequence**, which is frequently referred to as a **time series**. For example, if the output of a temperature sensor mounted outside a building is sampled once per hour as indicated in Fig. 3.1, the set of temperature readings obtained is a sample sequence or time series. It will be shown that this sample sequence can be expressed in terms of a **time-shift operator**. Second, an

69

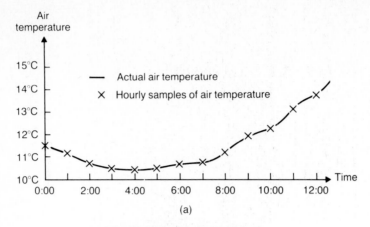

(a)

Sample	Hour [A.M.]	Temperature [°C]
	(midnight)	
0	0:00	11.5
1	1:00	11.2
2	2:00	10.8
3	03:00	10.5
4	4:00	10.4
5	5:00	10.5
6	6:00	10.7
7	7:00	10.8
8	8:00	11.3
9	9:00	11.9
10	10:00	12.3
11	11:00	13.2
12	12:00	13.8
	(noon)	

(b)

Figure 3.1 (a) Temperature of air surrounding a building, and
(b) sequence of hourly samples of air temperature.

expression representing the response of a system to changes in inputs can be
obtained directly using operator techniques in lieu of formulating and solving
difference equations or using the recursion method as described in Chapter 2.
Thus, using discrete transfer functions, we can obtain the response readily once
inputs are specified.

3.2
_____

SYSTEM INPUTS

The ability to describe time-varying signals in a mathematical form that is
analogous to the manner in which the computer handles data is an essential
ingredient in formulating a convenient method for analyzing system response.

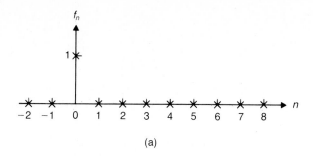

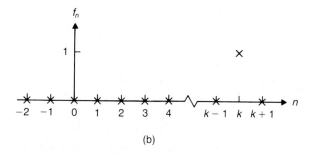

Figure 3.2 Unit sample sequence (a) $f_n = S_n$ and (b) $f_n = S_{n-k}$.

Discrete time sampling of a continuous function produces a sequence or series of numbers that is the only information about the function available to the computer. Consider a sample sequence that is unity at the origin in time ($t = 0$) and is zero elsewhere. As illustrated in Fig. 3.2(a), the **unit sample sequence** S_n is defined as a sequence of numbers such that

$$S_n = \begin{cases} 1 \text{ at } n = 0 \\ 0 \text{ at } n \neq 0 \end{cases} \tag{3.1}$$

where n is a normalized time variable, n is an integer, $n = t/T$, and T is the sample period.

The unit sample sequence may be shifted or translated in time to define a sample taken by the computer at any time $n = k$ by use of the **backward-shift operator** B. Figure 3.2(b) illustrates a sequence that describes a unit sample taken at $n = k$:[1]

$$S_{n-k} = B^k S_n \tag{3.2}$$

---

[1] Recall that a function $f(t)$ delayed by a time D can be represented by $f(t - D)$. Hence a sampled function f_n delayed by d samples where $d = D/T$ and d is an integer is represented by f_{n-d}.

where k is an integer and

$$S_{n-k} = \begin{cases} 1 \text{ at } n = k \\ 0 \text{ at } n \neq k \end{cases} \qquad (3.3)$$

Further, the result of sampling a function $f(t)$ can be described by summing translated and scaled unit sample sequences of the form of Eq. (3.3), which yields

$$f_n = \sum_{k=-\infty}^{\infty} f(kT)S_{n-k} = \sum_{k=\infty}^{\infty} f(kT)B^k S_n \qquad (3.4)$$

where $f(kT)$ is the value of function $f(t)$ at time $t = kT$.

By superimposing appropriately translated and scaled unit sample sequences, we can represent any continuous function in its sampled form. The following examples show how this is done for pulse, step, and ramp functions. These functions are commonly used as inputs in the study of system response.

Example 3.1 Pulse function   A pulse of duration $3T$ is shown in Fig. 3.3(a). If the pulse function is sampled at sample period T, the resulting sequence of samples is as shown in Fig. 3.3(b). We assume that once a sample is taken, the value of the function remains constant until the time of the next sample. Thus in Fig. 3.3(b) the sampled value at time $3T$ is zero, rather than A because the value of the function in the next interval is zero.

Figure 3.3 (a) Pulse of duration $3T$, and (b) sample sequence for pulse of duration $3T$.

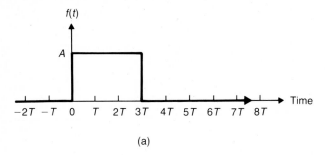

(a)

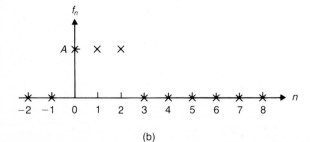

(b)

The sample sequence in Fig. 3.3(b) can be represented as

$$f_n = \cdots + 0S_{n+2} + 0S_{n+1} + AS_n + AS_{n-1} + AS_{n-2} + 0S_{n-3} + 0S_{n-4} + \cdots$$

which can be rewritten as

$$f_n = AS_n + ABS_n + AB^2S_n$$

or

$$f_n = A(1 + B + B^2)S_n$$

The foregoing equation can be referred to as a **generating function**. It is a function of the backward-shift operator B, and it operates on the unit sample sequence S_n to produce a series of samples representing a pulse of duration $3T$. □

Example 3.2 Step function Consider the step function illustrated in Fig. 3.4(a). The sample sequence illustrated in Fig. 3.4(b) is described by

$$f_n = \cdots + 0S_{n+2} + 0S_{n+1} + AS_n + AS_{n-1} + AS_{n-2} + AS_{n-3} + AS_{n-4} + \cdots$$

From Eq. (3.2),

$$f_n = AS_n + ABS_n + AB^2S_n + AB^3S_n + AB^4S_n + \cdots$$

and

$$f_n = A(1 + B + B^2 + B^3 + B^4 + \cdots)S_n$$

Figure 3.4 (a) Step function and (b) sample sequence for step function.

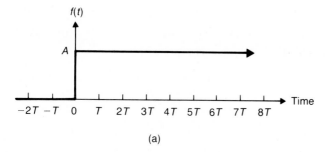

(a)

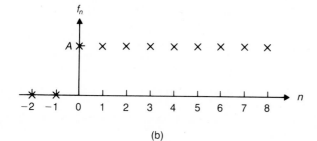

(b)

A more convenient form for expressing this series is

$$f_n = \frac{A}{1 - B} S_n$$

The equivalence of the infinite series $1 + B + B^2 + \cdots$ and its closed form $1/(1 - B)$ can be verified using long division:

$$
\begin{array}{r}
1 + B + B^2 + \cdots \\
1 - B\overline{)1} \\
\underline{1 - B} \\
B \\
\underline{B - B^2} \\
B^2 \\
\underline{B^2 - B^3} \\
B^3 \\
\vdots
\end{array}
$$

The quantity $\dfrac{A}{1 - B} S_n$ is the generating function for the step input. It represents a series of samples with value A for time $t \geq 0$ and value 0 for time $t < 0$. □

Example 3.3 Ramp function   Determine the generating function for a ramp function of slope A, beginning at $t = 0$. Here,

$$f(t) = \begin{Bmatrix} At, & t \geq 0 \\ 0, & t < 0 \end{Bmatrix}$$

Samples taken at integer intervals of time of $0, T, 2T, 3T, \ldots$ are equal to $0, AT, 2AT, 3AT, \ldots$. Thus the sample sequence is given by

$$f_n = 0S_n + ATS_{n-1} + 2ATS_{n-2} + 3ATS_{n-3} + \cdots$$

or

$$f_n = AT(0B^0 + 1B^1 + 2B^2 + 3B^3 + \cdots)S_n$$

The infinite series in the foregoing equation can be written in closed form as

$$\frac{B}{(1 - B)^2} = B + 2B^2 + 3B^3 + \cdots$$

Hence the generating function for the ramp function is

$$f_n = \frac{ATB}{(1 - B)^2} S_n$$

The validity of this closed-form representation of the infinite series can be shown using long division:

$$\begin{array}{r} ATB + 2ATB^2 + 3ATB^3 + \cdots \\ \hline 1 - 2B + B^2 \,\overline{\smash{\big)}\, ATB} \\ ATB - 2ATB^2 + ATB^3 \\ \hline 2ATB^2 - ATB^3 \\ 2ATB^2 - 4ATB^3 + 2ATB^4 \\ \hline 3ATB^3 - 2ATB^4 \\ \vdots \quad \square \end{array}$$

Table of Generating Functions

Table 3.1 lists generating functions for a number of commonly encountered continuous functions. These generating functions are closed-form expressions of algebraic series. Additional functions can be derived as desired for various continuous functions.

Table 3.1

Generating Functions for Selected Continuous Functions

CONTINUOUS FUNCTION, $f(t)$	DISCRETE FUNCTION, f_n, IN TERMS OF UNIT SAMPLE SEQUENCE S_n	GENERATING FUNCTION FOR f_n	GENERAL TERM OF SEQUENCE f_n
1. Unit Sample	$\begin{cases} 1 \text{ for } n = 0 \\ 0 \text{ for } n \neq 0 \end{cases}$	S_n	—
2. $f(t)$	$(f_0 + f_1 B + f_2 B^2 + \cdots)S_n$	$F(B)S_n$	f_n
3. $Kf(t)$	$K(f_0 + f_1 B + f_2 B^2 + \cdots)S_n$	$KF(B)S_n$	Kf_n
4. $u(t)$	$(1 + B + B^2 + \cdots)S_n$	$\left(\dfrac{1}{1-B}\right)S_n$	1
5. t	$T(B + 2B^2 + 3B^3 + \cdots)S_n$	$\left[\dfrac{TB}{(1-B)^2}\right]S_n$	nT
6. t^2	$T(B + 4B^2 + 9B^3 + \cdots)S_n$	$\left[\dfrac{T^2(1+B)B}{(1-B)^3}\right]S_n$	$(nT)^2$
7. e^{-at}	$(1 + AB + A^2B^2 + \cdots)S_n$ where $A = e^{-aT}$	$\left(\dfrac{1}{1-AB}\right)S_n$	A^n
8. $\sin \omega t$	—	$\dfrac{B \sin \omega T}{1 - 2B \cos \omega T + B^2}S_n$	—
9. $\cos \omega t$	—	$\dfrac{1 - B \cos \omega T}{1 - 2B \cos \omega T + B^2}S_n$	—

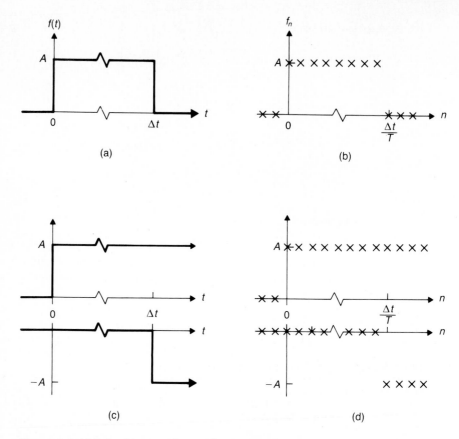

Figure 3.5 (a) Pulse function, (b) sample sequence for pulse function, (c) component step functions, and (d) component sample sequences.

Example 3.4 Pulse generating function   The generating function for the sample sequence of a pulse of length Δt and amplitude A can be obtained readily by using Table 3.1. A pulse is shown in Fig. 3.5(a), and the corresponding sample sequence is given in Fig. 3.5(b). The pulse may be thought of as being composed of two individual step functions: one at time $t = 0$ and an equal negative step function at $t = \Delta t$, as shown in Fig. 3.5(c). Figure 3.5(d) illustrates the two corresponding sample sequences. Thus

$$f_n = A(1 + B + B^2 + \cdots)S_n + (-A)B^\alpha(1 + B + B^2 + \cdots)S_n$$

where $\alpha = \Delta t/T$.

The generating function for a unit step can be found in Item 4 in Table 3.1. The generating function for the pulse can then be obtained by summing the generating functions of the two step functions shown in Fig. 3.5(c). The result is

$$f_n = A\left(\frac{1}{1-B}\right)S_n - AB^\alpha\left(\frac{1}{1-B}\right)S_n$$

or

$$f_n = A\left(\frac{1-B^\alpha}{1-B}\right)S_n$$

If $\alpha = 3$, then the foregoing equation can be simplified to

$$f_n = A(1 + B + B^2)S_n$$

which is identical to the result of Example 3.1. □

3.3

SYSTEM RESPONSE

The response of processes to specific computer manipulations or the response of complete closed-loop control systems to specific commands can be readily determined by using the concept of the generating function. Recall that the discrete transfer function relating the output of a process or system to discrete step changes of an input is

$$\frac{O_n}{I_n} = G(B) \tag{3.5}$$

where $G(B)$ is an algebraic expression in B. Thus the output is

$$O_n = G(B)I_n \tag{3.6}$$

If the nature of I_n is specified, its generating function is of the form

$$I_n = F(B)S_n \tag{3.7}$$

and substituting Eq. (3.7) into Eq. (3.6) yields

$$O_n = G(B)F(B)S_n \tag{3.8}$$

The quantity $G(B)F(B)S_n$ forms a new generating function for the output response O_n. The product $G(B)F(B)$ results in a new function

$$H(B) = G(B)F(B) \tag{3.9}$$

which is also an algebraic expression in B. By expanding $H(B)$ using long division, we can obtain the sample sequence of the output directly.

Example 3.5 Response of computer control system   Suppose we want to move the positioning table shown in Fig. 3.6(a) at a constant speed of 1 mm/s. To accomplish this, a computer is used to manipulate motor velocity in proportion to the error between commanded and actual motor position. If the motor can be approximately modeled by

$$\frac{dx(t)}{dt} = k_m v(t) \qquad \text{mm/s}$$

where $x(t)$ is the position of the table and $v(t)$ is the amplifier input voltage, then

Figure 3.6 (a) Positioning table and drive, (b) block diagram, (c) ramp command input to positioning system, and (d) response of positioning error to ramp input for Example 3.5 $(T = 0.005 \text{ s})$.

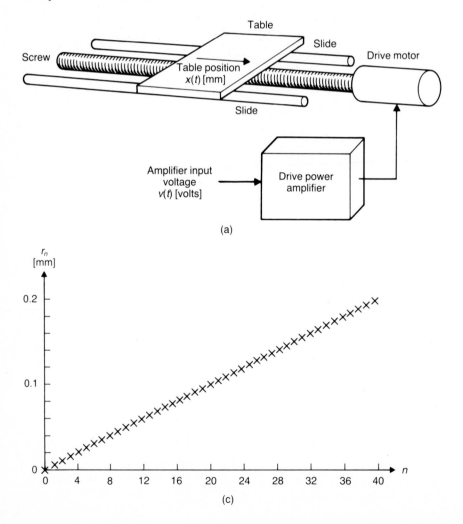

(a)

(c)

the discrete process transfer function from Table 2.1 is

$$\frac{x_n}{v_n} = \frac{k_m TB}{1 - B}$$

A block diagram for the system with proportional control is shown in Fig. 3.6(b). For a sample rate of 200 Hz ($T = 0.005$ s), process gain $k_m = 8.5$, and proportional controller gain $k_c = 2$, the closed-loop transfer function is

$$\frac{x_n}{r_n} = \frac{0.085B}{1 - 0.915B}$$

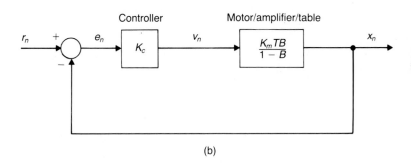

(b)

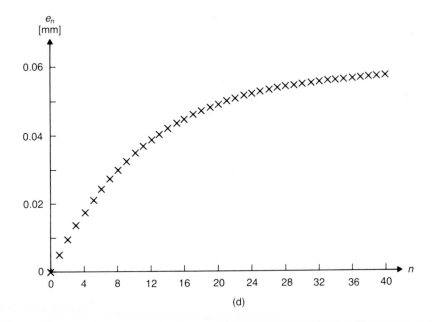

(d)

where r_n is the commanded table position. The transfer function for the error between actual position and commanded position is

$$\frac{e_n}{r_n} = \frac{1 - B}{1 - 0.915B}$$

If the table is to be commanded to move at a constant speed of 1 mm/s, the system input is the ramp function

$$r(t) = t \qquad mm$$

The sample sequence for this ramp function is shown in Fig. 3.6(c). Table 3.1 gives the generating function for this input as

$$r_n = \frac{0.005B}{(1 - B)^2} S_n$$

The response of the table to this ramp command input can be found from Eqs. (3.6) and (3.8), which yield

$$x_n = \frac{0.085B}{1 - 0.915B} r_n$$

and,

$$x_n = \left(\frac{0.085B}{1 - 0.915B} \right) \left(\frac{0.005B}{(1 - B)^2} S_n \right)$$

or

$$x_n = \frac{0.000425B^2}{1 - 2.915B + 2.830B^2 - 0.915B^3} S_n$$

This is the generating function for table position x_n in response to the ramp input. The response sequence can be calculated using long division:

$$
\begin{array}{r}
0.000425B^2 + 0.001239B^3 + 0.002409B^4 + 0.003904B^5 + \cdots \\
\hline
1 - 2.915B + 2.830B^2 - 0.915B^3 \overline{\smash{\big)}\, 0.000425B^2} \\
0.000425B^2 - 0.001239B^3 + 0.001203B^4 - 0.000389B^5 \\
\hline
0.001239B^3 - 0.001203B^4 + 0.000389B^5 \\
0.001239B^3 - 0.003611B^4 + 0.003506B^5 - \cdots \\
\hline
0.002409B^4 - 0.003117B^5 + \cdots \\
0.002409B^4 - 0.007021B^5 + \cdots \\
\hline
0.003904B^5 - \cdots \\
\vdots
\end{array}
$$

Hence, for the ramp input,

$$x_n = (0B^0 + 0B^1 + 0.000425B^2 + 0.001239B^3 + 0.002409B^4 + 0.003904B^5 + \cdots)S_n$$

It is convenient to use a computer program to carry out long division, and such a program is given as Program 3.1.

The error between actual and commanded table position can be found in a similar manner:

$$e_n = \frac{1 - B}{1 - 0.915B} \, r_n$$

$$= \left[\frac{1 - B}{1 - 0.915B} \right]\left[\frac{0.005B}{(1 - B)^2} \, S_n \right]$$

$$= \frac{0.005B}{(1 - 0.915B)(1 - B)} \, S_n$$

Error in response to the ramp input can be obtained from the above equation by long division:

$$
\begin{array}{r}
0.005B + 0.009575B^2 + 0.013761B^3 + 0.017591B^4 + \cdots \\
\hline
1 - 1.915B + 0.915B^2 \overline{)\, 0.005B} \\
0.005B - 0.009575B^2 + 0.004575B^3 \\
\hline
0.009575B^2 - 0.004575B^3 \\
0.009575B^2 - 0.018336B^3 + 0.008761B^4 \\
\hline
0.013761B^3 - 0.008761B^4 \\
0.013761B^3 - 0.026353B^4 + \cdots \\
\hline
0.017591B^4 - \cdots \\
\vdots
\end{array}
$$

Program 3.1 Long division of polynomials.

```
program Long_Division;
{                                    N[0] + N[1]B + D[2]B**2 + ... + N[5]B**5}
{Q[0] + Q[1]B + Q[2]B**2 + ...  =  ----------------------------------------------}
{                                    D[0] + D[1]B + D[2]B**2 + ... + D[5]B**5}
  var
    N, D : array[0..5] of real;                {numerator and denominator arrays}
    Q : real;
    i, j : integer;
begin                                          {input num. and denom. coefficients}
  readln(N[0], N[1], N[2], N[3], N[4], N[5]);
  readln(D[0], D[1], D[2], D[3], D[4], D[5]);
  for i := 0 to 20 do                          {calculate 20 terms of series}
    begin
      Q := N[0] / D[0];                        {quotient is next term in series}
      for j := 1 to 5 do
        N[j - 1] := N[j] - Q * D[j];           {numerator becomes remainder}
      N[5] := 0.0;
      writeln(i, Q)
    end
end.
```

Hence

$$e_n = (0B^0 + 0.005B^1 + 0.009575B^2 + 0.013761B^3 + 0.017591B^4 + \cdots)S_n$$

The coefficients of the series in B are the values of the response at corresponding values of n. The response of the system error for the sampled ramp command input is illustrated in Fig. 3.6(d). ☐

Partial Fraction Expansion

A function that is a ratio of two polynomials can be rewritten as a sum of partial fractions. This form can be useful in finding response sequences, in finding the final value of these sequences, and in reducing higher-order systems to a sum of lower-order systems. If the original function has the form

$$G(B) = \frac{N(B)B^d}{D(B)} \tag{3.10}$$

where

$$D(B) = 1 - a_1 B - a_2 B^2 - a_3 B^3 - \cdots - a_k B^k \tag{3.11}$$

and

$$N(B) = b_0 + b_1 B + b_2 B^2 + b_3 B^3 + \cdots + b_\ell B^\ell \tag{3.12}$$

with $b_0 \neq 0$ and $\ell < k$,[2] then the denominator $D(B)$ in Eq. (3.11) can be written in factored form with roots $r_1, r_2, r_3, \ldots,$ and r_k as

$$D(B) = \left(1 - \frac{1}{r_1}B\right)\left(1 - \frac{1}{r_2}B\right)\left(1 - \frac{1}{r_3}B\right)\cdots\left(1 - \frac{1}{r_k}B\right) \tag{3.13}$$

If there are no repeated roots,[3] then Eq. (3.10) can be rewritten as

$$H(B) = \left[\frac{c_1 B^d}{\left(1 - \frac{1}{r_1}B\right)} + \frac{c_2 B^d}{\left(1 - \frac{1}{r_2}B\right)} + \frac{c_3 B^d}{\left(1 - \frac{1}{r_3}B\right)} + \cdots + \frac{c_k B^d}{\left(1 - \frac{1}{r_k}B\right)}\right] \tag{3.14}$$

or

$$H(B) = \sum_{i=1}^{k} \frac{c_i B^d}{\left(1 - \frac{1}{r_i}B\right)} \tag{3.15}$$

---

[2] If $\ell \geq k$ then Eq. (3.10) can be rewritten as

$$H(B) = P(B) + N'(B)B^d/D(B)$$

where polynomials $P(B)$ and $N'(B)$ are the quotient and the remainder, respectively, obtained by long division of $N(B)$ by $D(B)$, and the order of $N'(B)$ is $\ell' < k$.

[3] A discussion of partial fraction expansion when there are repeated roots is included in Section 11.4, Eqs. (11.40) through (11.43).

where

$$c_i = \left(1 - \frac{1}{r_i}B\right)\frac{N(B)}{D(B)}\Bigg|_{B = r_i} \tag{3.16}$$

Example 3.6 Response from partial fraction expansion   It is noteworthy that, via partial fraction expansion, Table 3.1 can be used to obtain the response sequence directly, eliminating the need for long division. In Example 3.5 the generating function for e_n was determined to be

$$e_n = \frac{0.005B}{(1 - 0.915B)(1 - B)}S_n$$

Via partial fraction expansion,

$$e_n = \left(\frac{c_1 B}{1 - 0.915B} + \frac{c_2 B}{1 - B}\right)S_n$$

Next, we evaluate constants c_1 and c_2

$$c_1 = \frac{0.005}{1 - B}\Bigg|_{B = 1/0.915} = -0.05382$$

$$c_2 = \frac{0.005}{1 - 0.915B}\Bigg|_{B = 1} = 0.05882$$

Therefore,

$$e_n = \left(\frac{-0.05382B}{1 - 0.915B} + \frac{0.5882B}{1 - B}\right)S_n$$

or

$$e_{n+1} = -0.05382\left(\frac{1}{1 - 0.915B}\right)S_n + 0.05882\left(\frac{1}{1 - B}\right)S_n$$

Using the last two columns of Table 3.1, we find that the response is

$$e_{n+1} = 0.05882(1)^n - 0.05382(0.915)^n$$

From this equation, values of e_{n+1} can be found for $n = 0, 1, 2, \ldots$ as follows:

$$e_1 = 0.05882 - 0.05382(0.915)^0 = 0.005 \text{ mm}$$

$$e_2 = 0.05882 - 0.05382(0.915)^1 = 0.009575 \text{ mm}$$

$$e_3 = 0.05882 - 0.05382(0.915)^2 = 0.013761 \text{ mm}$$

$$e_4 = 0.05882 - 0.05382(0.915)^3 = 0.017591 \text{ mm}$$

$$\vdots$$

This result is identical to that which we obtained using long division in Example 3.5. We also observe that the value of e_n approaches 0.05882 mm as n approaches infinity. Thus 0.05882 can be referred to as the **final value** of the series e_n. In this example it can also be referred to as the **steady-state following error,** because the actual table position "follows," or lags behind, the commanded position by this amount when the table is moving at a steady velocity of 1 mm/s. This makes sense, given the control system shown in Fig. 3.6(b), because error is required to generate a change in position. $\square$

Final-Value Theorem

Frequently it is useful to be able to find the final or ultimate value of a response sequence directly from the generating function. Then it is not necessary to determine the sequence explicitly and look for a value of the function at time $t = \infty$ ($n = \infty$) to find the final value of the response.

In general, a generating function is of the form

$$y_n = H(B)S_n \tag{3.17}$$

where $H(B)$ is a ratio of two polynomials. Using Eq. (3.15) we can rewrite this as a sum of response series:

$$y_n = \sum_{i=1}^{k} \frac{c_i B^d}{\left(1 - \dfrac{1}{r_i} B\right)} S_n \tag{3.18}$$

where $r_1, r_2, r_3, \cdots, r_k$ are the roots of the denominator of $H(B)$. An analogy can be drawn between the concept of stability discussed in Section 2.6 and the final value of a generating function. If the absolute value of each of the roots $r_1, r_2, r_3, \cdots, r_k$ is greater than 1, then each of the series in Eq. (3.18) converges to zero and the final value of the response series y_n is zero. On the other hand, if the absolute value of any of the roots is less than 1, then the series in Eq. (3.18) diverges and the response series y_n diverges and has no final value.

The case where one root and *only one root* has a value $r_i = 1$ occurs often in generating functions. The result in this case is that all of the series except the series associated with root $r_i = 1$ converge to a final value of zero, whereas the series associated with root $r_i = 1$ has a constant value c_i. The final value of the response series y_n therefore is c_i, which can be found using Eq. (3.16). Alternatively, the final-value theorem can be stated as

$$y_\infty = \lim_{B \to 1} (1 - B)H(B) \tag{3.19}$$

which is valid as long as the system response defined by Eq. (3.17) is not unstable.

Example 3.7 Response using the final-value theorem In Example 3.5, the generating function for the error between the actual and the commanded position of a positioning-table control system subjected to a sampled-ramp input of 1 mm/s was

$$e_n = \frac{0.005B}{(1 - 0.915B)(1 - B)} S_n \qquad \text{mm}$$

Using Eq. (3.19) we find the final value of this response to be

$$e_\infty = \lim_{B \to 1} (1 - B) \frac{0.005B}{(1 - 0.915B)(1 - B)}$$

or

$$e_\infty = \frac{0.005}{(1 - 0.915)} = 0.05882 \qquad \text{mm}$$

The final (or steady-state) value of the error therefore is 0.05882 mm at a velocity of 1 mm/s. □

Response to a Unit Pulse Input

An input that is of unit magnitude and has a duration of one sample period can be called a **unit pulse** input. The unit pulse is shown in Fig. 3.7 and is described by the unit sample sequence generating function S_n. If the dynamic behavior of a system is characterized by the discrete transfer function

$$\frac{y_n}{x_n} = G(B) \qquad (3.20)$$

and the input x_n is a unit pulse, then

$$x_n = S_n \qquad (3.21)$$

and

$$y_n = G(B)S_n \qquad (3.22)$$

Figure 3.7 Unit pulse sample sequence $x_n = S_n$.

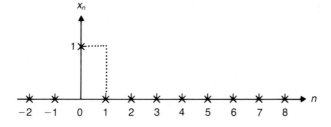

The response of a system to a unit pulse, therefore, is simply the sequence formed by the discrete transfer function operating on the unit sample. Although this may appear to be an obvious conclusion, the result is important in terms of analyzing discrete systems. The dynamic behavior of a system is described by its discrete transfer function, and expansion of that function into a power series in B that operates on the unit sample sequence S_n generates a unique response series that is characteristic of the system. This unit pulse response can be very useful in comparing system performance, estimating system parameters, and designing control systems.

Example 3.8 Response to a unit pulse   The dynamic characteristics of the holding tank shown in Fig. 3.8(a) can be observed by subjecting the system to a unit pulse in input flow rate. If the area of the tank is 3 square meters, the output

Figure 3.8 (a) Holding tank for Example 3.8, and (b) response of holding tank to unit pulse input ($T = 0.25$ min).

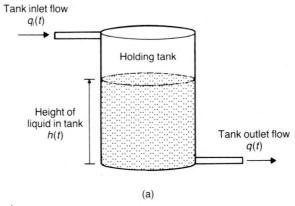

(a)

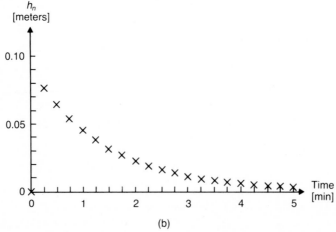

(b)

flow rate is 2 cubic meters per minute per meter of fluid height in the tank, and the sample period is 15 seconds (0.25 minutes), then a continuous model for the system is

$$1.5 \frac{dh(t)}{dt} + h(t) = 0.5 q_i(t) \qquad \text{meters}$$

where $q_i(t)$ is the input flow rate in cubic meters per minute. The corresponding discrete model from Table 2.1 is

$$\frac{h_n}{q_{i_n}} = \frac{0.0768B}{1 - 0.8465B}$$

This transfer function can be rewritten as the power series

$$\frac{h_n}{q_{i_n}} = 0B^0 + 0.0768B^1 + 0.0650B^2 + 0.0550B^3 + 0.0466B^4 + \cdots$$

which can be obtained using long division. If the input is a pulse of 1 m³/min for one sample period (0.25 min), then the input is described by

$$q_{i_n} = S_n \qquad \text{m}^3/\text{min}$$

The response to this input can be obtained directly from the power series form of the transfer function:

$$h_n = (0.0768B + 0.0650B^2 + 0.0550B^3 + 0.0466B^4 + \cdots)S_n$$

This response is plotted in Fig. 3.8(b). □

Example 3.9 Analysis of system with computer control As an example of the application of the principles discussed thus far, consider the analysis of a complete computer feedback control system, where it is desired to maintain the flow rate from a holding tank at a certain value. A schematic diagram for such a system is given in Fig. 3.9.

The computer performs the function of error detection and valve motor manipulation. The error detection function is given as

$$e_n = r_n - q_n$$

where flow rate from the tank q_n is obtained by sampling flowmeter voltage with an analog-to-digital (A/D) converter, and the manipulation m_n is output to the valve motor amplifier by a digital-to-analog (D/A) converter. For this example, proportional control action is chosen. Thus

$$m_n = K_c e_n$$

where K_c is the proportional control action gain. The inlet flow control valve is manipulated by a valve motor that produces a time rate of change of valve stem

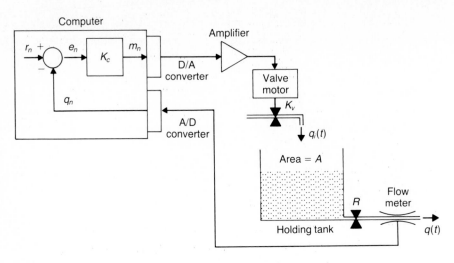

Figure 3.9 Flow control system with holding tank for Example 3.9.

position proportional to the manipulation voltage. Further, as a first approximation, consider that the flow rate through the valve is proportional to valve stem
position. Thus a manipulation voltage $m(t)$ produces a rate of change of inlet flow
$q_i(t)$ given by

$$\frac{dq_i(t)}{dt} = K_v m(t)$$

where K_v is the overall proportionality constant for the valve and actuator.
    The defining equation for the tank is

$$q_i(t) - q(t) = A \frac{dh(t)}{dt}$$

where $q(t)$ is the output flow rate, A is the tank cross-sectional area, and h is the
liquid level. The relationship between $q(t)$ and $h(t)$ is assumed to be

$$q(t) = \frac{h(t)}{R}$$

Thus

$$A \frac{dh(t)}{dt} + \frac{1}{R} h(t) = q_i(t)$$

    Differentiating this equation and combining it with the input-value equation
above gives the overall process model

$$AR \frac{d^2 h(t)}{dt^2} + \frac{dh(t)}{dt} = RK_v m(t)$$

The corresponding equation for output flow as a function of manipulation voltage is

$$\tau \frac{d^2q(t)}{dt^2} + \frac{dq(t)}{dt} = K_v m(t)$$

where the tank time constant is

$$\tau = AR$$

The discrete process transfer function is obtained from entry 5 in Table 2.1. It is

$$G_p(B) = \frac{q_n}{m_n} = \frac{(b_1 B + b_2 B^2)}{1 - a_1 B - a_2 B^2}$$

where

$$a_1 = 1 + e^{-T/\tau}$$

$$a_2 = -e^{-T/\tau}$$

$$b_1 = K_v[T - \tau(1 - e^{-T/\tau})]$$

$$b_2 = -K_v[Te^{-T/\tau} - \tau(1 - e^{-T/\tau})]$$

The transfer function for the controller is

$$G_c(B) = \frac{m_n}{e_n} = K_c$$

The closed-loop transfer function is obtained from

$$\frac{q_n}{r_n} = \frac{G_c(B)G_p(B)}{1 + G_c(B)G_p(B)}$$

or

$$\frac{q_n}{r_n} = \frac{K_c(b_1 B + b_2 B^2)}{1 + (K_c b_1 - a_1)B + (K_c b_2 - a_2)B^2}$$

This is the discrete transfer function of the overall closed-loop computer control system. The system characteristic equation, from which stability of the system can be determined, is

$$1 + (K_c b_1 - a_1)B + (K_c b_2 - a_2)B^2 = 0$$

The absolute values of the roots of the characteristic equation must be greater than unity for absolute stability. The system will exhibit limited stability with one root equal to unity and one root greater than unity.

The response of the system can be determined from the closed-loop transfer function. If the system is subjected to a unit step command input:

$$r_n = \frac{1}{1 - B} S_n$$

the generating function for the step response is

$$q_n = \frac{K_c(b_1 B + b_2 B^2)}{(1 - B)[1 + (K_c b_1 - a_1)B + (K_c b_2 - a_2)B^2]} S_n$$

If the system is subjected to a unit pulse command input, the response is

$$q_n = \frac{K_c(b_1 B + b_2 B^2)}{1 + (K_c b_1 - a_1)B + (K_c b_2 - a_2)B^2} S_n$$

With specific coefficient values substituted into these equations, we can expand them, using long division, to obtain the response sequence for the system. The actual nature of the performance varies with the gain constant setting K_c. Note that not all gain constants provide desirable response, even though they may result in a stable response. Low values of K_c tend to result in slow response, and high values of K_c tend to result in oscillatory response. □

Example 3.10 Computer control system stability   Consider the simplified flow control system without a holding tank shown schematically in Fig. 3.10. The valve produces a flow rate proportional to the integral of the manipulation voltage. The equations characterizing this system are

$$e_n = r_n - q_n$$

$$m_n = K_c e_n$$

$$\frac{dq(t)}{dt} = K_v m(t)$$

Figure 3.10 Flow control system without holding tank for Example 3.10.

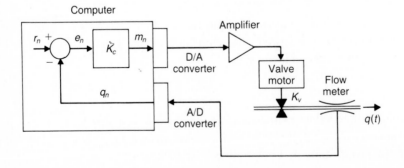

The discrete transfer function of the valve is

$$\frac{q_n}{m_n} = \frac{K_v TB}{1 - B}$$

When this equation is combined with the control equations, the overall closed-loop discrete transfer function becomes

$$\frac{q_n}{r_n} = \frac{K_c K_v TB}{1 - (1 - K_c K_v T)B}$$

for which the characteristic equation is

$$1 - (1 - K_c K_v T)B = 0$$

Thus the root is

$$r_1 = \frac{1}{1 - K_c K_v T}$$

To satisfy the criteria for absolute stability,

$$|r_1| > 1$$

or

$$\left| \frac{1}{1 - K_c K_v T} \right| > 1$$

which indicates that the system is stable for

$$|1 - K_c K_v T| < 1$$

or

$$-1 < 1 - K_c K_v T < 1$$

or

$$0 < K_c K_v T < 2$$

The actual response can now be analyzed for several choices of the gain-sample period product, $K_c K_v T$, for a unit step input $r_n = \dfrac{1}{1 - B} S_n$.

For $K_c K_v T = 1$,

$$q_n = \left(\frac{1}{1 - B} \right) \frac{B}{1} S_n$$

$$= \frac{B}{1 - B} S_n$$

$$= (0B^0 + B^1 + B^2 + B^3 + \cdots)S_n$$

The resulting step response is plotted in Fig. 3.11(a).

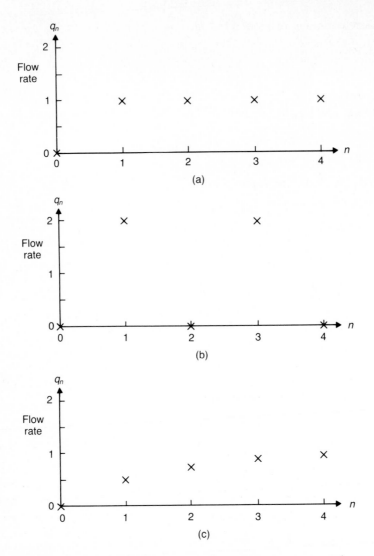

Figure 3.11 Response of flow control system to step command input. (a) $K_c K_v T = 1$; (b) $K_c K_v T = 2$; (c) $K_c K_v T = 0.5$.

For $K_c K_v T = 2$,

$$q_n = \left(\frac{1}{1 - B} \right) \frac{2B}{1 + B} S_n$$

$$= \frac{2B}{1 - B^2} S_n$$

$$= (0B^0 + 2B^1 + 0B^2 + 2B^3 + 0B^4 + 2B^5 + \cdots)S_n$$

The response is plotted in Fig. 3.11(b). As might be predicted by the choice of $K_cK_vT = 2$ rather than $K_cK_vT < 2$, the system possesses limited stability and is not absolutely stable.

For $K_cK_vT = 0.5$,

$$q_n = \left(\frac{1}{1-B}\right)\frac{0.5B}{1-0.5B}S_n$$

$$= \frac{0.5B}{1-1.5B+0.58B^2}$$

$$= (0B^0 + 0.5B^1 + 0.75B^2 + 0.875B^3 + 0.9375B^4 + \cdots)S_n$$

The response for this choice of system parameters is shown in Fig. 3.11(c). Comparing Figs. 3.11(a), (b), and (c) clearly shows that the nature of the response depends on the selection of system parameters, control algorithm gain, and sample period. The most desirable of the responses to the step input in many cases is that achieved with $K_cK_vT = 1$ and shown in Fig. 3.11(a), where the system is on target with zero error in one sample period. In other cases, particularly if the sample period is short and the step changes in input commands are large, this response specification may be too demanding for the system to achieve. Here, there is likely to be a maximum voltage that can be output by the D/A converter on the computer. That maximum voltage cannot be exceeded even if a large response is commanded to occur in a short time. □

Example 3.11 Machined part quality control   Quality control in machined part production has become increasingly important as pointed out in Chapter 1. For example, consider a "tool offset" controller constructed for a machine tool wherein each part produced is inspected by a laser measuring system. If an error in part dimension is detected, a correction (or offset) in tool position is applied before the next part is machined so that errors in the previous part are eliminated in subsequent parts.

Figure 3.12(a) shows the sequence of operations performed by the system in machining a part, of the type shown in Figure 3.12(b), on a lathe. The initial part is machined and its diameter is measured. A tool offset is computed for the next part based on current and past measurements. The tool position is then changed, offsetting it from its previous position, so that the diameter of the next part will be closer to the desired diameter, resulting in the production of higher quality parts.

The machine tool and the laser measuring system together constitute the process to be controlled. This process can be approximately defined by

$$L_n = 2Z_n + 2W_n$$

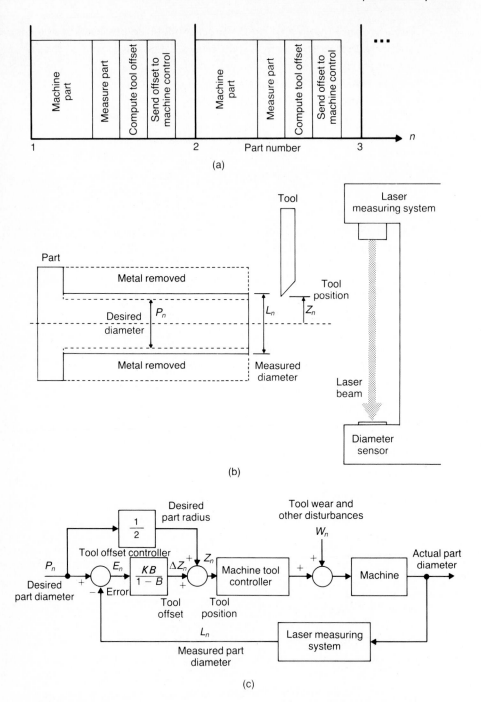

Figure 3.12 (a) Part machining and diameter error compensation sequence, (b) part machining and diameter measurement system, and (c) schematic block diagram of tool offset controller for Example 3.11.

where Z_n is the tool position, and W_n represents disturbances in the process (e.g., tool wear, machine deflection, etc.) that result in an unanticipated change in the diameter of the part, and L_n is the resulting part diameter measurement. Note that the process is discrete rather than continuous.

We chose a controller of the form

$$Z_n = \frac{P_n}{2} + K \sum_{i=0}^{n-1} (P_i - L_i)$$

where P_n is the desired part diameter and K is an integral control action gain. The $P_n/2$ term in the equation introduces a type of feedforward control, which is a form of open-loop control that will be discussed in Section 4.4. A schematic block diagram for the system is shown in Fig. 3.12(c).

The objective of the controller is to keep the measured diameter the same as the desired diameter in the presence of disturbances. Thus, the system is a regulator. Analytically, this can be observed in the closed-loop equation for the system, which is derived by substituting the control equation

$$Z_n = \frac{P_n}{2} + \frac{KB}{1-B}(P_n - L_n)$$

into the process equation

$$L_n = 2Z_n + 2W_n$$

to obtain the closed-loop equation

$$L_n = P_n + \frac{2(1-B)}{[1-(1-2K)B]} W_n$$

The controller gain K can be chosen to produce a desired rate of response to disturbance inputs. The choice $K = 0.5$ results in

$$L_n = P_n + 2(1-B)W_n$$

it will be seen that this choice of controller gain results in a very desirable system that completely compensates for step disturbance inputs, eliminating the error in the next part after the effect of the disturbance is detected in the part diameter measurement.

Typically, the desired part diameter is specified for a given batch of parts and remains constant. This can be represented by the generating function

$$P_n = \frac{D}{1-B} S_n$$

where D is the desired part diameter. The initial tool position would be set to one half this diameter ($Z_0 = D/2$), and the initial part ($n = 0$) would be manufactured. If there are no disturbances, then the measured diameter of the initial part and all subsequent parts will be D. This can be shown to be the case by substituting the foregoing generating function into the closed-loop equation and setting $W_n = 0$. The result is

$$L_n = \frac{D}{1 - B} S_n$$

or

$$L_n = (DB^0 + DB^1 + DB^2 + DB^3 + \cdots)S_n$$

The performance of the system also can be analyzed for various forms of process disturbances. An initial error in tool setup would be represented as a step input

$$W_n = \frac{U}{1 - B} S_n$$

where U is the tool setup error. If a cylindrical part is to be machined to a diameter of 20 mm and there is a setup error of 0.025 mm, then the initial part ($n = 0$) measures 20.05 mm. With $K = 0.5$, each subsequent part measures 20 mm. This is verified by substituting $D = 20$ mm and $U = 0.025$ mm into the foregoing generating functions and substituting the generating functions into the closed-loop equation. The result is

$$L_n = \left(\frac{20}{1 - B} + 2(0.025) \right) S_n$$

or

$$L_n = (20.05B^0 + 20B^1 + 20B^2 + \cdots)S_n$$

Table 3.2 shows the values of the system variables in Fig. 3.12(c) for this case.

A small chip on the tool point, occurring during the processing of part $n = j$, would be represented as a step disturbance by the generating function

$$W_n = \frac{CB^j}{1 - B} S_n$$

Table 3.2

Compensation for Tool Setup Error with $K = 0.5$

n	P_n	Z_n	W_n	L_n	E_n	ΔZ_{n+1}[mm]
0	20	10	0.025	20.05	−0.05	−0.025
1	20	9.975	0.025	20	0	0
2	20	9.975	0.025	20	0	0
3	20	9.975	0.025	20	0	0
etc.						

where C is the depth of the chip in the tool point. The response of the system to a chip of depth $C = 0.01$ mm while machining the fifth ($n = 4$) in a set of 20 mm diameter parts is

$$L_n = (20B^0 + 20B^1 + 20B^2 + 20B^3 + 20.02B^4 + 20B^5 + 20B^6 + \cdots)S_n$$

A constant rate of tool point wear would be represented by a ramp disturbance input. Without control, this would result in a ramp increase in part diameter. The generating function in this case has the form

$$W_n = \frac{RB}{(1 - B)^2} S_n$$

where R is the rate of tool wear. If the rate of tool wear is $R = 0.001$ mm/part, then for a 20 mm diameter part,

$$L_n = \left(\frac{20}{1 - B} + \frac{2(0.001)B}{1 - B} \right) S_n$$

or

$$L_n = (20B^0 + 20.002B^1 + 20.002B^2 + 20.002B^3 + \cdots)S_n$$

The values of the system variables for this case are given in Table 3.3. With the controller configuration and gain chosen, it should be noted that the tool offset moves at the same rate as the tool wear. However, there is a lag in the compensation of one part, resulting in a batch part with a dimensional offset of 0.002 mm.

Other disturbances may also be present in the system. Machine distortion, caused by temperature changes during warmup or by temperature fluctuations in the environment in which the machine is operated, can change the geometric relationship between the tool and the part. Cutting forces can result in deflection of the tool or part, making parts larger than desired. Many of these disturbances can be successfully compensated for by the controller used in this example.

Table 3.3

Compensation for Constant Rate of Tool Wear with $K = 0.5$

n	P_n	Z_n	W_n	L_n	E_n	ΔZ_{n+1}[mm]
0	20	10	0	20	0	0
1	20	10	0.001	20.002	−0.002	−0.001
2	20	9.999	0.002	20.002	−0.002	−0.002
3	20	9.998	0.003	20.002	−0.002	−0.003
4	20	9.997	0.004	20.002	−0.002	−0.004
etc.						

Chapter 4 shows that other controllers can be designed with the objective of improving performance in the presence of various types of disturbances. For example, a controller can be designed to correct for tool wear disturbances, even when the tool wear functions are parabolic or exponential in nature. □

3.4

SUMMARY

The major thrust of this chapter has been to develop the concept of sample sequences and a means of representing these sequences or time series mathematically so that they can be used in the design and analysis of computer control systems. The operator technique presented allows sample sequences for common control system inputs (such as pulse, step, ramp, and sine) to be conveniently represented as generating functions. They can then be used with the discrete transfer function of a dynamic system, which is found using the techniques described in Chapter 2, to obtain the response of the system to the particular input function. The method of partial fraction expansion was introduced, and a final-value theorem was developed that enables us to calculate the steady-state value of the system response.

The techniques presented in this chapter will be used in Chapter 4 in the design of discrete controllers that can be formulated to meet specific system response objectives. From the examples given at the end of the chapter, the reader should have recognized that using the presence of discrete sampling in closed-loop computer control systems introduces inherent stability problems. These problems must be addressed in the design process. Selection of the sample period to be used in a computer control system is particularly important. The interaction between sample period, stability, magnitude of manipulations generated, and speed of response to disturbances will be investigated in Chapter 4.

BIBLIOGRAPHY

Box, G. E. P., and G. M. Jenkins, *Time Series Analysis: Forecasting and Control.* San Francisco, CA: Holden-Day, 1970.

Cadzow, J. A., and H. R. Martens. *Discrete Time and Computer Control Systems.* Englewood Cliffs, NJ: Prentice-Hall, 1970.

Close, C. M., and D. K. Frederick. *Modeling and Analysis of Dynamic Systems.* Boston, MA: Houghton-Mifflin, 1978.

Doebelin, E. O. *Systems Modeling and Response.* New York, NY: Wiley, 1980.

Dorf, R. C. *Modern Control Systems, 4/e.* Reading, MA: Addison-Wesley, 1986.

Franklin, G. F., and J. D. Powell. *Feedback Control of Dynamic Systems.* Reading, MA: Addison-Wesley, 1986.

Harrison, H. L., and J. G. Bollinger. *Introduction to Automatic Controls.* New York, NY: Harper & Row, 1969.

Kuo, B. C. *Automatic Control Systems, 5/e.* Englewood Cliffs, NJ: Prentice-Hall, 1987.

Ogata, K. *Modern Control Engineering.* Englewood Cliffs, NJ: Prentice-Hall, 1970.

PROBLEMS

3.1 Use long division to calculate the sample sequence generated by each of the following generating functions.

a. $f_n = \dfrac{10}{1 - B} S_n$

b. $f_n = \dfrac{2(1 - B^3)}{(1 - B)^2} S_n$

c. $f_n = \dfrac{1}{1 - 1.99B + 0.99B^2} S_n$

d. $f_n = \dfrac{B^2}{1 - 0.95B} S_n$

e. $f_n = \dfrac{5(1 + B + B^2)}{1 - B^4} S_n$

f. $f_n = \dfrac{4.6B}{1 - 1.08B + B^2} S_n$

3.2 For each of the continuous functions shown in Fig. 3.13, develop the generating function f_n for the sequence obtained by sampling the function with a sample period of 1 second.

Figure 3.13 Functions for Problem 3.2 (to be sampled every 1 second).

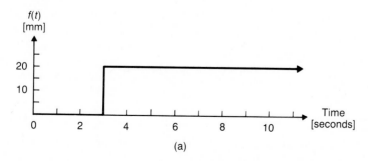

(a)

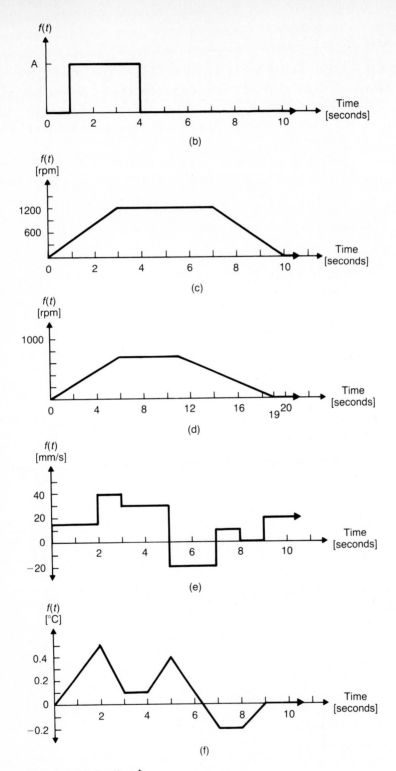

Figure 3.13 (*continued*)

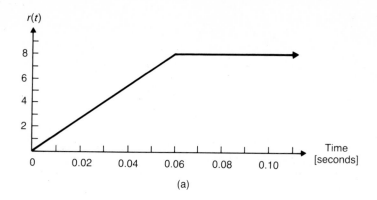

(a)

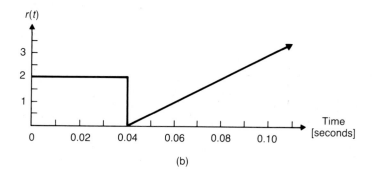

(b)

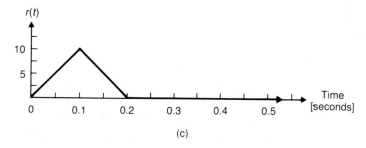

(c)

Figure 3.14 Functions for Problem 3.3 (to be sampled every 0.02 second).

3.3 For each of the functions shown in Fig. 3.14, develop the generating function f_n for the sequence obtained by sampling the function shown with a sample period of 0.02 second.

3.4 Give both the infinite series and a closed-form generating function for the sample sequence shown in Fig. 3.15.

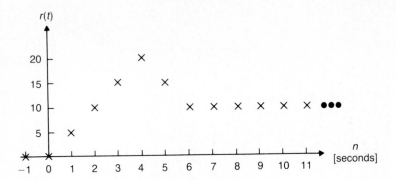

Figure 3.15 Sample sequence for Problem 3.4.

3.5 Develop the generating function for the sequence obtained by sampling $f(t) = At^2$ at instants T seconds apart.

3.6 Derive the generating function given in Table 3.1 for the sequence obtained by sampling $f(t) = \sin \omega t$ at intervals of T seconds.

3.7 For a sample rate of 200 Hz, develop generating functions for the periodic functions shown in Fig. 3.16. (*Hint:* The term $1/(1 - B^k)$ generates a sample sequence that has a nonzero term every k samples.)

Figure 3.16 Periodic functions for Problem 3.7.

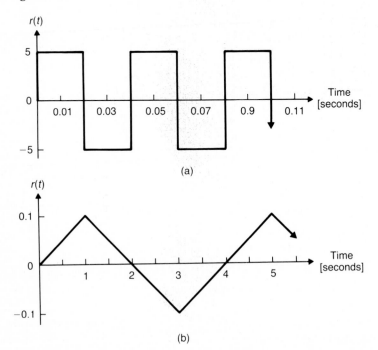

3.8 Expand the following generating functions or transfer functions into partial fractions.

a. $f_n = \dfrac{20B}{(1 - 0.9B)(1 - B)} S_n$

b. $f_n = \dfrac{K(1 - \delta)B}{(1 - \delta B)(1 - B)} S_n$

c. $f_n = \dfrac{B^2 - 1.95B^3 + 0.95B^4}{1 - 1.75B + 0.75B^2} S_n$

d. $\dfrac{c_n}{m_n} = \dfrac{0.21B + 0.17B^2}{(1 - 1.54B + 0.55B^2)(1 - B)}$

e. $\dfrac{c_n}{m_n} = \dfrac{0.55B^4 + 0.37B^5}{1 - 1.21B + 0.30B^2}$

f. $\dfrac{c_n}{m_n} = \dfrac{0.024B + 0.046B^2}{1 - 1.81B + 1.22B^2}$

3.9 In Example 3.5, the generating function for the response of a positioning system to a ramp input was found to be

$$x_n = \left[\frac{0.085B}{1 - 0.915B} \right] \left[\frac{0.005B}{(1 - B)^2} S_n \right]$$

Expand this function into partial fractions. (Note that the factor $(1 - B)$ has a multiplicity of two and represents a repeated root.)

3.10 The response sequence for a system is described by the generating function

$$c_n = \frac{9}{(1 - B)(1 - 0.1B)} S_n$$

a. Determine the response sequence using long division and plot the response.
b. Determine the response by employing partial fraction expansion and Table 3.1.
c. Determine the final value of the response using the final-value theorem.

3.11 Suppose that the positioning table shown in Fig. 3.6(a) and described in Example 3.5 is subjected to a step input of 1 mm.
a. Give the generating functions for table position and following error in response to the step input.
b. Calculate the steady-state table position and error using the final-value theorem.
c. What is the maximum value in the sequence of amplifier input voltages generated, v_n.

3.12 A floodgate for controlling tidal surges in a river is shown in Fig. 3.17. The gate must be rotated 270° in the counterclockwise direction to move it from its normal open position on the river bottom to its closed position in order to prevent an abnormally high tide from reaching a city upstream. The gate is driven by an electric motor, and gate position $\theta(t)$ is controlled by a computer that outputs a

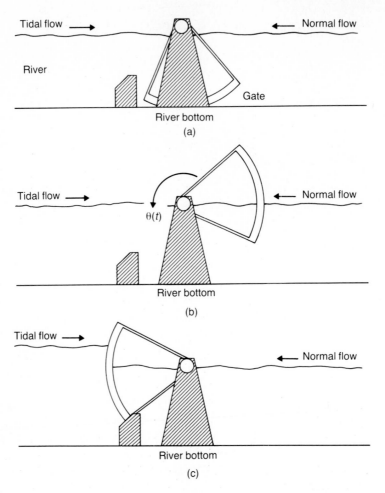

Figure 3.17 Flood tide control gate for Problem 3.12. (a) Gate open; (b) gate closing; (c) gate closed.

voltage $m(t)$ to the gate drive system. The speed of rotation of the gate that results is proportional to $m(t)$, and gate motion can be modeled using

$$\frac{d\theta(t)}{dt} = 2m(t) \qquad \text{deg/min}$$

The computer control equation used is

$$m_n = 0.1(\theta_{c_n} - \theta_n) \qquad \text{volts}$$

where $\theta_c(t)$ is the commanded gate position. The same period is 2.5 min.

a. Find the generating function for gate position that results from the gate being closed using a step input of 270°.

b. Find the generating function for gate position that results from the gate being closed using a ramp input of 11.4°/min, lasting 25 min.

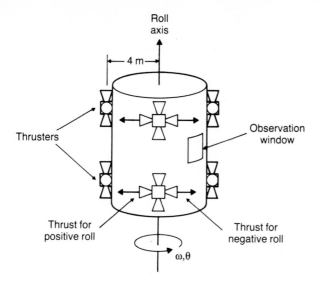

Figure 3.18 Manned space laboratory with eight groups of thrusters (two groups are hidden) for Problem 3.13.

    c. Plot and compare the responses θ_n that you obtained in parts (a) and (b).

    d. Plot and compare the sequences of manipulation voltages m_n generated in parts (a) and (b).

3.13 The manned space laboratory illustrated in Figure 3.18 has an inertia of 2000 N-m-s^2/rad. Roll of the spacecraft around the axis shown is produced by eight pairs of roll thrusters, each of which develops a thrust of 50 N in either the plus or the minus direction. A single variable $m(t)$ simultaneously controls all of the roll thrusters; $m(t)$ can have only the value $+1$, -1, or 0, and it is held constant over 0.2-second time intervals.

    a. Assume that the spacecraft roll rate $\omega(t)$ and position $\theta(t)$ are initially zero and that it is desired to roll the spacecraft approximately 30° to align an observation window with a new target. Develop a generating function for $m(t)$ that produces the desired roll in a minimum amount of time but keeps the roll rate under 0.5 revolution per minute.

    b. Give the sample sequences for m_n, ω_n, and θ_n produced by this generating function.

    c. Calculate the final value of θ_n using the final-value theorem. Why can't a roll of exactly 30° be achieved?

3.14 Show that the system in Example 3.9 is stable for the following process and controller parameters: $A = 10$ in^2, $R = 0.1$ s/in^2, $K_v = 1$ in^3/volt-s^2, $K_c = 10$, and $T = 0.1$ s.

3.15 Using the process and controller parameters given in Problem 3.14, determine the sequence representing the response of the system in Example 3.9 to a step input of 1 in^3/s. Plot the result.

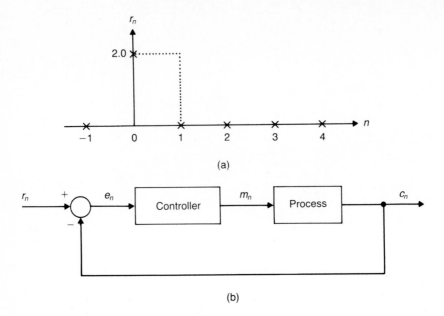

Figure 3.19 (a) Command input sequence and (b) system block diagram for Problem 3.16.

3.16 When the input r_n shown in Fig. 3.19(a) is applied to the system shown in Fig. 3.19(b), the response sequence is described by the following generating function:

$$c_n = \frac{4(1 - \delta)(1 - 0.5B)B}{(1 - \delta B)(1 - B) + 2(1 - \delta)B(1 - 0.5B)} S_n$$

where $\delta = e^{-1}$, the sample period is $T = 1$ s, and the discrete transfer function of the controller is

$$\frac{m_n}{e_n} = \frac{1 - 0.5B}{1 - B}$$

a. What is the closed-loop transfer function c_n/r_n?
b. Plot the response sequence c_n. Is the system stable?
c. Deduce the discrete process transfer function from the generating function for c_n as given above. What is the corresponding continuous process model? What type of control actions are being used? How does this explain the result of part (b)?

3.17 Each of the following discrete transfer functions represent a process that is to be subjected to a unit pulse input. For each process, plot the resulting response and characterize the nature of the process (first-order, second-order, underdamped, overdamped, and so on).

a. $\dfrac{c_n}{m_n} = \dfrac{6.32B}{1 - 0.368B}$

b. $\dfrac{c_n}{m_n} = \dfrac{20B^4}{1 - B}$

c. $\dfrac{c_n}{m_n} = \dfrac{100B}{1 - 0.9B}$

d. $\dfrac{c_n}{m_n} = \dfrac{23B^2}{1 - 1.08B + B^2}$

e. $\dfrac{c_n}{m_n} = \dfrac{82B + 242B^2}{1 - 0.79B + 0.37B^2}$

f. $\dfrac{c_n}{m_n} = \dfrac{0.21B + 0.17B^2}{1 - 1.54B + 0.55B^2}$

3.18 For each of the systems shown in Fig. 3.20, develop a discrete process model and plot the response of the system to a unit pulse input. For each system, use the

Figure 3.20 Systems to be modeled in Problem 3.18. (a) $T = 2$ s, $M = 12$ kg, $k = 0.15$ N/m, $c = 0.03$ N-s/m; (b) $T = 0.5$ min; (c) $T = 0.1$ s, $k = 5$ ton/ft, $c = 1.5$ ton-s/ft; (d) $T = 0.001$ s; (e) $T = 0.0001$ s.

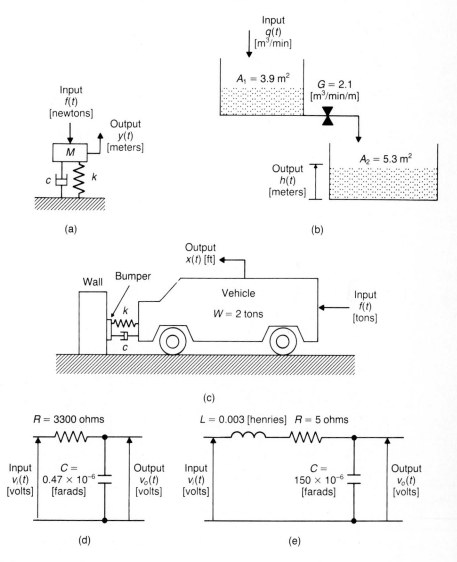

sample period, input, and output identified in the figure in developing the model. (Assume all initial conditions are zero.)

3.19 A solar water-heating system is shown in Fig. 3.21. The energy collected can be modeled using

$$Q_u(t) = F_R A_c \{S(t) - U_L[T_w(t) - T_a(t)]\}$$

and the storage tank can be modeled using

$$MC_p \frac{dT_w(t)}{dt} = Q_u(t) - L(t) - (UA)_T[T_w(t) - T_r(t)]$$

where

$F_R = 0.9$ (factor accounting for temperature gradients in collector)

$A_c = 5$ m² (collector surface area)

$U_L = 0.01$ MJ/m² − h-°C (collector heat-loss coefficient)

$M = 400$ kg (mass of water in system)

$C_p = 0.00418$ MJ/kg-°C (specific heat of water)

$(UA)_T = 0.054$ MJ/h-°C (tank heat-loss coefficient)

Figure 3.21 House with solar water-heating system for Problem 3.19.

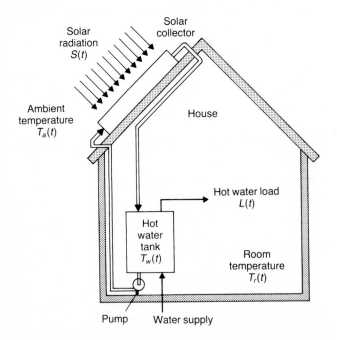

The variables are

$Q_u(t)$ = energy collected, in MJ/h

$S(t)$ = absorbed solar radiation per unit collector area, in MJ/m²-h

$T_a(t)$ = ambient temperature, in °C

$T_r(t)$ = temperature of room, in °C

$T_w(t)$ = temperature of water in tank, in °C

$L(t)$ = hot water load, in MJ/h

a. Develop a discrete model of the system for a sample period of 1 hour. The output should be tank water temperature, and the inputs should be solar radiation, ambient temperature, room temperature, and load.

b. Hourly values of solar radiation and ambient temperature are given in the following table. Derive a generating function for tank water temperature and calculate hourly tank water temperatures for the period 6 A.M. to 6 P.M., assuming zero hot water load, a constant room temperature of 20°C, and an initial tank water temperature of 40°C at 6 A.M. (*Hint*: Combine the inputs in a single generating function, and represent tank water temperature as $T_w = 40°C + \Delta T_w$.)

Time	Solar Radiation S [MJ/m²-h]	Ambient Temperature T_a [°C]	Water Temperature T_w [°C]	Energy Collected Q_u [MJ/h]	Collector Temperature T_o [°C]
6 A.M.	0.0	−6			
7 A.M.	0.0	−7			
8 A.M.	0.7	−5			
9 A.M.	1.8	−2			
10 A.M.	2.2	1			
11 A.M.	2.4	3			
12 Noon	3.1	3			
1 P.M.	3.0	4			
2 P.M.	2.5	5			
3 P.M.	2.0	5			
4 P.M.	1.3	4			
5 P.M.	0.3	0			
6 P.M.	0.0	−2			

c. The water temperature at the collector outlet, $T_o(t)$, can be found from

$$Q_u(t) = \dot{m}C_p[T_o(t) - T_w(t)]$$

where

$$\dot{m} = 270 \text{ kg/h (collector mass flow rate)}$$

Develop a discrete equation for energy collected and collector outlet temperature. Calculate the energy collected and collector outlet water temperature and complete the foregoing table, including the tank water temperature you found in part (b).

d. Identify any time in the table when the temperature of the collector output water either exceeds 100°C (boiling point) or is below 0°C (freezing point) or when the energy collected is negative (energy is lost rather than gained). Assume that the pump is turned off, the collector immediately drains, and $Q_{u}(t) = 0$ when any of these conditions is detected. Develop a pump-off difference equation for tank water temperature that reflects this pump-off condition. Use it, together with the pump-on difference equation obtained from the results of part (a), to successively calculate new values of water tank temperature. (Switch between difference equations depending on whether the pump should be on or off for any particular sample period.) Complete a modified table with the results from this modified system.

3.20 Reconsider Example 3.11 in which a discrete controller was used to correct tool position on a lathe to compensate for tool wear. In Table 3.3 the controller can be seen to track a constant rate of tool wear, maintaining the part within the error associated with machining one part. Suppose that the tool wear is characterized by the exponential function

$$W_{n} = 0.001e^{0.5n} \text{ mm}$$

Study the relationship between Z_{n} and E_{n} for this case. What conclusions can be drawn about the effectiveness of the controller in compensating for exponential disturbances? (Note that the wear rate after several parts is probably more rapid than is realistic.)

Chapter 4

Discrete Controller Design

INTRODUCTION

The formulation of computer control algorithms is both an art and a science. Clearly, the degree of success of a computer control installation can depend on the effectiveness of the algorithm chosen by the control engineer. Some of the basic concepts that can be applied to this selection process are developed in this chapter. Feedback control design for a desired closed-loop response, feedforward control algorithms, and noninteracting control algorithms are well-established approaches with a relatively straightforward methodology. Other approaches might be classified as inventive or common-sense control strategies that evolve from the designer's understanding of the process and exactly what the control is intended to accomplish.

Whatever the approach to controller design, it should be evident that a thorough knowledge of the process is essential for effective controller design. Knowledge of the process may take many different forms. An analysis of how designers have gained knowledge of process behavior reveals three broad approaches. First, the detailed laws governing the process may be analyzed and a mathematical model formulated. In many instances, this approach is adequate and straightforward. Second, a process may be extremely complex in mathematical analysis but relatively straightforward in response. In this case, formulation of a discrete process transfer function on the basis of test data from experiments with existing equipment may prove to be the most economical and effective approach. This is particularly true where processes are consistent in behavior and the control instrumentation produces good signal-to-noise ratios in the control range

111

of the variables. Third, it is not uncommon to encounter highly complex processes wherein the variables are corrupted by noise and disturbances. In this case, it may be necessary to draw upon statistically based techniques to eventually achieve a desired level of control.

This chapter focuses on the design of **control algorithms** for controlling well-behaved processes that can be adequately described by linear, discrete transfer functions and wherein the constants in these transfer functions are readily determined by computation or experiment. Extensive use will be made of techniques that have been developed for representing input sample sequences and calculating subsequent system response from the process transfer function. The important issue of selection of the sample period will also be discussed. Later chapters will deal with the experimental and analytical determination of process models and with the design of control algorithms using frequency domain and transformation techniques.

4.2
CONTROL ALGORITHMS FOR DESIRED CLOSED-LOOP RESPONSE

It has been shown that processes manipulated by a computer using discrete step changes in the process input variable can be represented by a discrete transfer function

$$G_p(B) = \frac{c_n}{m_n} \tag{4.1}$$

as illustrated by the block diagram in Fig. 4.1. The block diagram describes the process behavior at the sample (and control) instants $t = nT$, where $n = 0, 1, 2, \ldots$.

The controller and error detector that complete the closed-loop system are composed of a computer with associated computer software and interface hardware, as illustrated in Fig. 4.2. To implement direct computer control, the computer must accomplish the following tasks at each sample instant $t = nT$.

1. Sample the process output c_n and convert the signal to a number or digital word in the computer.

2. Use the command r_n to form the error

$$e_n = r_n - c_n \tag{4.2}$$

Figure 4.1 Block diagram for process to be controlled.

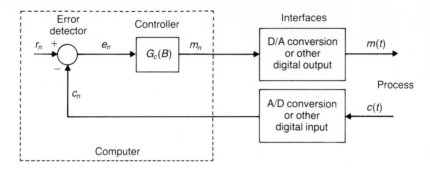

Figure 4.2 Control computer and interfaces to process.

3. Implement a software algorithm that solves the control equation in recursion form for the new value of, or the change in, the manipulated variable m_n. The control equation may be represented in equivalent discrete transfer function form as

$$G_c(B) = \frac{m_n}{e_n} \tag{4.3}$$

4. Output the newly calculated control manipulation m_n to the process. This step may involve either analog voltages or digital words, depending on the process control hardware.

This entire series of events must take a relatively insignificant amount of time relative to the sample time T and the process time constants in order for the theory advanced in this analysis to be applicable. Thus the computation time must be kept small to ensure that there is very little delay between the input of c_n and the output of m_n. These issues will be discussed in more detail in Chapter 5.

Keeping in mind the assumption just noted, we can study the closed-loop control system by using the block diagram in Fig. 4.3. From Eq. (2.27) the closed-loop transfer function that relates system output to system input is

$$G(B) = \frac{c_n}{r_n} = \frac{G_c(B)G_p(B)}{1 + G_c(B)G_p(B)} \tag{4.4}$$

Figure 4.3 Closed-loop block diagram for discrete control system.

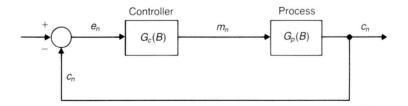

Control Algorithms for Desired Output–Input Relationships

One approach to designing a control algorithm is to determine the controller transfer function $G_c(B)$ required in order to have the overall system behave in some desirable manner. That is, the designer may wish to have the control system behave in accordance with a specified closed-loop transfer function $G_{dr}(B)$. Then,

$$G_{dr}(B) = \frac{c_n}{r_n} = \frac{G_c(B)G_p(B)}{1 + G_c(B)G_p(B)} \tag{4.5}$$

Equation (4.5) can be solved for the required controller discrete transfer function in terms of the plant, $G_p(B)$, and the desired overall characteristics, $G_{dr}(B)$. Solving Eq. (4.5) for $G_c(B)$ yields

$$G_c(B) = \frac{1}{G_p(B)} \frac{G_{dr}(B)}{[1 - G_{dr}(B)]} \tag{4.6}$$

Thus, given a process and the characteristics we wish to obtain in the overall closed-loop system at the sample instants, we can specify a controller in terms of a discrete transfer function. Furthermore, a recursion equation can be obtained from Eq. 4.3 that completely specifies the computer control algorithm.

Unfortunately, however, it is possible to ask for too much. The m_n computed from $G_c(B)$ may not be physically obtainable, or it may be too costly to achieve, in which case the designer must modify his or her demands This is illustrated in the following example.

Example 4.1 Design for dead-beat response   In Example 3.10, we examined a flow control system in which the valve shown in Fig. 3.10 produces a flow rate proportional to the integral of the valve manipulation voltage. The process was defined using

$$\frac{dq(t)}{dt} = K_v m(t)$$

where $q(t)$ is the flow rate through the valve, $m(t)$ is the valve manipulation voltage, and K_v is the proportionality constant of the valve. From Table 2.1 we learn that the discrete transfer function for the process, describing its behavior at the sample instants, is

$$G_p(B) = \frac{q_n}{m_n} = \frac{K_v TB}{1 - B}$$

Suppose that we want to manipulate the process with the computer so that, at the sample instants, the actual flow $q(t)$ will always equal the commanded flow $r(t)$. The desired closed-loop transfer function is therefore

$$G_{dr}(B) = \frac{c_n}{r_n} = 1$$

From Eq. (4.6), the controller required must have the characteristics described by

$$G_c(B) = \frac{1}{G_p(B)} \frac{G_{dr}(B)}{[1 - G_{dr}(B)]} = \frac{1}{\dfrac{K_v TB}{1 - B}(1 - 1)} = \infty$$

Clearly, this is not a physically realizable controller. Also, the closed-loop transfer function chosen defines $c_n = r_n$ and implies $e_n = 0$. This is not realistic in this example; some error is required between actual and commanded flow rate in order to generate process manipulations.

As an alternative, suppose that the specification is eased so as to accept a response that equals the command one sample period later. This condition can be stated mathematically as

$$q_{n+1} = r_n$$

A response that equals the command one sample period later is a **dead-beat response**, which can be more generally defined as a response that exactly reaches its final value after a finite number of sample periods. In this case, the desired closed-loop transfer function is

$$G_{dr}(B) = \frac{q_n}{r_n} = B$$

Then

$$G_c(B) = \frac{B}{\dfrac{K_v TB}{1 - B}(1 - B)} = \frac{1}{K_v T}$$

The control equation, then, is

$$m_n = \frac{1}{K_v T} e_n$$

Thus a simple proportional controller with gain

$$K_c = \frac{1}{K_v T}$$

will accomplish the objective.

Suppose that a 5-volt manipulation signal delivered for 2 seconds to the valve results in a change in flow rate of 180 gallons per minute (gpm). The valve constant, then, is

$$K_v = \frac{180 \text{ gpm}}{(5 \text{ volts})(2 \text{ s})} = \frac{18 \text{ gpm}}{\text{volt-s}}$$

If the sample period is $T = 0.5$ s and the units of feedback from the flow meter are gpm, then the discrete transfer function for the valve is

$$\frac{q_n}{m_n} = \frac{9B}{1 - B} \qquad \text{gpm/volt}$$

For dead-beat response in one sample period, the proportional control gain is

$$K_c = \frac{1}{18(0.5)} = 0.111 \text{ volt/gpm}$$

The flow rate and manipulation voltage in response to a step command input of 50 gpm are plotted in Fig. 4.4. □

Figure 4.4 Flow control process with controller gain $K_c = 0.111$ (a) output, and (b) manipulation.

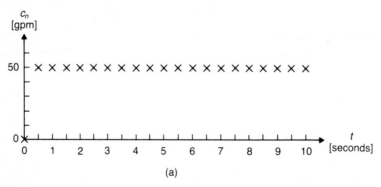

(a)

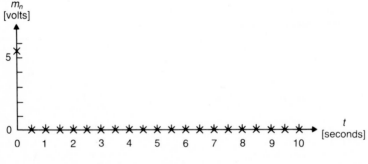

(b)

Example 4.2 Design for first-order response   Dead-beat response can be too demanding, particularly when the sample period is short compared to process time constants or when responses to large step inputs are required. Suppose that the response specification for Example 4.1 is eased so that the desired closed-loop transfer function corresponds to a first-order continuous system with a time constant τ_{dr}. From Table 2.1, the desired closed-loop transfer function is

$$G_{dr}(B) = \frac{K(1 - \delta)B}{1 - \delta B}$$

where $\delta = e^{-T/\tau_{dr}}$ and $K = 1$, because the final value of the response should equal the input when the input is constant. From Eq. (4.6), the controller required to achieve the specified closed-loop transfer function is

$$G_c(B) = \frac{\dfrac{(1 - \delta)B}{1 - \delta B}}{\dfrac{K_v TB}{(1 - B)}\left[1 - \dfrac{(1 - \delta)B}{1 - \delta B}\right]} = \frac{(1 - \delta)}{K_v T}$$

The control equation is then

$$m_n = \frac{(1 - \delta)}{K_v T} e_n$$

This is a proportional controller with gain $K_c = (1 - \delta)/K_v T$.

When we compare this result to that obtained for dead-beat response in Example 4.1, we see that for positive non-zero K_v, T, and τ_{dr},

$$\frac{1 - \delta}{K_v T} < \frac{1}{K_v T}$$

Hence the proportional controller gain is lower for first-order response than for dead-beat response, a result which is to be expected because first-order response is less demanding. Also, first-order response is less dependent than dead-beat response on the choice of sample period, because a time constant is specified rather than response in one sample period.

For a desired closed-loop time constant of $\tau_{dr} = 2$ s and sample period $T = 0.5$ s, as used in Example 4.1, the desired closed-loop transfer function is found from

$$\delta = e^{-0.5/2} = 0.779$$

and

$$\frac{q_n}{r_n} = \frac{(1 - 0.779)B}{1 - 0.779B} = \frac{0.221B}{1 - 0.779B}$$

The proportional controller gain that achieves this response with the valve constant

$$K_v = 18 \frac{\text{gpm}}{\text{volt-s}}$$

used in Example 4.1 is

$$K_c = \frac{1 - 0.779}{18(0.5)} = 0.025 \text{ volt/gpm}$$

This value of K_c is significantly less than the value 0.111 volt/gpm required for dead-beat response in Example 4.1.

The system can be simulated using Program 4.1, and the responses of flow rate and manipulation voltage to a step command input of 50 gpm are plotted in Fig. 4.5(a) and (b). Comparing Fig. 4.5(b) to Fig. 4.4(b) reveals that the maximum manipulation voltage in this case is less than one-fourth of that in Fig. 4.4(b). This is due to the reduction in gain from $K_c = 0.111$ to $K_c = 0.025$ in reducing response requirements from *complete* response in 0.5 s to achieving a *nearly complete* response in 10 s. □

Program 4.1 Simulation of flow control system.

```
program Flow_Control_Simulation;
  const
    T = 0.5;                              {sample period - seconds}
    Kp = 18.0;                            {process gain - gpm/s-volt}
    Kc = 0.025;                           {controller gain - volts/gpm}
  var
    n : integer;
    Qn, Rn, En, Mn, time : real;
  begin
    Qn := 0.0;                            {initial flow - gpm}
    time := 0.0;
    Rn := 50.0;                           {step command - gpm}
    for n := 0 to 20 do
      begin
        En := Rn - Qn;                    {calculate error}
        Mn := Kc * En;                    {calculate manipulation}
        writeln(time, Rn, Qn, En, Mn);
        Qn := Qn + Kp * T * Mn;           {calculate next process output}
        time := time + T
      end
  end.
```

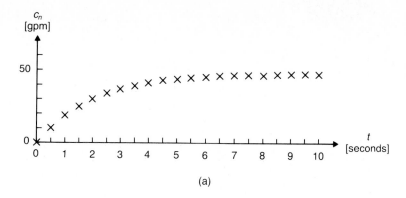

(a)

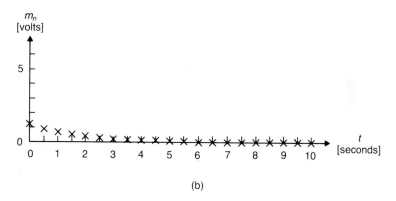

(b)

Figure 4.5 Flow control process with controller gain $K_c = 0.025$
(a) output, and (b) manipulation.

Example 4.3 Design of vehicle control The velocity of a vehicle as a function of throttle position can be approximately modeled using

$$\tau_p \frac{dv(t)}{dt} + v(t) = K_p m(t) \qquad km/h$$

where $v(t)$ is the vehicle velocity in km/h, $m(t)$ is the throttle position in percent of full actuation, $\tau_p = 15$ s, and $K_p = 2$ km/h per percent throttle actuation. The object is to design a velocity control with velocity command input $r(t)$[1] that achieves

$$v_{n+1} = r_n$$

with a velocity sample period of 0.1 s. The system and its control are illustrated in Figure 4.6.

[1] A velocity regulator or "cruise" control will be designed in Example 4.5.

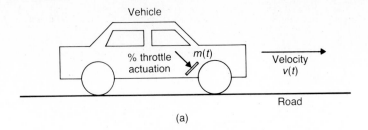

(a)

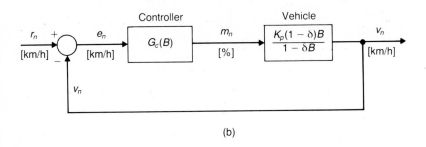

(b)

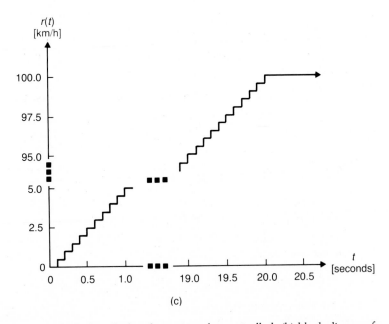

(c)

Figure 4.6 (a) Vehicle for which velocity is to be controlled; (b) block diagram for vehicle velocity control system; and (c) ramp input to smoothly accelerate vehicle to 100 km/h.

From Table 2.1, the discrete transfer function for the process is

$$\frac{v_n}{m_n} = \frac{K_p(1 - \delta_p)B}{1 - \delta_p B} \qquad \frac{\text{km/h}}{\%}$$

where

$$\delta_p = e^{-T/\tau_p} = e^{-0.1/15} = 0.9934$$

For the desired control,

$$G_{dr}(B) = \frac{v_n}{r_n} = B$$

From Eq. (4.6),

$$G_c(B) = \frac{\dfrac{B}{K_p(1 - \delta_p)B(1 - B)}}{1 - \delta_p B} \qquad \frac{\%}{\text{km/h}}$$

or

$$G_c(B) = \frac{1 - \delta_p B}{K_p(1 - \delta_p)(1 - B)}$$

The required control equations are therefore

$$m_n = m_{n-1} + \frac{1}{K_p(1 - \delta_p)} e_n - \frac{\delta_p}{K_p(1 - \delta_p)} e_{n-1} \qquad \%$$

and

$$e_n = r_n - v_n \qquad \text{km/h}$$

Substituting the parameter values given into the above control equation results in

$$m_n = m_{n-1} + 75.25 e_n - 74.75 e_{n-1} \qquad \%$$

There are usually limitations on the magnitudes of manipulations that can be applied to a process. With this control, the relationship between the commanded velocity and throttle actuation is

$$m_n = \frac{1 - \delta_p B}{K_p(1 - \delta_p)} r_n \qquad \%$$

For a unit step velocity command input of 1 km/h and the vehicle initially at rest,

$$r_n = \frac{1}{1 - B} S_n \qquad \text{km/h}$$

and

$$m_n = \frac{1 - \delta_p B}{K_p(1 - \delta_p)(1 - B)} S_n \qquad \%$$

or

$$m_n = \frac{1}{K_p}\left(\frac{1}{(1 - \delta_p)} B^0 + B^1 + B^2 + B^3 + \cdots \right) S_n \qquad \%$$

For the parameter values given, the maximum throttle actuation is

$$m_{max} = \frac{1}{K_p(1 - \delta_p)} = \frac{1}{2(1 - 0.9934)} = 75.25\%$$

Noting that the throttle actuation limit is 100% leads to the conclusion that the maximum step input amplitude for which the desired output−input relationship can be achieved is

$$100\%\left(\frac{1 \text{ km/h}}{75.25\%}\right) = 1.33 \text{ km/h}$$

A step input of greater than 1.33 km/h causes the controller to demand a throttle actuation exceeding the 100% limit.

If it is desired to accelerate the vehicle up to a speed of 100 km/h, then a ramp input 20 s in duration and a slope (acceleration) of 5 km/h/s would be a more reasonable choice than a step input of 100 km/h. As can be seen in Fig. 4.6(c), the ramp consists of 200 individual steps of 0.5 km/h. In this case, the closed-loop transfer function for which the control equation was designed remains valid because the actuation limit is not exceeded. As designed, the vehicle velocity lags behind this command input by only one sample period, 0.1 s. □

Control Algorithms for Desired Error−Input Relationships

A control algorithm can be formulated on the basis of a desired error−input relationship. If the desired error transfer function is given as $G_{de}(B)$, then referring to Fig. 4.3,

$$G_{de}(B) = \frac{e_n}{r_n} = \frac{1}{1 + G_c(B)G_p(B)} \qquad (4.7)$$

and solving for the controller transfer function yields

$$G_c(B) = \frac{m_n}{e_n} = \frac{1}{G_p(B)} \frac{[1 - G_{de}(B)]}{G_{de}(B)} \qquad (4.8)$$

Two points should be recognized concerning the selection of $G_{de}(B)$. First, in Eq. (4.7) it is unlikely that a specification demanding that $e_n = 0$ is consistent with the control geometry of Fig. 4.3. If there is never any error at the sample instants, there can be no change in process manipulation, and the process output remains unchanged. Second, the steady-state value of the error that results from a particular input sequence r_n can be found from the final-value theorem in Eq. (3.19) as

$$e_\infty = \lim_{B \to 1} \left[(1 - B)G_{de}(B) \frac{r_n}{S_n} \right] \tag{4.9}$$

Equation (4.9) is useful for studying the error–input relationship and specifying $G_{de}(B)$.

Example 4.4 Design for minimum error Suppose that the goal is to design a controller that causes a process to follow a ramp command and reach zero steady-state error as rapidly as possible. The process is described by $G_p(B)$, and the command is $r_n = AnT$. Figure 4.7 illustrates a desired response sequence. Clearly, there can be no response at $t = 0$. Also, there can be no response at $t = T$, because this is the first sample instant at which the computer controller detects the existence of an error. From $t = T$, the computer can implement a control action on the basis of the slope of the ramp command and arrive on target at time $t = 2T$. If the command input remains at constant slope, the process response can proceed with zero error.

For zero steady-state error, the error sequence is given by

$$e_n = [0B^0 + ATB + 0B^2 + \cdots]S_n$$

and

$$e_n = (ATB)S_n$$

Figure 4.7 Desired response to ramp input $r_n = AnT$ for Example 4.4.

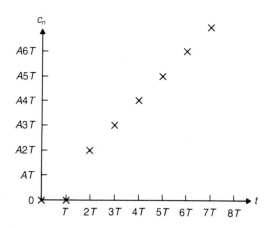

Referring to Table 3.1, we find that the generating function for the ramp input is

$$r_n = \frac{ATB}{(1 - B)^2} S_n$$

But

$$G_{de}(B) = \frac{e_n}{r_n}$$

Thus,

$$G_{de}(B) = \frac{(ATB)S_n}{(ATB)S_n}(1 - B)^2$$

or

$$G_{de}(B) = (1 - B)^2$$

The required controller transfer function can then be computed from Eq. (4.8) as

$$G_c(B) = \frac{1 - (1 - B)^2}{G_p(B)(1 - B)^2}$$

or

$$G_c(B) = \frac{(2 - B)B}{G_p(B)(1 - B)^2}$$

This example could also be solved by defining the desired sequence for c_n and then describing $G_{dr}(B)$, the desired discrete transfer function for c_n/r_n. For this approach,

$$c_n = (0B^0 + 0B^1 + A2TB^2 + A3TB^3 + \cdots + AnTB^n)S_n$$

The process response can be expressed as

$$c_n = \frac{ATB}{(1 - B)^2} S_n - ATBS_n$$

Thus

$$c_n = r_n - (1 - B)^2 r_n$$

from which the desired output−input relationship is

$$G_{dr}(B) = \frac{c_n}{r_n} = (2 - B)B$$

Equation (4.6) then can be used in place of Eq. (4.8) to find the required controller transfer function. □

Control Algorithms for Desired Output–Disturbance Relationships

Process control systems are commonly required to maintain an output variable at some desired level despite disturbances entering the system. These systems are often called **regulators** and are designed to respond to changes in load forces, line pressure, ambient temperature, speed, and the like. One schematic representation of a system acted on by disturbances is given in Fig. 4.8(a). In order to treat the effect of disturbance inputs, we must make one basic assumption: that the disturbance enters the process in the form of step changes occurring at the sample instants. Clearly, this is not really the case. However, without this assumption it is difficult to treat disturbance effects analytically. And in many cases it is a good approximation, particularly if the process control sample rate is high relative to the rate of change of the disturbance.

Figure 4.8 (a) Block diagram for discrete control system with disturbance input, and (b) reduced block diagram for $r_n = 0$.

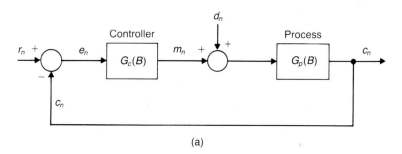

(a)

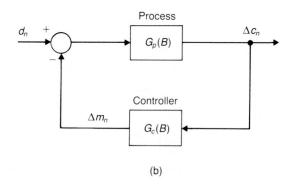

(b)

Figure 4.8(b) is a reduced block diagram that can be used to study the effect of disturbances. The command has been considered zero for the disturbance study. This poses no difficulty because, for linear systems, a total response may be computed by superimposing the individual response sequences from multiple sources. In the case of a disturbance, the output sequence is redefined as Δc_n to indicate a deviation of the process output from the target value. A sign change at the disturbance is the result of elimination of the error-summing junction. From Fig. 4.8(b),

$$\frac{\Delta c_n}{d_n} = \frac{G_p(B)}{1 + G_c(B)G_p(B)} \tag{4.10}$$

If we wanted to have the deviation of the output behave in a specific fashion, we might specify that in this case

$$G_{dd}(B) = \frac{\Delta c_n}{d_n} \tag{4.11}$$

Then

$$G_{dd}(B) = \frac{G_p(B)}{1 + G_c(B)G_p(B)} \tag{4.12}$$

and the desired controller could be solved for as follows:

$$G_c(B) = \frac{1}{G_p(B)} \frac{[G_p(B) - G_{dd}(B)]}{G_{dd}(B)} \tag{4.13}$$

One conclusion we reach from studying Eq. (4.13) is that with the configuration of Fig. 4.8(a), it is not possible to design a controller $G_c(B)$ such that Δc_n is zero for all disturbances d_n. This implies $G_{dd}(B) = 0$, and $G_c(B) = \infty$. To actually decide on $G_{dd}(B)$, it is necessary to consider both an acceptable sequence for Δc_n and the nature of the sequence approximating the disturbance. As one might expect, the more that is known about the nature of a disturbance, the better the control that can be implemented.

Example 4.5 Design for disturbance rejection In Example 4.3, a velocity controller was designed for a vehicle that achieved complete response to velocity command inputs in one 0.1-s sample period. However, it may be more desirable to design a velocity ("cruise") control that maintains a nearly constant vehicle velocity in the presence of disturbances—most significantly, changes in the slope of the roadway as indicated by θ in Fig. 4.9(a). The object is to design a controller that minimizes the deviation of vehicle velocity from a constant command velocity setting when a change in roadway slope occurs.

If it is known that a slope of $20°$ slows the vehicle with an effect similar to reducing the throttle actuation by 10%, then a block diagram for the system,

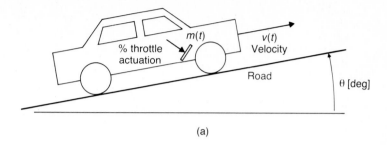

(a)

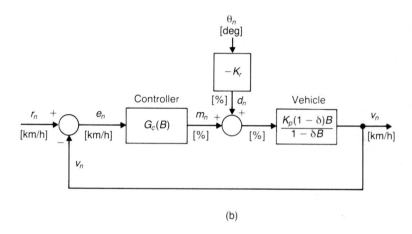

(b)

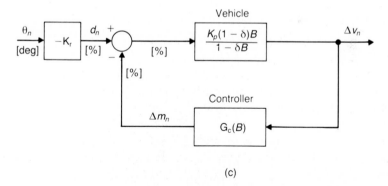

(c)

Figure 4.9 (a) Vehicle to be controlled on sloping roadway, (b) vehicle velocity control system with road slope disturbance input, (c) block diagram for vehicle with road slope input (command input $r_n = 0$).

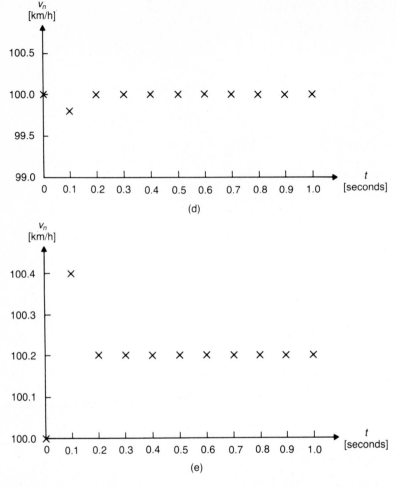

Figure 4.9 (d) response of vehicle speed to roadway slope change of 30° with controller designed in Example 4.5, and (e) response of vehicle speed to step command of 0.2 km/h with controller designed in Example 4.5.

including the disturbance input d_n, can be drawn as shown in Fig. 4.9(b), where

$$K_r = \frac{10}{20} = 0.5 \quad \%/\text{deg}$$

and

$$d_n = -K_r \theta_n \quad \%$$

Because the system is assumed to be linear, the effects of the two inputs r_n and θ_n can be separated and the output defined as

$$v_n = v_n' + \Delta v_n$$

where v_n' is affected only by r_n' and Δv_n is affected only by θ_n. Similarly, the manipulation generated by the controller can be represented as

$$m_n = m_n' + \Delta m_n$$

where m_n' is the result of changes in r_n' and Δm_n is the result of changes in θ_n. A block diagram for Δv_n is shown in Fig. 4.9(c).

When a step D in roadway slope occurs, no error is detected by the system in Fig. 4.9(c) until the first sample instant after the step. Therefore, the best performance that can be hoped for is that the velocity will be on target at the sample instant after the error is detected (two sample periods after the step). The desired sequence for the deviation in velocity, then, is

$$\Delta v_n = (0B^0 + AB^1 + 0B^2 + \cdots)S_n$$

where A is the magnitude of the velocity deviation at the sample period beginning at $t = T$.

The magnitude A may be computed from the relationship

$$\Delta v_n = \frac{K_p(1 - \delta)}{1 - \delta B}(d_n - \Delta m_n)$$

The step in roadway slope is a step disturbance to the system that can be modeled using

$$d_n = \frac{D}{1 - B} S_n$$

or

$$d_n = (DB^0 + DB^1 + DB^2 + DB^3 \cdots)S_n$$

No manipulation is output by the controller until time $t = T$, when an error is detected. Therefore

$$\Delta m_n = (0B^0 + M_1 B^1 + M_2 B^2 + M_3 B^3 \cdots)S_n$$

where $M_1, M_2, M_3, \ldots$ represents the sequence of manipulations generated. Thus

$$\Delta v_n = \frac{K_p(1 - \delta_p)B}{1 - \delta_p B}[DB^0 + (D - M_1)B^1 + (D - M_2)B^2 + \cdots]S_n$$

or, after division by $1 - \delta_p B$,

$$\Delta v_n = K_p(1 - \delta)\{0B^0 + DB^1 + [(1 + \delta_p)D - M_1]B^2$$
$$+ [(1 + \delta_p + \delta_p{}^2)D - M_1 - M_2]B^3 + \cdots\}S_n$$

Comparing coefficients with the desired sequence yields

$$A = K_p(1 - \delta_p)D$$

Thus the response sequence Δv_n is

$$\Delta v_n = (0B^0 + K_p(1 - \delta_p)DB^1 + 0B^2 + 0B^3 + \cdots)S_n$$

Substituting into Eq. (4.11), we get

$$G_{dd}(B) = \frac{\Delta v_n}{d_n} = \frac{K_p(1 - \delta_p)DB}{\dfrac{D}{1 - B}S_n}S_n$$

or

$$G_{dd}(B) = K_p(1 - \delta_p)B(1 - B)$$

The controller discrete transfer function is then computed via Eq. (4.13) as

$$G_c(B) = \frac{\dfrac{K_p(1 - \delta_p)B}{1 - \delta_p B} - K_p(1 - \delta_p)B(1 - B)}{\dfrac{K_p(1 - \delta_p)B}{1 - \delta_p B}[K_p(1 - \delta_p)B(1 - B)]}$$

or

$$G_c(B) = \frac{(1 + \delta_p) - \delta_p B}{K_p(1 - \delta_p)(1 - B)}$$

from which the control equation is

$$m_n = m_{n-1} + \frac{(1 + \delta_p)}{K_p(1 - \delta_p)}e_n - \frac{\delta_p}{K_p(1 - \delta_p)}e_{n-1}$$

Substituting the parameter values given into this equation with the sample period of 0.1 s used in Example 4.3 results in

$$m_n = m_{n-1} + 150.00e_n - 74.75e_{n-1}$$

Using this control equation results in dead-beat response to a disturbance input (road slope) of $30°$ in two sample periods, as shown in Fig. 4.9(d). This can be calculated using Program 4.2. Note the limits $0 \leq m_n \leq 100\%$ applied in the program.

In Example 4.3, the control equation

$$m_n = m_{n-1} + 75.25e_n - 74.75e_{n-1}$$

was developed for this system with the objective of achieving dead-beat response to command inputs rather than disturbance inputs. Unfortunately, both objectives cannot be accomplished with this control formulation, and the response to a step change of 0.2 km/hr in the command input of the system designed for disturbance response is shown in Fig. 4.9(e). The effect of the increased gain in the e_n term of the control equation can be clearly seen in the

```
program Vehicle_Speed_Control;
  const
    V0 = 100.0;                        {100-km/h initial velocity}
    Rn = 100.0;                        {100-km/h command input}
    THETAn = 30.0;                     {30-deg step in roadway slope}
    T = 0.1;                           {0.1-s sample period}
    TAUp = 15.0;                       {15-s vehicle time constant}
    Kp = 2.0;                          {2-km/h-% throttle actuation gain}
    Kr = 0.5;                          {0.5-%/degree distubance gain}
    K0 = 150.00;                       {controller gains from Ex. 4.5}
    K1 = -74.75;
  var
    n : integer;
    time, Vn, Dn : real;
    En, En1 : real;                    {error history}
    Mn, Mn1 : real;                    {manipulation history}
    A1, B1 : real;
  begin
    Vn := V0;                          {initialize vehicle velocity}
    En1 := 0.0;                        {initialize history}
    Mn1 := V0 / Kp;
    A1 := exp(-T / TAUp);              {constants for process model}
    B1 := Kp * (1.0 - A1);
    time := 0.0;                       {initialize time}
    for n := 0 to 10 do                {simulate 10 sample periods}
      begin
        Dn := -Kr * THETAn;            {calculate disturbance input}
        En := Rn - Vn;                 {calculate error}
        Mn := Mn1 + K0 * En + K1 * En1;  {calculate manipulation}
        writeln(time, Rn, Dn, Vn, En, Mn);  {print current variable values}
        Vn := A1 * Vn + B1 * (Mn + Dn);  {calc. next velocity}
        En1 := En;                     {update history}
        Mn1 := Mn;
        time := time + T               {update time}
      end
end.
```

Program 4.2 Simulation of vehicle with speed control for Example 4.5.

overshoot in the step command response. However, the response does settle quickly to the desired level, and for this application, good speed regulation in response to disturbances is likely to be more important than response to changes in speed setting. □

Controller Realizability

In the foregoing sections, we designed discrete controllers by forcing a closed-loop computer control system to a desired transient and/or steady-state response. It should be emphasized that the closed-loop transfer functions used in designing controllers must be carefully selected to ensure that desired responses make sense with respect to the physical nature of the process and control implementation. An example is the specification of response to a disturbance input in Example 4.5, where the effect of the disturbance on the process was calculated so that a reasonable response could be specified. Once a controller has been designed, it is necessary to determine whether the control hardware can deliver, and the process can accept, the manipulations called for by the controller for the expected range of system inputs. If this is not the case, the design specifications must be modified to reduce the amount of control power called for in manipulating the process. Also, though the principles involved appear straightforward, two critical design restrictions should be observed. The first restriction involves controller realizability; the second, controller stability.

In forcing closed-loop system response to some desired form, it is apparent that it is possible to require information about future errors. The controller calculated is

$$G_c(B) = \frac{m_n}{e_n} \tag{4.14}$$

or

$$m_n = G_c(B)e_n \tag{4.15}$$

which can be expanded by long division into the form

$$m_n = \cdots + K_{-3}e_{n+3} + K_{-2}e_{n+2} + K_{-1}e_{n+1} + K_0 e_n + K_1 e_{n-1} \\ + K_2 e_{n-2} + K_3 e_{n-3} + \cdots \tag{4.16}$$

Any nonzero $K_{-1}, K_{-2}, K_{-3}, \ldots$ terms imply a knowledge of future errors.

Clearly, demanding that a calculation be carried out with data that lie in the future is not physically possible with the techniques at hand. Such a controller must therefore be considered physically unrealizable. This condition generally results from establishing performance requirements that are unrealistic for the physical elements of the system. The problem can usually be overcome by relaxing some degree of performance, such as speed of response. It should be recognized that some form of estimation based on statistics might also be useful.

Example 4.6 An unrealizable controller   Suppose that a process can be modeled as an integration with a delay of two sample periods. The discrete process model, then, is

$$G_p(B) = \frac{K_v TB}{1 - B} B^2$$

where K_v is the process gain and T is the sample period. If a desired closed-loop transfer function is specified as

$$G_{dr}(B) = B$$

as was the case in Example 4.1, then the required controller transfer function obtained using Eq. (4.6) is

$$G_c(B) = \frac{B}{\dfrac{K_v TB^3}{(1-B)}(1-B)} = \frac{1}{K_v TB^2}$$

The corresponding controller equation is

$$m_{n-2} = \frac{1}{K_v T} e_n$$

or

$$m_n = \frac{1}{K_v T} e_{n+2}$$

This control law is not realizable because it requires knowledge of the error two sample periods into the future in order to calculate the present manipulation. □

Controller Stability

A closed-loop control system must be stable. For practical implementation of a digital controller, it is also important to recognize that the controller itself must not be unstable as an independent element. This condition is satisfied by ensuring that the magnitude of all the roots of the characteristic equation of the controller are greater than 1, with the exception that one root may have a value $r = 1$. This often occurs in controllers with integral control actions as defined by Eq. (2.16), and it is illustrated by the results of Examples 4.3 and 4.5.

When the controller $G_c(B)$ of Eq. (4.6), (4.8), or (4.13) is in a closed-loop system as shown in Fig. 4.3, the $G_p(B)$ term in the denominator of the control equation cancels the process transfer function $G_p(B)$. This can lead to two problems. First, terms in the numerator of the process transfer function tend to become terms in the denominator of the controller transfer function. The values of the roots of the numerator of the process transfer function therefore can affect the stability of controllers designed using these equations. Sometimes selecting a shorter sample period can improve the values of these roots—and hence the stability of the controller. Second, although these terms may appear to cancel, this may not truly happen. In practice, the real process is likely to vary somewhat from the model used to calculate the controller. Thus the cancellation does not exactly occur, and the result may be an actual closed-loop transfer function that varies significantly from the desired transfer function.

Example 4.7 Controller instability Stability problems that we may encounter when directly calculating controller transfer functions without considering the nature of the process to be controlled or the characteristics of the control equation obtained can be illustrated by the result of deriving a controller to obtain dead-beat response for a double-integration process. From Table 2.1, the discrete transfer function for a process defined by

$$\frac{d^2c(t)}{dt^2} = Km(t)$$

is

$$G_p(B) = \frac{KT^2(1 + B)B}{2(1 - B)^2}$$

This process is not stable because the two roots of its characteristic equation are $r_1, r_2 = 1$. Nevertheless, a controller transfer function can be calculated via Eq. (4.6) that achieves a stable desired closed-loop transfer function of

$$G_{dr}(B) = B$$

The result is

$$G_c(B) = \frac{B}{\dfrac{KT^2(1 + B)B}{2(1 - B)^2}(1 - B)} = \frac{2(1 - B)}{KT^2(1 + B)}$$

This controller is not stable because of the root at $r = -1$.

Well-behaved closed-loop response is predicted by the closed-loop transfer function. For example, the response to a unit pulse input

$$r_n = S_n$$

is predicted to be

$$c_n = BS_n$$

and the error for the same input is

$$e_n = (1 - B)S_n$$

However, the manipulation generated by this controller is not well-behaved and, for this input, can be shown to be

$$m_n = \frac{2(1-B)^2}{KT^2(1+B)} S_n$$

or

$$m_n = \frac{2}{KT^2} (1B^0 - 3B^1 + 4B^2 - 4B^3 + 4B^4 - 4B^5 + \cdots)S_n$$

The response of variables r_n, c_n, e_n, and m_n are plotted for the unit pulse input in Fig. 4.10 and can be obtained by simulating the system using Program 4.3.

Figure 4.10 Response of system variables with the controller designed in Example 4.7.

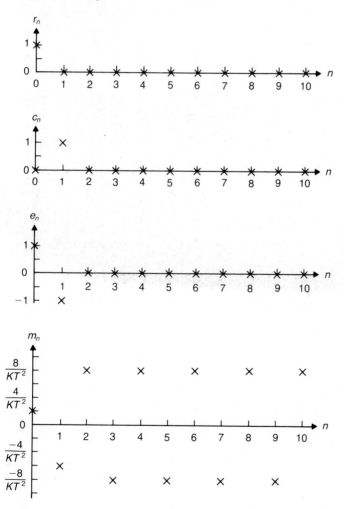

```
program Oscillating_Controller;
  const
    T = 1.0;                                  {1-s sample period}
    K = 2.0;                                  {double integration process gain}
  var
    n : integer;
    time : real;
    Rn : real;                                {command}
    Cn, Cn1, Cn2 : real;                      {process output history}
    En, En1 : real;                           {error history}
    Mn, Mn1, Mn2 : real;                      {manipulation history}
    Kp : real;
  begin
    Kp := K * sqr(T) / 2.0;                   {constant for process model}
    Cn := 0.0;                                {initialize process output history}
    Cn1 := 0.0;
    En1 := 0.0;                               {initialize error history}
    Mn1 := 0.0;                               {initialize mnipulation history}
    time := 0.0;                              {initialize time}
    Rn := 1.0;                                {unit pulse command}
    for n := 0 to 10 do                       {simulate 10 sample periods}
      begin
        En := Rn - Cn;                        {calculate error and manipulation}
        Mn := -Mn1 + (En - En1) / Kp;
        writeln(time, Rn, Cn, En, Mn);        {print current variable values}
        En1 := En;                            {update error and manip. history}
        Mn2 := Mn1;
        Mn1 := Mn;
        Cn2 := Cn1;                           {update process output history}
        Cn1 := Cn;
        Cn := 2.0 * Cn1 - Cn2 + Kp * (Mn1 + Mn2);   {calculate next process output}
        time := time + T;                     {update time}
        Rn := 0.0                             {Rn = 0 after time = T}
      end
  end.
```

Program 4.3 Double-integration process control simulation for Example 4.7.

Although the output of the controller is bounded, its characteristic of oscillation at one-half of the sample frequency is unlikely to be acceptable. The process input $m(t)$ and the process output $c(t)$ that result are plotted in Fig. 4.11, and the undesirable oscillation is not detected by sampling the process because the frequency is exactly one-half of the sample frequency. This phenomenon is a manifestation of **aliasing**, which will be discussed in Chapters 6 and 11. □

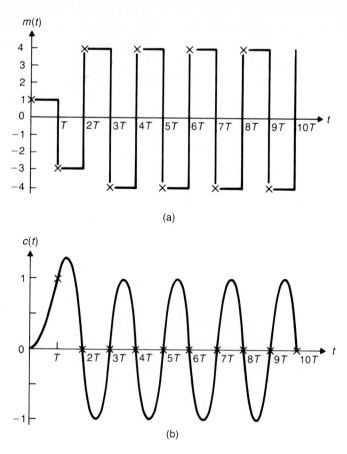

(a)

(b)

Figure 4.11 (a) Continuous process input (controller output) for Example 4.7, and (b) continuous process output for Example 4.7, showing oscillation not detected by sampling (aliasing).

4.3

SAMPLE-PERIOD SELECTION

The sample period chosen for a computer control system can have a significant impact on its performance. The sample period can affect the stability of the system and how rapidly disturbances and command inputs are responded to. The sample period should be chosen with care at the time the discrete controller is designed so that favorable system performance is achieved. It is usually desirable to make the sample period as short as possible in order to achieve rapid response and minimize the adverse effects of the sample period on stability. On the other hand, calculations must be performed by the control computer every sample period, and the shorter the sample period, the greater the computational load placed on the control computer. Ways of programming control computations efficiently will be discussed in Chapter 5; however, there usually is a practical limit on how short the sample period can be made. It is therefore useful to know what the limitations are on how long the sample period can be made with respect to speed of response to inputs, the size of process deviations that result from disturbances, and system stability.

Speed of Response to Inputs

In general, the sample period chosen for a discrete control system must be equal to or shorter than the time in which it is desired that the system respond to inputs. The simple case of dead-beat response to a unit step input in one sample period is illustrated in Fig. 4.12(a). The sample period chosen directly governs the speed of response, which is completed in exactly one sample period T. A somewhat different effect of sample period on response can be seen in Fig.

Figure 4.12 (a) Dead-beat response in one sample period, and (b) step response of first-order system with various values of T/τ.

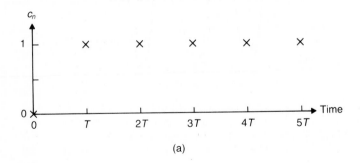

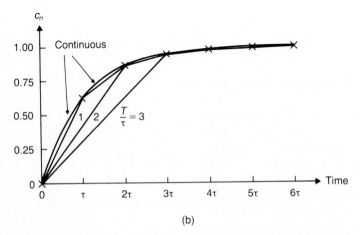

4.12(b), which shows step response, for various choices of T/τ, of a system designed so that its closed-loop transfer function is

$$G(B) = \frac{(1 - e^{-T/\tau})B}{1 - e^{-T/\tau}B} \qquad (4.17)$$

For purposes of comparison, step response of a first-order continuous system with a time constant τ is also shown. The choice of

$$T \leq \tau \qquad (4.18)$$

results in a response that resembles that of the continuous system. For many systems, Eq. (4.18) can be used as a general rule of thumb for selecting the sample period.

Rejection of Disturbances

The magnitude of the response of a process output to a disturbance before the system controller detects and compensates for the disturbance is dependent on the sample period. If the sample period is short, disturbances are detected quickly and compensated for before they can significantly affect the process output. If the sample period is too long, the disturbance affects the process for a relatively long time before compensation is applied. In this case, the magnitude of deviations of the process output from the desired level can become unacceptably large. It is therefore necessary to consider what disturbances can occur in a process and to choose a sample period that is short enough so they can be rejected by the controller before they adversely affect system performance.

Example 4.8 Sample-period selection for disturbance rejection The motor shown in Fig. 4.13(a) is occasionally subjected to load torque disturbance $D(t)$. If motor shaft position $\theta(t)$ can be approximately modeled using

$$\frac{d\theta(t)}{dt} = K_m v(t) + K_t D(t)$$

where $v(t)$ is a computer-controlled input voltage, then the corresponding process difference equation is

$$\theta_n - \theta_{n-1} = K_m T v_{n-1} + K_t T d_{n-1}$$

A closed-loop control system for this process is shown in Fig. 4.13(b). The disturbance can be present for an entire sample period before the controller has a chance to detect an error and modify the manipulation voltage. The change in motor shaft position due to the disturbance during that sample period is

$$\Delta\theta = K_t TD$$

where D is the magnitude of the load torque.

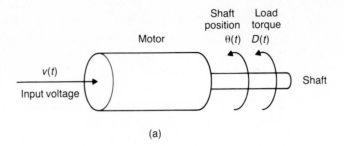

(a)

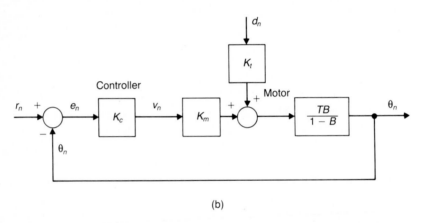

(b)

Figure 4.13 (a) Motor with load torque disturbance, and (b) motor and controller block diagram for Example 4.8.

If load torques of 200 in.-lb$_f$ are expected and it is required that deviations in shaft position due to these torques be limited to 0.01 revolutions, then for $K_t = 0.6$ rpm/in.-lb$_f$, the sample period is limited to

$$T \leq \frac{0.01 \text{ rev } (60 \text{ s/min})}{(0.6 \text{ rpm/in.-lb}_f)(200 \text{ in.-lb}_f)}$$

or

$$T \leq 0.005 \text{ s} \quad \square$$

Stability

A maximum sample period for a given control system can be determined from its closed-loop characteristic equation. The roots of the characteristic equation determine the system stability, and the value of these roots is a function of the sample period T in most cases. If T is made too large, the magnitude of one or more of the roots can be less than 1 and the resulting system is unstable. Keep in mind, however, that a system can be stable but still possess undesirable performance characteristics such as highly oscillatory responses and large overshoots before reaching a final value. For this reason, analysis of stability can establish a limit on the sample period, but it cannot necessarily find the shorter sample period required for desirable results.

Example 4.9 Sample-period selection for stability   Suppose it is known that an output of $v(t) = 5$ volts from the controller in Fig. 4.13(b) produces a motor velocity of 1000 rpm when no load torque is applied to the motor in Fig. 4.13(a). The motor gain is therefore

$$K_m = \frac{1000 \text{ rpm}}{5 \text{ volts}} = 200 \text{ rpm/volt}$$

If it is desired that the closed-loop system response be similar to that of a continuous system with a time constant of 0.05 s, then from Eq. (4.18), the sample period for the controller should be chosen such that

$$T \le 0.05 \text{ s}$$

Using Eq. (4.6) and a sample period of 0.05 s, we can determine that the proportional controller gain required is

$$K_c = \frac{(1 - e^{-T/\tau})}{K_m T} = 3.793 \text{ volts/rev}$$

The closed-loop transfer function for command inputs is

$$\frac{\theta_n}{r_n} = \frac{K_c K_m TB}{1 - (1 - k_c K_m T)B}$$

and the characteristic equation of the system is therefore

$$1 - (1 - K_c K_m T)B = 0$$

If the system is to be stable, then

$$0 < K_c K_m T < 2$$

and the sample period is limited to the range

$$0 < T < \frac{2}{K_c K_m}$$

From the values given,

$$0 < T < \frac{2(60 \text{ s/min})}{(200 \text{ rpm/volt})(3.793 \text{ volts/rev})}$$

or
$$0 < T < 0.158 \text{ s}$$

Even though the system in this example is stable for sample periods up to 0.158 s, the choice of a sample period of 0.05 s is likely to be more reasonable (response to a step command input then will correspond to the plot for $T/\tau = 1$ in Fig. 4.12b). However, even this sample period may be too long if the result of Example 4.8 is considered. There we found that a sample period of 0.005 s or less was required to limit the effect of load torque disturbances of 200 in.-lb$_f$ to deviations of 0.01 revolutions. Response to disturbances may well be the overriding consideration in an application such as this, and the sample period must be short enough so that disturbances can be rejected by the controller before deviations in the process reach unacceptable magnitudes.

No attempt has been made in this analysis to represent the time required by the control computer to sample the motor shaft position, calculate the error and manipulation, and output the manipulation voltage to the motor. We have assumed that the time delay between position sampling and manipulation output is insignificant compared to the sample period. From a practical point of view, this analysis is likely to be valid if these actions are carried out by the computer within one-tenth of the sample period. Even though this can be a relatively short period of time (500 μs for a 0.005-s sample period), it is easily achievable in this example because the simple proportional controller that has been designed requires very few computations: only one subtraction to compute the error and one multiplication to compute the manipulation. □

4.4
FEEDFORWARD CONTROL

Feedforward control is an important and widely used process control technique for providing effective elimination of undesirable response to various types of inputs. It is applicable whenever a process disturbance is measurable and its effect on the process is predictable. It can also be used to anticipate the effects of a command input on process output and to modify manipulations generated by a closed-loop control system in such a way that error in the process output is reduced. The principle is to feed the disturbance or command input signal forward through a feedforward controller and to combine it with manipulations generated by a feedback controller.

Disturbance Feedforward Control

Figure 4.14 illustrates the concept of feedforward control of disturbances. If a disturbance input d_n can be sampled by the computer, the feedforward controller transfer function $G_f(B)$ can be used to eliminate or at least reduce the effects of disturbances on the system. If $F(B)$ is the disturbance input transfer function, then process input m_n' is

$$m_n' = m_n + G_f(B)d_n + F(B)d_n \qquad (4.19)$$

Clearly, the most desirable situation is to design the feedforward control such that

$$f_n \approx -F(B)d_n \qquad (4.20)$$

or

$$G_f(B) \approx -F(B) \qquad (4.21)$$

If this is the case, then

$$m_n' \approx m_n \qquad (4.22)$$

Although Eq.(4.21) is analytically straightforward, practical shortcomings arise. The computer must treat the disturbance signal as a sampled variable, even though the disturbance changes continuously and the process responds continuously to the changing disturbance. Calculations using $G_f(B)$ are based on a discrete analysis, treating the disturbance as a series of step inputs entering the process at the sample times. Ideally, the feedforward control algorithm $G_f(B)$ should completely compensate for the disturbance, but in reality, this goal can only be approximated. The principle of feedforward control action must therefore be combined with feedback control, as illustrated in Fig. 4.14. If the feedforward control is successful, the portion of the disturbance input that is uncompensated

Figure 4.14 Feedforward control of disturbances.

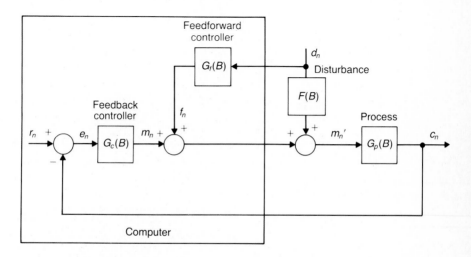

for is small, eliminating the disturbance as a major factor affecting system performance.

Example 4.10 Feedforward control of disturbances   Consider the vertical axis (z axis) drive of the robot arm in Fig. 4.15(a), which is shown carrying a load of W kg. A model that ignores the dynamic effects of changes in load W during motion of the robot arm, but approximately describes the effect of the load carried on the position of the arm when at rest, is shown in Fig. 4.15(b). Assume that a proportional controller with a gain of 140 already has been designed for the system.

Figure 4.15 (a) Robot arm to be controlled in Example 4.10, and (b) feedforward control of load on robot arm.

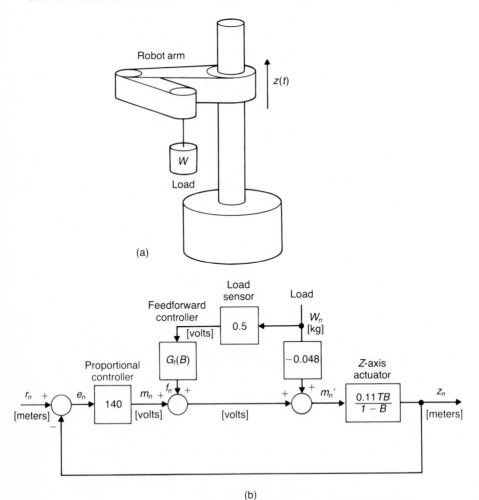

(a)

(b)

The effect of the addition of a 5-kg load on the system can be determined by modeling it as a step input:

$$W_n = \frac{5}{1 - B} S_n \quad \text{kg}$$

The change in position that results can be found by using the final-value theorem:

$$\Delta z_\infty = \lim_{B \to 1} (1 - B) \left[-0.048 \frac{\dfrac{0.11 TB}{1 - B}}{1 + 140 \dfrac{0.11 TB}{1 - B}} \right] \frac{5}{1 - B} = -1.7 \quad \text{mm}$$

This error can be eliminated through load sensing and feedforward control.

When a load sensor is provided that produces an output of 5 volts for a load of 10 kg, **disturbance feedforward control** can be incorporated as shown in Fig. 4.15(b). The feedforward transfer function required is

$$0.5 G_f(B) = 0.048$$

or

$$G_f(B) = 0.096$$

With this value of $G_f(B)$, the error is zero when the arm is at rest.

Note that the addition of feedforward disturbance control does not affect the closed-loop position control transfer function z_n/r_n. Rejection of disturbance inputs therefore has been achieved without affecting either the stability of the system or the nature of its response to changes in the position command input r_n. This would not have been the case if errors in position due to load disturbances were reduced either by increasing the gain of the feedback controller or by adding integral control action. ◻

Command Feedforward Control

Figure 4.16(a) illustrates the use of **command feedforward control**. The objective is to modify the manipulation input to the process in such a way that the performance of the system is improved by reducing or eliminating errors between the command input and the process output. We want to find $G_f(B)$ such that

$$c_n = r_n \tag{4.23}$$

It can be observed from Fig. 4.16(a) that the system transfer function is

$$\frac{c_n}{r_n} = \left(1 + \frac{G_f(B)}{G_c(B)} \right) \frac{G_c(B) G_p(B)}{1 + G_c(B) G_p(B)} \tag{4.24}$$

If $G_f(B)$ is chosen such that

$$G_f(B) = \frac{1}{G_p(B)} \tag{4.25}$$

then substituting into Eq. (4.24) yields

$$\frac{c_n}{r_n} = 1 \tag{4.26}$$

which is the result required by Eq. (4.23).

The effect of command feedforward control is perhaps better illustrated by redrawing the block diagram given in Fig. 4.16(a) as shown in Fig. 4.16(b), where it can be more clearly seen that if

$$G_f(B) \approx \frac{1}{G_p(B)} \tag{4.27}$$

then

$$e_n \approx 0 \tag{4.28}$$

$$m_n \approx 0 \tag{4.29}$$

and

$$c_n \approx r_n \tag{4.30}$$

$G_f(B)$ can be thought of as implementing open-loop control of the process output c_n. (The output of the feedforward controller is not affected by c_n.) Because feedforward control is essentially an open-loop control, the accuracy of control depends on the accuracy of knowledge of the process transfer function. Any inaccuracies in this control are compensated for by the closed-loop controller $G_c(B)$.

It is interesting to note that implementation of command feedforward control as shown in the block diagram in Fig. 4.16(a) can be simplified by using the modified block diagram shown in Fig. 4.16(c). The modified command input to the feedback controller is

$$r_n' = \left(1 + \frac{G_f(B)}{G_c(B)}\right) r_n \tag{4.31}$$

This formulation can be particularly useful when an entire sequence of command inputs r_n is known in advance. It is then possible to precalculate and store the entire modified command sequence, and r_n' rather than r_n can be used as the command input for each sample period. This approach eliminates the need to calculate f_n at each sample period, and it speeds up control computations.

Regardless of what transfer function is chosen for the controller $G_c(B)$, a

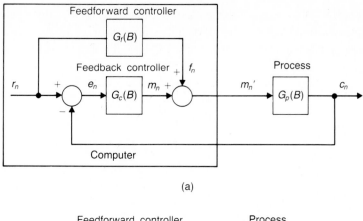

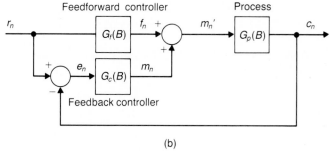

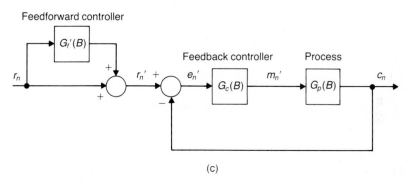

Figure 4.16 (a) System with command feedforward control, (b) block diagram redrawn to emphasize the open-loop nature of feedforward control, and (c) feedforward control using modified command input.

feedforward controller $G_f(B)$ can be designed that tends to eliminate system error. Hence a lower-performance feedback control loop can be implemented than would be needed without feedforward control—for example, a low-gain proportional controller often can be used. One must often take care to avoid large step inputs when using feedforward control, because overly large manipulations may be needed to achieve the required response.

Example 4.11 Again consider the robot arm in Fig. 4.15(a). Suppose that it is now desired to eliminate the following errors that occur as the command input r_n is changed in Fig. 4.15(b). Assuming $W_n = 0$, the error transfer function is

$$\frac{e_n}{r_n} = \frac{1}{1 + 140\dfrac{0.11TB}{1 - B}} = \frac{1 - B}{1 - (1 - 15.4T)B}$$

For a ramp input of 0.1 m/s,

$$r_n = \frac{0.1TB}{(1 - B)^2} S_n$$

and the final value of the error is

$$e_\infty = \lim_{B \to 1} (1 - B) \left[\frac{1 - B}{1 - (1 - 15.4T)B} \right] \frac{0.1TB}{(1 - B)^2} = 6.5 \qquad \text{mm}$$

These errors can be eliminated by using command feedforward control as illustrated in Fig. 4.17. From Eq. (4.25), the required feedforward controller is

$$G_f(B) = \frac{1}{\dfrac{0.11TB}{1 - B}} = \frac{1 - B}{0.11TB}$$

The corresponding feedforward control equation is

$$f_{n-1} = \frac{1}{0.11T}(r_n - r_{n-1})$$

or

$$f_n = \frac{1}{0.11T} r_{n+1} - \frac{1}{0.11T} r_n$$

Figure 4.17 Command feedforward control for robot arm.

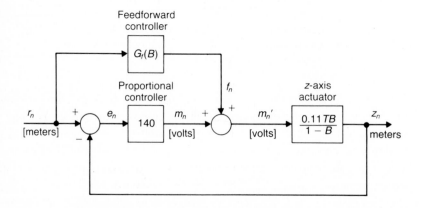

Generally, this result is unrealizable because it requires future knowledge of r_{n+1}; however, in many applications of the robot arm shown in Fig. 4.15(a), motion commands are prestored in a data-base in the control computer. In this case, the entire sequence of commands r_n is known and the foregoing feedforward control equation can be implemented. If the formulation for command feedforward control shown in Fig. 4.16(c) is used, the modified command sequence r_n' can be precalculated via Eq. (4.31) as

$$r_n' = \left(1 + \frac{\dfrac{1-B}{0.11TB}}{140}\right) r_n$$

or

$$r_n' = \frac{1}{15.4T} r_{n+1} + \left(1 - \frac{1}{15.4T}\right) r_n$$

In other applications of the robot arm, the command r_n must be generated in real time, perhaps using outputs from task-oriented sensors on the robot. In this case, a modified feedforward equation must be used that is realizable. If it is known that

$$r_{n+1} - r_n \approx r_n - r_{n-1}$$

for most of the time the robot is operating, then a logical choice is

$$f_n = \frac{1}{0.11T} r_n - \frac{1}{0.11T} r_{n-1}$$

and

$$G_f(B) = \frac{1-B}{0.11T}$$

With this feedforward control transfer function, the error transfer function can be determined to be

$$\frac{e_n}{r_n} = \frac{(1-B)^2}{1 - (1 - 15.4T)B}$$

For the ramp input of 0.1 m/s, the final value of the error is

$$e_\infty = \lim_{B \to 1} (1 - B) \left[\frac{(1-B)^2}{1 - (1 - 15.4T)B} \right] \frac{0.1TB}{(1-B)^2} = 0$$

Therefore, the modified command feedforward controller both is realizable and eliminates steady-state following errors. ▫

4.5

CASCADE CONTROL

Cascade control refers to the use of multiple feedback loops to control a multistage process. The result of the use of cascade control usually is improved performance, particularly because the effects of disturbances can be minimized by the addition of inner loops.

Cascade control may prove useful whenever a process has intermediate response variables that can be measured. The process shown in Figure 4.18(a) has an output $c(t)$ and an intermediate output $c'(t)$, both of which can be sampled. As we noted in Section 2.4, discrete transfer functions must be individually derived for each of these process outputs. The intermediate output is therefore related to the manipulation through the discrete transfer function $G_p'(B)$, whereas the complete process can be represented by $G_p(B)$.

Figure 4.18 (a) Process with intermediate output, and (b) block diagram for cascade control.

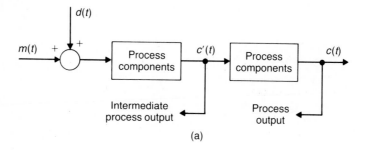

(a)

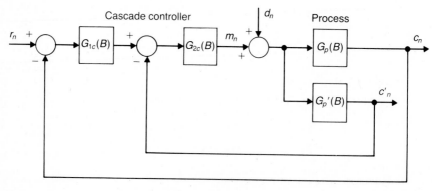

(b)

Implementation of cascade control is illustrated by the complete block diagram of Fig. 4.18(b). Two control algorithms, $G_{1c}(B)$ and $G_{2c}(B)$, give the designer the flexibility of specifying two criteria—for example, command input response and disturbance response. When we take the command input as zero ($r_n = 0$), the following equations describe the system in Fig. 4.18(b).

$$\Delta m_n = -G_{1c}(B)G_{2c}(B)\Delta c_n - G_{2c}(B)\Delta c_n' \tag{4.32}$$

$$\Delta c_n' = G_p'(B)(\Delta m_n + d_n) \tag{4.33}$$

and

$$\Delta c_n = G_p(B)(\Delta m_n + d_n) \tag{4.34}$$

Combining these three equations yields

$$\frac{\Delta c_n}{d_n} = \frac{G_p(B)}{1 + G_{1c}(B)G_{2c}(B)G_p(B) + G_{2c}(B)G_p'(B)} \tag{4.35}$$

Equation (4.35) can be set equal to some desired relationship between response and an anticipated type of disturbance. Thus,

$$G_{dd}(B) = \frac{\Delta c_n}{d_n} \tag{4.36}$$

Considering the response to input commands with zero disturbance ($d_n = 0$), the following equations describe the system in Fig. 4.18(b):

$$m_n = G_{1c}(B)G_{2c}(B)(r_n - c_n) - G_{2c}(B)c_n' \tag{4.37}$$

$$c_n' = G_p'(B)m_n \tag{4.38}$$

$$c_n = G_p(B)m_n \tag{4.39}$$

And the overall discrete transfer function for the closed-loop system becomes

$$\frac{c_n}{r_n} = \frac{G_{1c}(B)G_{2c}(B)G_p(B)}{1 + G_{1c}(B)G_{2c}(B)G_p(B) + G_{2c}(B)G_p'(B)} \tag{4.40}$$

Equation (4.40) can be equated to a desired output–input relationship:

$$G_{dr}(B) = \frac{c_n}{r_n} \tag{4.41}$$

Equations (4.35) and (4.36) can be combined to produce

$$\frac{G_p(B)}{G_{dd}(B)} = 1 + G_{1c}(B)G_{2c}(B)G_p(B) + G_{2c}(B)G_p'(B) \tag{4.42}$$

and Eqs. (4.40) and (4.41) can be combined to produce

$$\frac{G_{1c}(B)G_{2c}(B)G_p(B)}{G_{dr}(B)} = 1 + G_{1c}(B)G_{2c}(B)G_p(B) + G_{2c}(B)G_p'(B) \tag{4.43}$$

Using the principle of linear superposition, Eqs. (4.42) and (4.43) can be combined to produce

$$G_{1_c}(B)G_{2_c}(B) = \frac{G_{dr}(B)}{G_{dd}(B)} \qquad (4.44)$$

Substituting from Eq. (4.44) into Eq. (4.42) yields the inner-loop controller

$$G_{2_c}(B) = \frac{G_p(B)[1 - G_{dr}(B)] - G_{dd}(B)}{G_{dd}(B)G_p'(B)} \qquad (4.45)$$

The outer-loop controller then has the transfer function

$$G_{1_c}(B) = \frac{G_{dr}(B)G_p'(B)}{G_p(B)[1 - G_{dr}(B)] - G_{dd}(B)} \qquad (4.46)$$

Example 4.12 Cascade control   The elevator shown in Fig. 4.19(a) is suspended by a cable, and its position is determined by controlling the position of a drum upon which the cable is wrapped as the elevator rises. The drum is driven by an electric motor, the speed of which is manipulated using a voltage that is output by a computer. The elevator control computer measures drum position and generates elevator position commands based on an algorithm that considers the elevator's present position and which floors have been requested as stops. Disturbances are introduced into the system by the weight of people entering and leaving the elevator, which applies varying load torque to the drum. The object is to design a control system that accurately controls elevator position in response to both command and disturbance inputs.

Assume that the drum can be modeled using

$$\omega(t) = 0.3v(t - 0.018) - 0.15D(t - 0.018)$$

$$\frac{d\theta(t)}{dt} = \omega(t)$$

$$z(t) = 1.57\theta(t)$$

where $v(t)$ is the manipulation voltage that is output by the computer, in volts.

$\omega(t)$ is the drum velocity, in rev/s.

$\theta(t)$ is the drum position, in rev.

$z(t)$ is the elevator position in m.

For a sample rate of 50 Hz ($T = 0.02$ s), the process transfer functions for drum speed and position are obtained from Table 2.1 as

$$G_p'(B) = 0.3B$$

$$G_p(B) = \frac{0.006B^2}{1 - B}$$

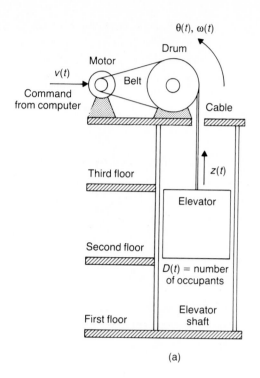

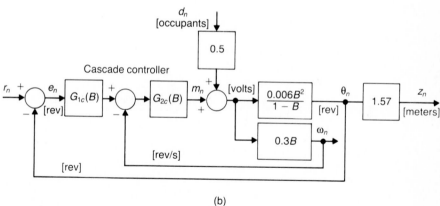

Figure 4.19 (a) Elevator system for Example 4.12 and (b) cascade control for elevator system.

Also

$$\frac{z_n}{\theta_n} = 1.57$$

The delay of 0.018 s has been represented by an integer number of sample periods ($d = 1$ in Table 2.1).

In order to specify desired response to both command and disturbance inputs, cascade control can be used as illustrated in Fig. 2.19(b). Because the complete process transfer function includes the factor B^2, the most desirable closed-loop transfer function that can be specified for command inputs is

$$G_{dr}(B) = \frac{c_n}{r_n} = B^2$$

For the unit step disturbance caused by one person entering the elevator,

$$d_n = \frac{1}{1 - B} S_n$$

the best response that can be expected with the process transfer function is

$$\Delta c_n = (0B^0 + 0B^1 + AB^2 + 0B^3 + 0B^4 + \cdots)S_n$$

where A is the amplitude of the response caused by the disturbance. The closed-loop transfer function specified for disturbance inputs is therefore

$$G_{dd}(B) = \frac{\Delta c_n}{d_n} = \frac{AB^2 S_n}{\dfrac{1}{1 - B} S_n} = AB^2(1 - B)$$

Using Eq. (4.45), the controller transfer function $G_{2c}(B)$ is

$$G_{2c}(B) = \frac{\dfrac{0.006B^2}{1 - B}(1 - B^2) - AB^2(1 - B)}{AB^2(1 - B)(0.3)B} = \frac{0.006B^2 + 0.006B^3 - AB^2 + AB^3}{AB^2(1 - B)(0.3)B}$$

Picking $A = 0.006$ results in the controller

$$G_{2c}(B) = \frac{6.667}{1 - B}$$

Similarly, $G_{1c}(B)$ can be found from Eq. (4.46) as

$$G_{1c}(B) = \frac{B^2(0.3)B}{\dfrac{0.006B^2}{1 - B}(1 - B^2) - 0.006B^2(1 - B)} = 25$$

Together, these controllers result in complete response to command inputs and complete rejection of disturbance inputs, both in two sample periods (0.04 s). □

4.6
NONINTERACTING CONTROL OF
INTERACTING SYSTEMS

Many processes with more than a single input and a single output exhibit the property of interaction between the manipulation and output variables. The general block diagram configuration of a two-variable interacting system is shown in Fig. 4.20, wherein the plant, or process, is represented by four transfer functions, $g_{11}, g_{21}, g_{12},$ and g_{22}.[2] The objective of the control design is to select a set of controller transfer functions $d_{11}, d_{21}, d_{12},$ and d_{22} that eliminate the interaction effects and also ensure that the overall closed-loop response of each loop follows some desired relationship.

The total plant of Fig. 4.20 may be represented by the two simultaneous equations

$$c_1 = g_{11}m_1 + g_{12}m_2 \tag{4.47a}$$

$$c_2 = g_{21}m_1 + g_{22}m_2 \tag{4.47b}$$

From Fig. 4.20,

$$m_1 = d_{11}e_1 + d_{12}e_2 \tag{4.48a}$$

$$m_2 = d_{21}e_1 + d_{22}e_2 \tag{4.48b}$$

Figure 4.20 Configuration for noninteracting control of interacting process.

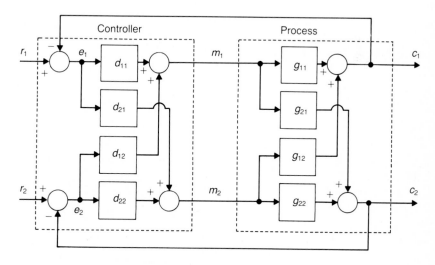

[2] Lower-case g is used here to represent the process discrete transfer functions, and lower-case d is used to represent the controller discrete transfer functions. Lower-case q will be used to represent the desired process transfer functions.

Substituting Eqs. (4.48) into Eqs. (4.47) yields

$$c_1 = g_{11}(d_{11}e_1 + d_{12}e_2) + g_{12}(d_{21}e_1 + d_{22}e_2) \tag{4.49a}$$

$$c_2 = g_{21}(d_{11}e_1 + d_{12}e_2) + g_{22}(d_{21}e_1 + d_{22}e_2) \tag{4.49b}$$

or

$$c_1 = (g_{11}d_{11} + g_{12}d_{21})e_1 + (g_{11}d_{12} + g_{12}d_{22})e_2 \tag{4.50a}$$

$$c_2 = (g_{21}d_{11} + g_{22}d_{21})e_1 + (g_{21}d_{12} + g_{22}d_{22})e_2 \tag{4.50b}$$

For noninteraction, it is necessary for c_1 to be influenced only by e_1 and for c_2 to be influenced only by e_2. Then in Eqs. (4.50),

$$g_{11}d_{12} + g_{12}d_{22} = 0 \tag{4.51a}$$

$$g_{21}d_{11} + g_{22}d_{21} = 0 \tag{4.51b}$$

If q_{11} and q_{22} are the desired closed-loop transfer functions, then

$$c_1 = q_{11}r_1 \tag{4.52a}$$

$$c_2 = q_{22}r_2 \tag{4.52b}$$

Because

$$e_1 = r_1 - c_1 \tag{4.53a}$$

$$e_2 = r_2 - c_2 \tag{4.53b}$$

then

$$c_1 = \frac{q_{11}}{1 - q_{11}} e_1 \tag{4.54a}$$

$$c_2 = \frac{q_{22}}{1 - q_{22}} e_2 \tag{4.54b}$$

Substitution of Eqs. (4.51) into Eqs. (4.50) and comparison with Eqs. (4.54) yield the result

$$\frac{q_{11}}{1 - q_{11}} = g_{11}d_{11} + g_{12}d_{21} \tag{4.55a}$$

$$\frac{q_{22}}{1 - q_{22}} = g_{21}d_{12} + g_{22}d_{22} \tag{4.55b}$$

Solving Eqs. (4.51) and (4.55) for the control transfer functions d_{11}, d_{12}, d_{21}, and d_{22} yields

$$d_{11} = \frac{g_{22}q_{11}}{\Delta(1 - q_{11})} \tag{4.56a}$$

$$d_{12} = \frac{-g_{12}q_{22}}{\Delta(1 - q_{22})} \tag{4.56b}$$

$$d_{21} = \frac{-g_{21}q_{11}}{\Delta(1 - q_{11})} \tag{4.56c}$$

$$d_{22} = \frac{g_{11}q_{22}}{\Delta(1 - q_{22})} \tag{4.56d}$$

where

$$\Delta = g_{11}g_{22} - g_{12}g_{21} \tag{4.57}$$

It is possible to take a more general approach for multivariable interacting systems by using matrix notation. For example, Eqs. (4.47) can be rewritten as

$$\begin{bmatrix} c_1 \\ c_2 \end{bmatrix} = \begin{bmatrix} g_{11} & g_{12} \\ g_{21} & g_{22} \end{bmatrix} \begin{bmatrix} m_1 \\ m_2 \end{bmatrix} \tag{4.58}$$

or, more generally,

$$\mathbf{c} = \mathbf{Gm} \tag{4.59}$$

where **c** is an $n \times 1$ process output vector, **G** is an $n \times n$ matrix of process transfer functions, **m** is an $n \times 1$ process input manipulation vector, and n is the number of output variables. Eqs. (4.48) can be rewritten as

$$\begin{bmatrix} m_1 \\ m_2 \end{bmatrix} = \begin{bmatrix} d_{11} & d_{12} \\ d_{21} & d_{22} \end{bmatrix} \begin{bmatrix} e_1 \\ e_2 \end{bmatrix} \tag{4.60}$$

or

$$\mathbf{m} = \mathbf{De} \tag{4.61}$$

where **D** is an $n \times n$ matrix of controller transfer functions to be determined and **e** is an $n \times 1$ error vector.

Equations (4.53) can be rewritten as

$$\begin{bmatrix} e_1 \\ e_2 \end{bmatrix} = \begin{bmatrix} r_1 \\ r_2 \end{bmatrix} - \begin{bmatrix} c_1 \\ c_2 \end{bmatrix} \tag{4.62}$$

or

$$\mathbf{e} = \mathbf{r} - \mathbf{c} \tag{4.63}$$

where $\mathbf{r}$ is an $n \times l$ vector of command inputs. Substituting Eq. (4.63) into Eq. (4.61) gives

$$\mathbf{m} = \mathbf{D}(\mathbf{r} - \mathbf{c}) \tag{4.64}$$

Substituting Eq. (4.64) into Eq. (4.59) yields

$$\mathbf{c} = \mathbf{GD}(\mathbf{r} - \mathbf{c}) \tag{4.65}$$

and solving this equation for $\mathbf{c}$ yields

$$\mathbf{c} = (\mathbf{I} + \mathbf{GD})^{-1}\mathbf{GDr} \tag{4.66}$$

where $\mathbf{I}$ is the $n \times n$ identity matrix.

For the system shown in Figure 4.20, the desired transfer functions can be expressed generally as

$$\begin{bmatrix} c_1 \\ c_2 \end{bmatrix} = \begin{bmatrix} q_{11} & q_{12} \\ q_{21} & q_{22} \end{bmatrix} \begin{bmatrix} r_1 \\ r_2 \end{bmatrix} \tag{4.67}$$

$\mathbf{Q}$ can be defined as an $n \times n$ matrix of desired transfer functions where

$$\mathbf{c} = \mathbf{Qr} \tag{4.68}$$

Comparing Eq. (4.68) with Eq. (4.66) yields

$$\mathbf{Q} = (\mathbf{I} + \mathbf{GD})^{-1}\mathbf{GD} \tag{4.69}$$

The matrix of controller transfer functions then can be found by solving Eq. (4.69) for $\mathbf{D}$. The result is

$$\mathbf{D} = \mathbf{G}^{-1}\mathbf{Q}(\mathbf{I} - \mathbf{Q})^{-1} \tag{4.70}$$

Example 4.13 Elimination of interaction   For the system shown in Fig. 4.20, the design objective is eliminating the interaction effects. The following set of equations must therefore be satisfied:

$$c_1 = q_{11}r_1 + 0r_2$$
$$c_2 = 0r_1 + q_{22}r_2$$

or

$$\mathbf{Q} = \begin{bmatrix} q_{11} & 0 \\ 0 & q_{22} \end{bmatrix}$$

In the case of the two-dimensional systems of Fig. 4.20,

$$\mathbf{G} = \begin{bmatrix} g_{11} & g_{12} \\ g_{21} & g_{22} \end{bmatrix}$$

If the inverse matrices in Eq. (4.70) exist, then

$$
D = \frac{1}{\Delta}
\begin{bmatrix}
g_{22} & -g_{12} \\
-g_{21} & g_{11}
\end{bmatrix}
\begin{bmatrix}
\dfrac{q_{11}}{1 - q_{11}} & 0 \\
0 & \dfrac{q_{22}}{1 - q_{22}}
\end{bmatrix}
$$

where $\Delta = g_{11}g_{22} - g_{12}g_{21}$. Thus,

$$
D = \frac{1}{\Delta}
\begin{bmatrix}
\dfrac{g_{22}q_{11}}{1 - q_{11}} & \dfrac{-g_{12}q_{22}}{1 - q_{22}} \\
\dfrac{-g_{21}q_{11}}{1 - q_{11}} & \dfrac{g_{11}q_{22}}{1 - q_{22}}
\end{bmatrix}
$$

from which the individual controller transfer functions can be written as

$$
d_{11} = \frac{g_{22}q_{11}}{\Delta(1 - q_{11})}
$$

$$
d_{12} = \frac{-g_{12}q_{22}}{\Delta(1 - q_{22})}
$$

$$
d_{21} = \frac{-g_{21}q_{11}}{\Delta(1 - q_{11})}
$$

$$
d_{22} = \frac{g_{11}q_{22}}{\Delta(1 - q_{22})}
$$

Note that this result, which has been obtained by matrix methods, is the same as that obtained explicitly in Eqs. (4.56). ◻

Example 4.14 Two-input/Two-output system   An artificial tropical climate is created in the room shown in Fig. 4.21(a) by adding steam to raise the humidity and temperature of cold outside air entering the room. An electric heater is used to further raise the room temperature. Steam valve position $\theta(t)$ and heater voltage $v(t)$ are manipulated by a control computer to control room temperature $c(t)$ and room humidity $h(t)$. It is desired that room humidity and temperature be individually controllable to simulate real climatic fluctuations. A control formulation that decouples room temperature from humidity is required, because steam valve position affects both variables.

Figure 4.21(b) shows a control configuration that achieves noninteracting control of this process. If the room humidity can be modeled using

$$
\tau_\theta \frac{dh(t)}{dt} + h(t) = K_h \theta(t)
$$

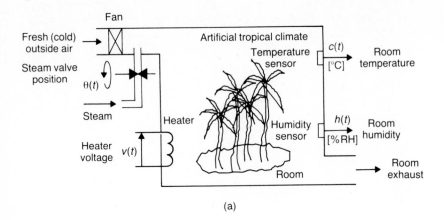

(a)

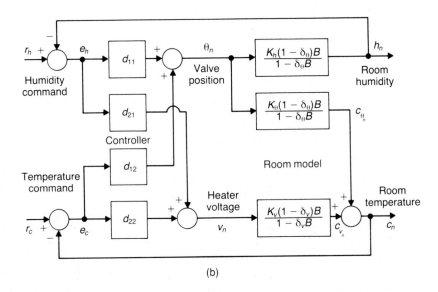

(b)

Figure 4.21 (a) Room with artificial climate to be controlled in Example 4.14, and (b) controller for noninteracting control of room temperature and humidity.

where τ_θ is the time constant associated with the effects of steam valve position in the room, and room temperature can be modeled using

$$\tau_\theta \frac{dc_\theta(t)}{dt} + c_\theta(t) = K_\theta \theta(t)$$

$$\tau_v \frac{dc_v(t)}{dt} + c_v(t) = K_v v(t)$$

$$c(t) = c_\theta(t) + c_v(t)$$

where τ_v is the time constant associated with the effect of electric heater voltage on the room, then (see Fig. 4.20) the individual process transfer functions are

$$g_{11} = \frac{K_h(1 - \delta_\theta)B}{1 - \delta_\theta B}$$

$$g_{21} = \frac{K_\theta(1 - \delta_\theta)B}{1 - \delta_\theta B}$$

$$g_{12} = 0$$

$$g_{22} = \frac{K_v(1 - \delta_v)B}{1 - \delta_v B}$$

where

$$\delta_\theta = e^{-T/\tau_\theta}$$

$$\delta_v = e^{-T/\tau_v}$$

If the desired time constant for climatic changes in the room is τ_d, then the desired room humidity and temperature transfer functions are

$$q_{11} = q_{22} = \frac{(1 - \delta_d)B}{(1 - \delta_d B)}$$

where

$$\delta_d = e^{-T/\tau_d}$$

The individual controller transfer functions can be found from Eqs. (4.56) and (4.57) as

$$d_{11} = \frac{\dfrac{(1 - \delta_d)B}{(1 - \delta_d B)}}{\dfrac{K_h(1 - \delta_\theta)B}{(1 - \delta_\theta B)}\left[1 - \dfrac{(1 - \delta_d)B}{(1 - \delta_d B)}\right]} = \frac{(1 - \delta_d)(1 - \delta_\theta B)}{K_h(1 - \delta_\theta)(1 - B)}$$

$$d_{21} = \frac{\dfrac{-K_\theta(1 - \delta_\theta)B}{(1 - \delta_\theta B)}\dfrac{(1 - \delta_d)B}{(1 - \delta_d B)}}{\dfrac{K_h(1 - \delta_\theta)B}{(1 - \delta_\theta B)}\dfrac{K_v(1 - \delta_v)B}{(1 - \delta_v B)}\left[1 - \dfrac{(1 - \delta_d)B}{(1 - \delta_d B)}\right]}$$

$$= \frac{-K_\theta(1 - \delta_d)(1 - \delta_v B)}{K_h K_v(1 - \delta_v)(1 - B)}$$

$$d_{12} = 0$$

$$d_{22} = \frac{\dfrac{(1 - \delta_d)B}{(1 - \delta_d B)}}{\dfrac{K_v(1 - \delta_v)B}{(1 - \delta_v B)}\left[1 - \dfrac{(1 - \delta_d)B}{(1 - \delta_d B)}\right]} = \frac{(1 - \delta_d)(1 - \delta_v B)}{K_v(1 - \delta_v)(1 - B)} \quad \square$$

4.7
SUMMARY

The synthesis of effective control algorithms is limited only by the designer's imagination and ability in algebra. However, successful performance depends on realistic process models and a good control environment.

The inherent advantage of discrete digital computer control should be apparent. With continuous-process controllers, the analog circuits and hardware required to implement complex control functions may be difficult to conceive. For computer control, the algorithms are directly obtained by forming recursion equations from the discrete controller transfer functions. Thus the designer is limited only by the realizability and the stability of the control algorithm. The designer must, of course, also be responsible for selecting a reasonable sample period, limiting the magnitude of manipulations that are output by a controller in response to command and disturbance inputs, and considering their effect on the process.

Little has been said here about the practical aspects of implementing a control algorithm on a given control computer and the task of interfacing the computer to the process to be controlled. These subjects will be discussed in detail in Chapters 5, 6, and 7. The subject of generating reasonable command inputs for closed-loop systems will be addressed in Chapter 8. This is an important topic, because the magnitude of changes in command inputs from sample period to sample period can greatly affect the success of implementation of the control strategies that have been discussed in this chapter. We will return to the subject of designing discrete control systems in Chapter 12, where we will discuss such techniques as pole placement using root locus, design using frequency response, and design of continuous controllers from which equivalent discrete controllers can be derived.

BIBLIOGRAPHY

Astrom, K. J., and B. Wittenmark. *Computer Controlled Systems: Theory and Design.* Englewood Cliffs, NJ: Prentice-Hall, 1984.

Dorf, R. C. *Modern Control Systems, 4/e.* Reading, MA: Addison-Wesley, 1986.

Franklin, G. F., and J. D. Powell. *Digital Control of Dynamic Systems.* Reading, MA: Addison-Wesley, 1981.

Kuo, B. C. *Automatic Control Systems, 5/e.* Englewood Cliffs, NJ: Prentice-Hall, 1987.

Kuo, B. C. *Digital Control Systems.* New York, NY: Holt, Reinhart and Winston, 1980.

Ogata, K. *Discrete-Time Control Systems.* Englewood Cliffs, NJ: Prentice-Hall, 1987.

Ogata, K. *Modern Control Engineering*. Englewood Cliffs, NJ: Prentice-Hall, 1970.

Ragazzini, J. R., and G. F. Franklin. *Sampled-Data Control Systems*. New York, NY: McGraw-Hill, 1958.

PROBLEMS

4.1 For a process modeled by

$$\frac{dc(t)}{dt} = 5m(t)$$

design a discrete controller to give approximately first-order system response to a step input with a time constant of 0.5 s and zero steady-state error. The sample period is to be 0.1 s.

4.2 A process is described by the discrete transfer function

$$\frac{c_n}{m_n} = \frac{K(1 + B)B}{1 - 2B + B^2}$$

Formulate a control equation such that the closed-loop system behaves as follows:

$$\frac{c_n}{r_n} = \frac{\alpha B}{1 - \gamma B}$$

4.3 A first-order continuous process is to be controlled as shown in Fig. 4.22 with a computer using *proportional control*. The sample period is to be 1 s, and the process model is

$$1.443 \frac{dc(t)}{dt} + c(t) = 10m(t)$$

a. Can "dead-beat" response be achieved via proportional control without steady-state error?

b. Calculate the proportional controller gain K_c that achieves dead-beat response.

c. What is the largest value of K_c for a stable system?

d. What steady-state output error would exist for a unit step input corresponding to your answer in part (c)?

Figure 4.22 System with proportional control for Problem 4.3.

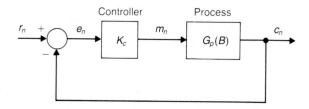

4.4   A process is described mathematically by a pure integration such that the discrete transfer function is

$$G_p(B) = \frac{K_p TB}{1 - B}$$

The discrete controller transfer function chosen is

$$G_c(B) = \frac{1}{K_p T}$$

with sample period $= T$.

a. Plot the system response for a step input $r(t) = R$ for $t \geq 0$. Plot r_n, e_n, m_n, c_n, and $c(t)$ versus t/T.

b. Plot the response for a ramp input $r(t) = At$ for $t \geq 0$. What conclusions can be drawn concerning the error in the system, e_n, when the command is a ramp?

c. Formulate a control algorithm that eliminates the steady-state error found in part (b) when the system is required to follow a ramp input $r(t) = At$. Accomplish the best control possible for the ramp command.

d. For the system with the control algorithm designed in part (c), plot the response to a step input $r(t) = R$. Compare this response with that which you obtained in part (a), and comment on which is better from the standpoint of step command changes.

4.5   A process is characterized as first-order with dead time D.

$$\tau_p \frac{dc(t)}{dt} + c(t) = K_p m(t - D)$$

Find a control equation that represents the best dead-beat controller possible for step changes in command in view of the presence of dead time.

4.6   A process is described by

$$\tau_p \frac{dc(t)}{dt} + c(t) = K_p m(t)$$

Formulate a control equation that forces the closed-loop system to behave as the discrete equivalent of a first-order system defined by

$$\tau_o \frac{dc(t)}{dt} + c(t) = r(t)$$

4.7   A machine drive is equipped with a linear potentiometer for position feedback with a gain of 0.5 volt/cm. The drive is known to behave approximately according to the following equations:

$$0.00002 \frac{d^2\omega(t)}{dt^2} + 0.012 \frac{d\omega(t)}{dt} + \omega(t) = 5v(t)$$

$$\frac{dx(t)}{dt} = 0.1\omega(t)$$

where $v(t)$ is the drive input voltage in volts, $\omega(t)$ is the rotational velocity of the drive in rev/s, and $x(t)$ is the machine position in cm.

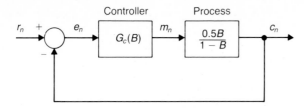

Figure 4.23 Block diagram for Problem 4.9.

a. Develop a discrete control equation for the system that obtains approximately first-order position response to a step input with a time constant of 0.06 s and a sample period of 0.004 s.

b. Repeat part (a) using a simplified process model obtained by eliminating the first and second derivatives in the equation for $\omega(t)$.

c. Compare the control equation you obtained in part (a) with that which you obtained in part (b). Plot and compare the response to a step unit obtained by combining each of the controllers with the complete process model.

4.8 The discrete transfer function for a process is

$$G_p(B) = \frac{B}{1 - 0.5B}$$

Determine the best controller possible to minimize error for the ramp input $r(t) = 2t$. The sampling time is 0.5 s.

4.9 Given the block diagram for a discrete control system shown in Fig. 4.23, find a control equation for m_n such that, if the system is subjected to an input

$$r_n = \frac{2(1 - B^3)}{1 - B} S_n$$

then the error is,

$$e_n = 2(1 - B^3)S_n$$

4.10 A flow control system is subjected to live pressure disturbances p_n. The system block diagram is shown in Fig. 4.24. Select a value of K_c that results in dead-beat response of q_n to step changes in p_n.

Figure 4.24 System with disturbance for Problem 4.10.

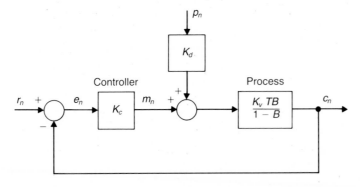

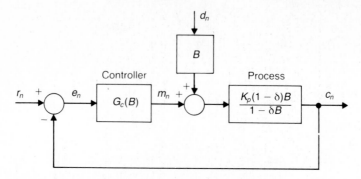

Figure 4.25 System with disturbance for Problem 4.11.

4.11 A control system schematic diagram is shown in Fig. 4.25. It is suspected that disturbances will appear from time to time at the sample instants as step changes of magnitude D. Develop the simplest possible control equation for m_n that will force the deviation in the output c_n to be

$$\Delta c_n = \alpha B^2 S_n$$

4.12 The tank in the liquid-level system shown in Fig. 4.26 has a surface area of 5 m². Flow into the tank can be modeled using

$$q(t) = 0.01\theta(t) \qquad \text{m}^3/\text{s}$$

where $\theta(t)$ is the valve position in degrees. Design a liquid-level controller transfer function for the system that achieves dead-beat rejection of disturbance inputs with a sample period of 10 s.

Figure 4.26 Liquid-level process with disturbance for Problem 4.12.

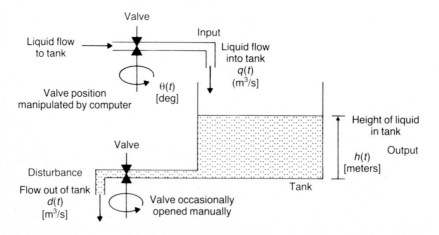

4.13 Plot the response of the system in Problem 4.12 to a step disturbance that occurs midway through a sample period. Explain any deviation from dead-beat response.

4.14 A float gauge that has been used to measure the liquid level in the system in Fig. 4.26 is to be replaced by a servo-controlled tank level gauge that positions a weight at the surface of the liquid and produces a more accurate output. The gauge system is shown in Fig. 4.27(a). Motor current is to be controlled as indicated in the block diagram in Fig. 4.27(b). Design a current controller for the system that produces no steady-state error for step disturbance inputs and achieves 98 % of its response in 0.4 s. The sample period is T seconds.

Figure 4.27 (a) Servo-controlled tank level gauge and (b) block diagram for control of motor current in Problem 4.14.

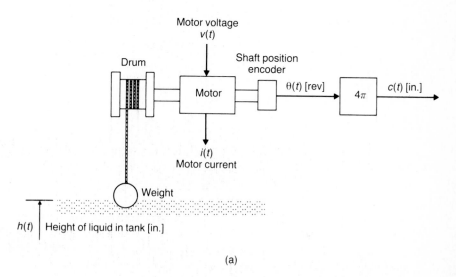

(a)

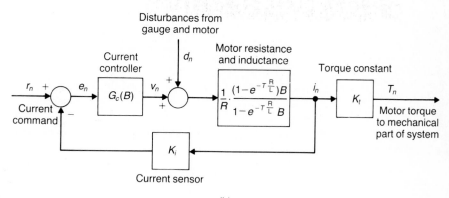

(b)

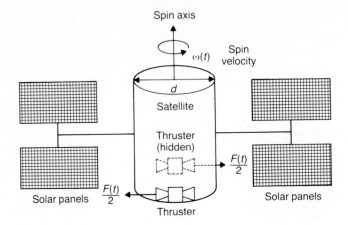

Figure 4.28 Satellite with spin control thrusters for Problem 4.16.

4.15 A process has been estimated to have the following transfer function:

$$\frac{c_n}{m_n} = \frac{K_p(1 - \delta)B}{1 - \delta B}$$

     a. Design a controller that achieves dead-beat response to step command inputs.

     b. It is anticipated that, in the course of operating the process, the process gain could vary $\pm 50\%$ from the estimated value of K_p. Would the closed-loop control system remain stable under the worst conditions?

4.16 The orbiting satellite shown in Fig. 4.28 is spin-velocity-stabilized by a computer control system. The inertia of the satellite is $J = 1000$ kg-m-s^2/rad, its diameter is $d = 1$ m, and its maximum thrust force is $F = 100$ N.

     a. The maximum possible change in spin velocity between samples is to be 0.05 rad/s. Compute the minimum allowable sampling frequency in hertz (Hz).

     b. The discrete transfer function of the spinning satellite is

$$G_p(B) = \frac{\omega_n}{F_n} = \frac{\dfrac{d}{2J} TB}{1 - B}$$

and the sample period is 1 s. Solve for the computer control algorithm that will give a zero-error dead-beat response to a step input as long as the thrusters are not called upon to deliver more than maximum force.

     c. What is the maximum amplitude of step input that will not call for more than the available thruster force?

4.17 A computer controller with sample period T is to operate as shown in Fig. 4.29. The significant time constants in the process are $\tau_1 = 0.02$ s and $\tau_2 = 0.001$ s. Pick an appropriate sample period T.

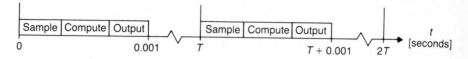

Figure 4.29 Timing diagram for computer controller in Problem 4.17.

4.18 Production control of a factory is described by the block diagram in Fig. 4.30. Factory output is currently measured weekly ($T = 1$ week).
   a. It has been proposed that record-keeping costs would be lowered if factory output were sampled monthly. Will the system be stable if this change is made?
   b. An alternative proposal is to reduce inventory costs by improving the acquisition of production data so that a 1-day sample period can be achieved. Redesign the feedback control for optimal response with this sample period.
   c. Show how feedforward control can be added to the system to help minimize the number of unfilled orders at any given time.

4.19 Modify the block diagram in Fig. 4.24 to include feedforward control that attempts to eliminate deviation in process output due to disturbances. Determine the feedforward control equation required when a sensor is used that directly measures p_n.

4.20 At regular intervals the solar panels on the satellite shown in Fig. 4.28 are reoriented. The effect of the reorientation can be treated as a disturbance $D(t)$ with an effect on spin velocity $\omega(t)$ similar to the thrust force $F(t)$.
   a. Draw a block diagram for the system with closed-loop control showing the disturbance input. (Assume that the units of $D(t)$ and $F(t)$ are identical.)
   b. Show how feedforward control can be added to nullify the effect of reorientation.

4.21 Show how command feedforward control would be added to the system shown in Fig. 4.31. Include the feedforward transfer function to be used.

4.22 The robot arm shown in Fig. 4.32(a) carries a load of W kg. The load being carried is known at all times. Show how command feedforward control can be added to the block diagram in Fig. 4.32(b) to compensate for the static (steady-state) errors in

Figure 4.30 Factory production model for Problem 4.18.

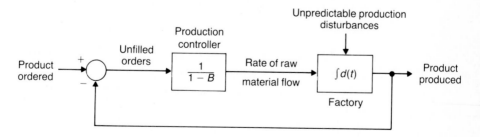

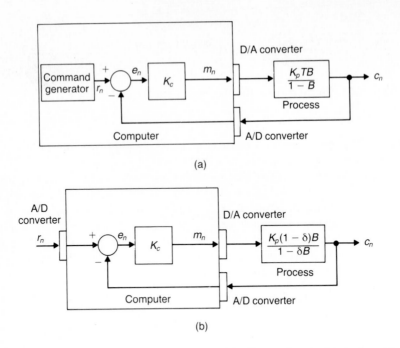

Figure 4.31 Systems to which command feedforward control is to be added in Problem 4.21.

arm position caused by the load. (Note that the feedforward transfer function must be nonlinear, because the torque generated by the load is a nonlinear function of arm position.)

4.23 Repeat Problem 4.22, but use a linear feedforward transfer function that is correct at $\theta = -90°$, $\theta = 0°$, and $\theta = +90°$. Plot the steady-state error in position over the range $-135° < \theta < +135°$. Compare the result with that obtained with no feedforward control (with feedback control only).

4.24 A simplified model for a DC motor/amplifier is shown in Fig. 4.33. Friction and inductance are ignored, and feedback has been added as shown to cancel the back-emf k_e. R is the armature resistance, k_t is the torque constant, J is the inertia of the motor and load, K_{tach} is the tachometer gain, k_{pot} is the potentiometer gain, $m(t)$ is the amplifier input, $d(t)$ is the load torque, $\omega(t)$ is the shaft velocity, $\theta(t)$ is the shaft position, $v(t)$ is the tachometer voltage, and $p(t)$ is the potentiometer voltage.

a. Design a controller to reject step load torque inputs. For a constant velocity command input, $\omega(t)$ should return to its original value two sample periods after a step in $d(t)$.

b. Determine what the sample period must be to limit the position disturbance to $\Delta\theta$ one sample period after a unit step in load torque.

c. Derive the equations for a cascade controller that satisfies part (a) and also achieves a dead-beat response to a position command input $r(t)$. Draw a system block diagram.

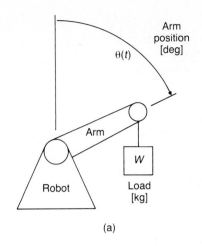

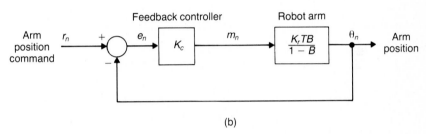

(b)

Figure 4.32 (a) Robot arm and load, and (b) block for control of robot arm in Problem 4.22.

Figure 4.33 Motor/amplifier system to be controlled in Problem 4.24.

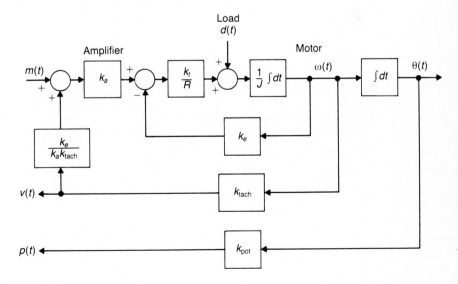

4.25 Draw a block diagram for the system in Fig. 4.33 that includes proportional velocity and position control as well as command and disturbance feedforward control.

4.26 Prove that for the special case of no interaction (g_{12} and $g_{21} = 0$),

$$d_{11} = \frac{q_{11}}{g_{11}(1 - q_{11})}$$

$$d_{22} = \frac{q_{22}}{g_{22}(1 - q_{22})}$$

and

$$d_{21} = d_{12} = 0$$

4.27 A chemical process has a discrete model of the form

$$g_{11} = \frac{a_{11}B}{1 - \gamma_{11}B} \qquad g_{21} = \frac{a_{21}B}{1 - \gamma_{21}B}$$

$$g_{12} = \frac{a_{12}B}{1 - \gamma_{12}B} \qquad g_{22} = \frac{a_{22}B}{1 - \gamma_{22}B}$$

where

$$a_{ij} = K_{ij}(1 - \gamma_{ij})$$

Formulate a set of discrete control equations such that the process interactions are eliminated and

$$q_{11} = \frac{a_d B}{1 - \gamma_d B}$$

$$q_{22} = \frac{a_d B}{1 - \gamma_d B}$$

4.28 Consider Example 3.11 in which a tool offset control system was used to track and compensate for a constant rate of tool wear on a lathe. In Problem 3.20 an exponential tool wear function was introduced, and system performance was found to deteriorate. Design a controller that achieves significantly improved performance over that found in Problem 3.20. What will be some of the theoretical and practical problems with controllers that attempt to compensate for disturbances that grow rapidly? (Note that as the tool wears in a real system, the surface finish deteriorates and the tool must eventually be replaced. The disturbance, therefore, will not grow without bounds.)

Chapter 5
Control Computers

INTRODUCTION

The availability of low-cost computers places a powerful tool in the hands of control engineers for a variety of applications that could not be justified in the past because of the cost of digital control. As a result, implementation of the control theory discussed in the earlier chapters can be accomplished using a wide range of computer configurations. Although the general category of control computers includes large-scale digital computer systems, this chapter focuses on the microcomputer. In order for the microcomputer to be utilized to its maximum capacity and effectiveness, the control engineer must understand its operation and the nature of instruction sets and data storage.

The objective of this chapter is to present important aspects of control computer architecture and operation in a way that is independent of computer manufacturer or model. The chapter begins with a discussion of basic computer organization and the main components in a control computer: the central processing unit, memory, and input/output devices. The concept of binary logic is crucial in all computer operations, and this topic is covered briefly via discussions of logic gates, flip-flops, adders, registers, and so on. For those readers who have not had previous exposure to the binary arithmetic used in computers, the subject is reviewed in Appendix B.

The next topics covered are the concepts of instructions and data: the operations that computers are capable of performing and the operands that the computer is capable of manipulating. This most basic level of computer programming is often referred to as the (computing) machine level. Each

173

computer manufacturer has developed its own machine languages used for programming at this level; thus a simplified set of instructions will be presented here that can be extended to most machine languages.

The concepts of program-driven and interrupt-driven input/output are discussed next. These represent the two main techniques for synchronizing control algorithms with hardware clocks that are used to establish a sample period. The use of high-level languages in programming control algorithms is advantageous in software development and maintenance, and a set of procedures (subroutines) is defined that allows analog voltage input/output and logic input/output as well as interrupt handling. These procedures enable us to more clearly describe various options in controller software organization.

The chapter concludes with a discussion of how closed-loop control functions can be organized on a control computer. Consideration is given to synchronization with clocks to establish the sample period, interrupts, control software organization, time delays, reduction of computation time, and so on. The chapter is organized in such a way that basic concepts of computer technology such as binary logic are presented first, followed by machine language, high-level language, and controller organization. This sequence reflects the various levels of abstraction in which designers of computer control systems must carry out their work.

5.2

BASIC COMPUTER ORGANIZATION

The details of interaction and flow of information between the various components of a control computer are strongly dependent on the architecture of the specific computer being considered. However, the main functional components of a computer are a central processing unit (CPU), memory, and input-output (I/O) devices (see Fig. 5.1). The basic functions performed by these components are described in the following paragraphs.

Figure 5.1 Basic computer components.

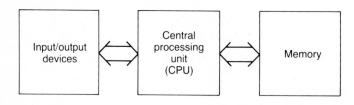

Central Processing Unit (CPU)

The **central processing unit** (**CPU**) of the computer transfers data and instructions from one part of the computer to another, controls the order in which processing instructions are executed, decodes instructions, and executes the instructions by using adders, shifters, comparators, and so on. Instructions and data are taken in and out of memory under the direction of the CPU. Data are also transferred to and from I/O devices by the CPU. This latter function is fundamental to real-time computer control and will be dealt with extensively in later sections.

A **program counter** is used to keep track of the next instruction to be executed by the computer. In normal operation, instructions are stored and executed sequentially; thus the program counter is advanced every time an instruction is executed. An **instruction decoder** determines the action that has to be initiated as a result of the next instruction. For example, the instruction may be decoded and found to be an output instruction. The CPU then initiates the transfer of data from the CPU to an output device.

The CPU contains logic circuits that perform arithmetic and logical operations. **Registers** are provided in the CPU where information can be stored and processed. Accumulators are the main operational registers; they get their name from the fact that they are often used to accumulate partial sums during program execution. Data transfers in and out of the CPU may also be performed via accumulators. There are usually indexing registers that can be used in various modes for addressing locations in memory. Sometimes, general-purpose registers are provided that combine the functions of accumulators and index registers.

A state generator controls the **state** of the CPU. For example, the CPU may be fetching an instruction from memory, it may be executing an instruction, or it may be transferring data to or from memory. Special states may also be provided to stop the CPU, respond to interrupts from external devices, or allow transfers of information directly from I/O devices to memory without intervention by the CPU.

Memory

In most cases, the computer **memory** is of the random access, read/write (RAM) type, but parts of it may be of the read-only (ROM) type. Addresses specifying locations of data in memory are selected by the CPU as a function of the program counter, addresses fetched with instructions, or addresses computed by the CPU. Once a memory address has been selected by the CPU, data can be either transferred from the CPU to that address by a write operation or transferred to the CPU from that address by a read operation.

Input/Output Devices

Among the most basic computer **input/output (I/O) devices** are monitors for alphanumeric display and keyboards for data entry. These allow the user to load programs and data into the computer. The programs can then be executed and the results displayed. Other I/O devices include printers, magnetic tapes, and disks. Control devices such as digital-to-analog converters, analog-to-digital converters, push-button switch inputs, and relay and lamp-driver outputs are also considered I/O devices.

The central processing unit, memory, and I/O devices are components that are basic to most computers, and their varying forms and degrees of sophistication influence the computer's performance and capabilities. Their functions will become more clear in the remainder of this chapter and in Chapter 6. First, however, it is convenient to review a concept that is fundamental to the operation of digital computers: binary logic.

5.3

BINARY LOGIC

Binary logic is used in computer hardware to carry out operations specified by programmed instructions stored in memory. These basic logic operations include AND, OR, NOT, and so on, and the term binary refers to the fact that the result of these operations and their operands can have only one of two possible values: 0 or 1. These values correspond to false and true, low voltage and high voltage, off and on, and so on. Logic operations are performed in a computer by solid-state electronic hardware called **gates**. Although there are a number of solid-state technologies that can be used to make the gates discussed in this section, we will utilize voltages of 0 volts and $+5$ volts as nominal values for logic 0 and logic 1, respectively. Gates react on the order of nanoseconds (a nanosecond is 10^{-9} second) to changes in inputs, and this results in the high-speed operation of computers.

AND Operation

The operation A AND B is often written $A \cdot B$. The value of $A \cdot B$ is determined by the value of A and the value of B as indicated by Table 5.1. It can be seen from the table that $A \cdot B$ has the value "logic 1" only when both variable A and variable B have the value logic 1. Otherwise, $A \cdot B$ has the value logic 0. A typical graphic symbol for an AND gate is shown in Fig. 5.2.

Table 5.1		
AND Logic		
A	B	$A \cdot B$
0	0	0
0	1	0
1	0	0
1	1	1

Figure 5.2 Two-input AND gate.

OR Operation

The operation A OR B is often written $A + B$. The value of the expression $A + B$ is determined by the values of A and B as indicated in Table 5.2. Note that the value of $A + B$ is logic 1 unless both A and B have the value logic 0. Fig. 5.3 is a typical graphic symbol for an OR gate.

Table 5.2		
OR Logic		
A	B	$A + B$
0	0	0
0	1	1
1	0	1
1	1	1

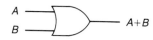

Figure 5.3 Two-input OR gate.

NOT Operation

The operation NOT A is often written $\bar{A}$. Table 5.3 indicates that the value of $\bar{A}$ is logic 1 when A is logic 0; it is logic 0 when A is logic 1. $\bar{A}$ is often called the complement or inverse of A. The NOT gate (Fig. 5.4) is often referred to as an inverter.

Table 5.3	
NOT Logic	
A	$\bar{A}$
0	1
1	0

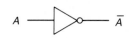

Figure 5.4 NOT gate (inverter).

Other Operations

Several other logic operations are commonly found in computer hardware, including EOR (EXCLUSIVE OR), NAND (NOT AND), and NOR (NOT OR). The result of these operations and the symbols for the gates that implement them are shown in Tables 5.4 through 5.6 and Figs 5.5 through 5.7, respectively. The NAND gate shown in Fig. 5.6 has three inputs, illustrating the fact that more than two inputs can be present. (Two, three, four, and eight input gates can often be found for NAND, NOR, AND, and OR functions.)

Table 5.4

EOR Logic

A	B	$A \oplus B$
0	0	0
0	1	1
1	0	1
1	1	0

Figure 5.5 EOR gate.

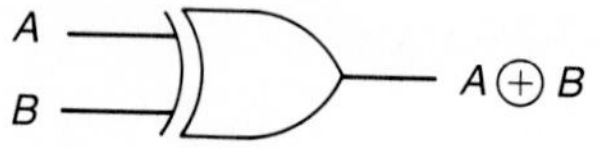

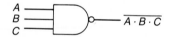

Table 5.5

NAND Logic

A	B	C	$A \cdot B \cdot C$
0	0	0	1
0	0	1	1
0	1	0	1
0	1	1	1
1	0	0	1
1	0	1	1
1	1	0	1
1	1	1	0

Figure 5.6 Three-input NAND gate.

Table 5.6

NOR Logic

A	B	$\overline{A + B}$
0	0	1
0	1	0
1	0	0
1	1	0

Figure 5.7 Two-input NOR gate.

Two other logic devices are among those commonly used. These are **drivers** (Table 5.7 and Figure 5.8) and **bus drivers** (Table 5.8 and Figure 5.9), both of which are used when it is necesary to amplify or strengthen a logic signal so that it can be used in subsequent logic operations. The bus driver includes an enable input that allows the output to be disconnected from the input and placed in a high-impedance state. This effectively creates an open circuit that is useful when multiple outputs are to be connected to a single circuit in the "bus" structures (described in the next chapter) used for communication between different computer components.

Table 5.7

Driver Logic

A	B
0	0
1	1

Figure 5.8 Driver.

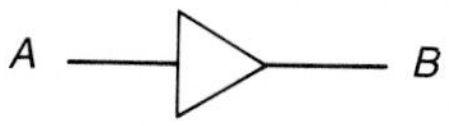

Table 5.8

**Bus Driver Logic
(E = enable input)**

E	A	B
0	0	no connection
0	1	no connection
1	0	0
1	1	1

Figure 5.9 Bus driver.

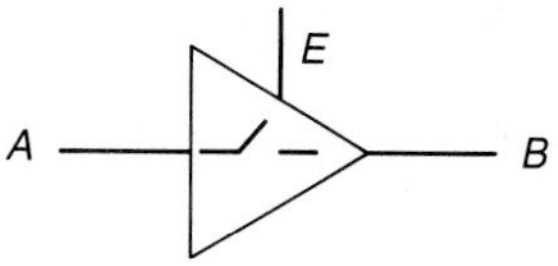

$\bar{R}-\bar{S}$ Flip-Flop

Gates are combined into successively more complex logic circuits to produce the logic functions necessary to perform computer operations. A simple gate combination called the $\bar{R}-\bar{S}$ (RESET–SET) flip-flop is used extensively as a device to implement "memory" functions. Table 5.9 describes its operation. A circuit implementing the flip-flop with two NAND gates is shown in Fig. 5.10. It can be seen that the outputs of the flip-flop, Q and $\bar{Q}$, remain unchanged as long as both inputs are logic 1. A change in output is achieved by setting either input $\bar{R}$ or input $\bar{S}$ briefly to logic 0. The input can then be returned to logic 1. Effectively, the flip-flop "remembers" which input was last set to logic 0. The $\bar{R}-\bar{S}$ flip-flop is often represented using the symbol shown in Fig. 5.11.

Table 5.9

$R-S$ Flip-Flop Logic

R	S	Q	$\bar{Q}$
0	0	1	1
0	1	0	1
1	0	1	0
1	1	Q	$\bar{Q}$

Figure 5.10 $R-S$ flip-flop implemented with NAND gates.

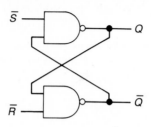

Figure 5.11 $R-S$ flip-flop.

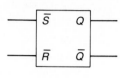

D Flip-Flop

Additional logic can be added to the $R-S$ flip-flop to modify its operation so that it can be used more conveniently in computer operations. Analysis of the D flip-flop circuit shown in Fig. 5.12 reveals that the outputs of this flip-flop remain unchanged while the input C is at logic 0. When C is changed to logic 1, the output Q takes on the value of input D and will follow any changes in D until C is returned to logic 0, at which point output Q will be the value of D when C changed from 1 to 0. As indicated in Table 5.10, the flip-flop can be viewed as sampling D while C is at logic 1 and holding the value of D when C goes from logic 1 to logic 0. D is often referred to as the data input, and C is referred to as the clock input. The D flip-flop is often represented using the symbol shown in Fig. 5.13.

Table 5.10

D Flip-Flop Logic

C	D	Q	$\bar{Q}$
0	0	Q	$\bar{Q}$
0	1	Q	$\bar{Q}$
1	0	0	1
1	1	1	0

Figure 5.12 D flip-flop implemented with NAND and NOT gates.

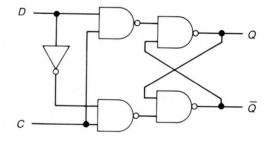

Figure 5.13 D flip-flop.

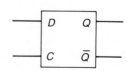

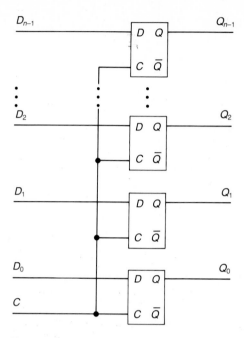

Figure 5.14 An *n*-bit register or buffer
constructed with *D* flip-flops.

Register or Buffer

D flip-flops are particularly useful in computer hardware because registers, buffers, and other groups of memory functions can be readily constructed using them. The register shown in Fig. 5.14 stores *n* bits of data input at the command of a single signal *C*.

Adder

Many addition functions need to be performed in a computer, including the calculation of addresses in memory and the excecution of addition instructions in computer programs. An adder, capable of adding input *A* to input *B* and of generating sum *S* and carry *C* outputs can be built from gates as shown in Fig. 5.15. This is often called a half-adder, because it is not capable of handling a

Figure 5.15 Half-adder circuit.

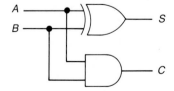

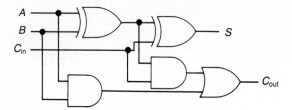

Figure 5.16 Full-adder circuit.

carry input as generally required for binary addition. A full-adder with carry input C_{in} is shown in Fig. 5.16. It should be noted that the full-adder shown is constructed from two half-adders.

Ripple-Carry Adder

Full-adders can be cascaded to form an n-bit ripple-carry adder as shown in Fig. 5.17. This is referred to as a ripple-carry adder, because carries generated in

Figure 5.17 An n-bit ripple-carry adder constructed with n full-adders.

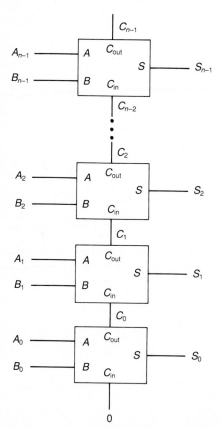

less significant bits propagate toward the most significant bit. Other adder circuits can be constructed with improved performance and implementation characteristics.

5.4

INSTRUCTIONS AND DATA

Computers operate only on binary sequences; therefore, these sequences must represent both data and instructions. A binary sequence stored in memory can be interpreted as an instruction to perform operations on data stored somewhere else in the computer memory, or it can be interpreted as data in the form of binary numbers to be operated on by an instruction. The programmer puts both data and instructions in appropriate locations in memory and determines which binary sequences the computer is to execute as instructions and which binary sequences form the data. Each computer has its own design, and the same binary sequence may mean different instructions to different computers. The language of binary numbers that the computer understands is called its **machine language**.

It is desirable to allow the programmer to write in a notation other than the binary sequences of a machine language. This capability is provided by a program called the assembler. An **assembler** recognizes a set of combinations of alphanumeric characters and symbols that form its language, and this set of symbols and character combinations is called the **assembly language**. Assembly language programming for a specific computer involves a set of alphanumeric symbols called **mnemonics**, each of which represents an instruction that can be performed by the computer. The assembler establishes a correspondence between these mnemonic symbols and the binary instructions that the computer understands. Features are also provided in an assembly language to define data conveniently, such as in the form of decimal numbers.

A computer generally can perform only one instruction at a time, and more complex functions such as arithmetic statements have to be performed by a set of instructions. For example, the statement "set F to the value of $X + Y - Z$" can be written

$$F \leftarrow X + Y - Z \qquad (5.1)$$

and could be programmed using instructions such as

Load X into a register.

Add Y to the contents of the register.

Subtract Z from the result in the register.

Store register contents in F.

In assembly language, each of these instructions is represented by the mnemonic that corresponds to that particular instruction. An assembly-language program for a very simple computer with one register could be

```
LOAD       X
ADD        Y
SUBTRACT   Z
STORE      F
```

The variables X, Y, Z, and F would be assigned to particular locations in the computer's memory by other portions of the assembly-language program. An assembler program would then be used to convert the above program into binary machine-language instructions and data.

With most computers, programs can also be written in higher-level languages such as BASIC, FORTRAN, Pascal, Ada, and "C." Arithmetic statements such as Eq. (5.1) can be entered and then translated into a set of machine-language instructions to perform the desired operations. The program that does the translation is called a **compiler**. Hence a program written in the FORTRAN language would be processed by a FORTRAN compiler before the desired operations were performed by a computer.

Computers are generally organized using fixed-length binary sequences for machine-language instructions and data. Such a sequence contains a fixed number of binary bits (0 or 1) of information and is usually referred to as a **word**. More than one word may be required to store an instruction or data on some computers, particularly those with short word lengths. As shown in the example given in Fig. 5.18, part of the binary sequence representing an instruction indicates the type of instruction that the sequence specifies. This set of bits is called the **instruction code**. The remaining part of the instruction gives more information to indicate the **operand** to be operated on. Every computer design has its own architecture and its own instruction set. The discussion here will focus on the various types of instructions that are common to most computers.

Typical types of computer instructions are listed in Table 5.11. In normal operation, the processor executes instructions stored sequentially in memory unless the instruction itself specifies otherwise. Such instructions include "jump," "go to," and "subroutine call" instructions, among others. The instructions specified by the instruction code are detected in the CPU by the instruction

Figure 5.18 Instruction consisting of an instruction code and an operand.

Instruction code	Operand
0 0 0 1 0 1 1 1	1 0 1 1 1 0 1 0

Table 5.11

Typical Computer Operations

Load
Store
Add
Subtract
Multiply
Divide
Increment or decrement
Negate
Complement
And
Or
Exclusive or
Transfer from register to register
Shift right or left
Jump or branch
Subroutine, function, or procedure call
Input
Output

decoder, which then initiates the operation. The operand is specified in a way unique to the computer design. In most computers, there are several modes of addressing memory locations. These modes include direct addressing, wherein the operand contains the memory address of the data to be operated on; indexed addressing, wherein an index register contains the address of the data to be operated on; indirect addressing, wherein the memory location specified by the operand contains the address of the data to be operated on; and immediate addressing, wherein the operand constitutes the actual data to be operated on rather than the address of the data.

Load and Store Instructions

Most operations in a computer are performed in registers rather than in memory. This is because the registers are internal to the CPU, and the CPU contains the logic circuits necessary to perform the operations. It is necessary to be able to **load** these registers with data stored in memory and to be able later to **store** the data in the register back into the memory, perhaps after the data has been modified by other operations.

Many computers have LOAD and STORE instructions to perform these operations. As an example, consider a simple computer with a single register for data manipulation. If it is desired to move the data stored in memory location 6 to memory location 14, then the assembly-language program to do so might be

```
LOAD    6
STORE   14
```

The first instruction loads the data stored in memory location 6 into the register. The second instruction then stores the data in the register into memory location 14. No operations were performed on the data while it was in the register, so it is unchanged.

Some computers provide a general purpose MOVE instruction that can be used to move data from memory location to memory location. For such a computer, the assembly-language instruction

```
MOVE  6,14
```

might be used to move the data in memory location 6 to memory location 14.

Arithmetic and Logic Instructions

Arithmetic and **logic** instructions involve arithmetic or logic operations between a word in a register and a word in a memory location. Typically, data to be operated upon is loaded into a register in the CPU prior to performing arithmetic or logic operations. Suppose again that only a single register is available for data manipulation. If Eq. (5.1) is to be implemented, for example, it is first necessary to know where the variables X, Y, Z, and F are stored in the computer memory. If they are stored in memory locations 100, 101, 102, and 103, respectively, the program becomes

```
LOAD      100     (get X in register)
ADD       101     (add Y to register)
SUBTRACT  102     (subtract Z from register)
STORE     103     (store result in F)
```

To make programs more readable and easier to write, most assemblers allow programs such as this to be written in the form

```
X   EQUALS    100
Y   EQUALS    101
Z   EQUALS    102
F   EQUALS    103

    LOAD      X
    ADD       Y
    SUBTRACT  Z
    STORE     F
```

where the EQUALS instructions are actually **pseudo-instructions** whose only purpose is to tell the assembler the meaning (location where data is stored) of the characters X, Y, Z, and F. They are not instructions to be excecuted by the computer.

Besides ADD and SUBTRACT instructions, other possible arithmetic instructions include NEGATE, MULTIPLY, and DIVIDE. If the object is to perform logic operations on data, logic instructions such as AND, OR, and NOT are normally available. For example, the logic statement

$$C \leftarrow A + \bar{B} \tag{5.2}$$

could be programmed as

```
LOAD    B     (get B in register)
NOT           (complement B)
OR      A     (or result with A)
STORE   C     (store result in C)
```

The immediate addressing mode is often convenient when one is programming arithmetic and logic operations. For example, a variable C representing a temperature in the Celsius (°C) scale can be converted to a variable F in the Fahrenheit (°F) scale by using

```
LOAD      C
MULTIPLY  #9
DIVIDE    #5
ADD       #32
STORE     F
```

where the "#" symbol in the operands #9, #5, and #32 signifies that the number that follows is the actual data to be used in the operation rather than the address of the data in memory. An ADD instruction with immediate addressing would have a different instruction code than an ADD instruction with memory addressing.

A program to set all bits in a variable B to zero except the four right-most bits, which are to be left unchanged, is

```
LOAD   B
AND    #15
STORE  B
```

The binary representation of the decimal number 15 is $0 \cdots 001111$. When each bit in this pattern is ANDed with the corresponding bit in the register, all bits ANDed with zero become zero regardless of their previous value, whereas all bits ANDed with 1 retain their previous value (see Table 5.1).

Branching Instructions

Instructions are performed sequentially by a computer unless otherwise specified by **branching instructions**. Two common branching instructions are JUMP and procedure (subroutine) CALL instructions. These instructions change the order of instruction execution and transfer control to other sets of instructions stored elsewhere in the memory. The operand of a JUMP instruction specifies the location of the next instruction to be executed. Consider the program

```
        ORIGIN  200
200  LOAD     6
201  JUMP     300
202  _____

        ORIGIN  300
300  STORE    14
301  _____
```

Here the ORIGIN pseudo-instruction is used to define where instructions are to be stored in the computer memory, and a JUMP instruction is used to transfer the sequence of execution from one set of instructions to another. The effect of the program is again to move the data in memory location 6 to memory location 14. However, the program is now split into two parts. One part begins at memory address 200 and consists of the LOAD instruction in location 200 and a JUMP instruction in location 201. The STORE instruction now does not directly follow the LOAD instruction but instead has been placed in the second part of the program at address 300, as defined by the second origin pseudo-instruction.

When the program is executed by the computer beginning at memory location 200, the LOAD instruction is executed first. The next instruction is the JUMP instruction, which, when executed, causes the STORE instruction at location 300 to be executed next rather than whatever instruction was stored in location 202. After the STORE instruction is executed, the instruction in location 301 is executed. This example illustrates the important point that both data and instructions are stored in computer memory. In this example, data is stored in locations 6 and 14, and instructions are stored in locations 200, 201, and 300.

The procedure CALL instruction differs from the JUMP instruction in that the location from which the CALL is made is automatically saved. This can then be used to return to the next instruction after the CALL after the instructions in the procedure are executed. For example, the previous program could be rewritten as

```
        ORIGIN  200
200  CALL     300   (CALL procedure)
201  _____
```

```
          ORIGIN  300
   300  LOAD     6      (LOAD/STORE procedure)
   301  STORE   14
   302  RETURN          (RETURN from procedure)
   303  _____
```

In this case, execution of the CALL instruction in location 200 causes the LOAD instruction in location 300 to be executed next. However, the location after the CALL instruction is saved by the computer so that when the RETURN instruction is executed, control is transferred back to location 201 rather than continuing at 303.

Other branching instructions conditionally transfer execution to a new location on the basis of the result of previous operations. For example, a JUMP IF POSITIVE instruction causes a jump to occur only if the number stored in the register of the simple computer system being described here is positive. This is illustrated in the following program for obtaining the absolute value of a number.

```
       ORIGIN              200
  200  LOAD                6
  201  JUMP IF POSITIVE    203    (skip negate instruction if positive)
  202  NEGATE
  203  STORE               14
```

In this program, the number stored in location 6 is loaded into the register. If the number, now in the register, is positive, the JUMP IF POSITIVE instruction causes execution to be transferred to the STORE instruction, storing the positive number in location 14. If the number in the register is negative, however, the JUMP IF POSITIVE instruction does not cause a jump. The NEGATE instruction is executed in this case, the result of which is positive and is subsequently stored in location 14.

Other conditioned branching instructions could include

```
JUMP  IF  ZERO
JUMP  IF  NOT ZERO
JUMP  IF  NEGATIVE
JUMP  IF  CARRY/BORROW
JUMP  IF  OVERFLOW
```

These instructions play a very important role in decision making and logical operations.

Shifting and Rotating Instructions

When programming algebraic functions, it often becomes necessary to shift or rotate numbers in registers. Rotating a word involves shifting the bits "end around," with all bits being shifted right or left. The least significant bit then is

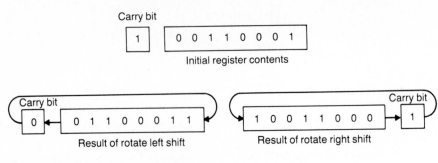

Figure 5.19 Examples of rotate shifts.

shifted into a carry bit, and the carry bit is shifted into the most significant bit for a right rotate (or vice versa for a left rotate). Examples of right and left **rotate shifts** are shown in Fig. 5.19.

Another important set of shifting instructions are the **arithmetic shifts** illustrated in Fig. 5.20. In the case of an arithmetic right shift, the sign bit[1] is propagated toward the right. This has the effect of dividing the number in a register by 2. An arithmetic left shift has the effect of multiplying the number by 2. If an arithmetic left shift is performed on a word and its sign bit is changed, an overflow error has occurred. A program to multiply A by 10 and store the result in B could be

```
LOAD    A                      (get A)
ARITHMETIC LEFT SHIFT          (times 2)
ARITHMETIC LEFT SHIFT          (times 2)
ADD     A                      (plus A is now times 5)
ARITHMETIC LEFT SHIFT          (times 2)
STORE   B                      (result is times 10)
```

Figure 5.20 Examples of arithmetic shifts.

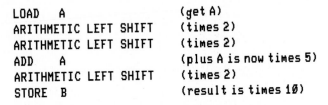

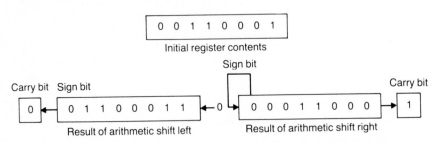

[1] The sign bit is the left-most bit of a 2's-complement binary word. This topic is reviewed in Appendix B.

Input/Output Instructions

Information transfers to input/output (I/O) devices are often programmed with special sets of INPUT and OUTPUT instructions. I/O operations are necessary to communicate with peripherals, be they digital-to-analog convertors, printers, or other devices.

One class of I/O operations is used to check the **status** of a device. Most peripheral devices have a means of telling the computer their status—indicating whether or not they are ready to transmit or receive binary data. Before any information transfer is initiated, the device status is usually checked, and if the device is not ready, the program waits until it is. Another type of I/O operation initiates an action in the device. Examples of such operations are stopping a tape reader and starting an analog-to-digital conversion. There are also, of course, I/O operations that initiate transfer of a word of data from a device to the computer and corresponding operations that transfer data from the computer to a device.

Suppose that a simple computer system has one register and has three I/O devices. Devices 1 and 2 are binary logic inputs, and device 3 is a binary logic output. If it is desired to make the output of device 3 logic 1 when either but not both of the inputs of devices 1 and 2 are logic 1, then the program could be as follows:

```
TEMP   EQUALS        100       (temporary storage location)
       ORIGIN        200
200    INPUT         1         (get first logic input)
201    STORE         TEMP      (temporarily save it)
202    INPUT         2         (get second logic input)
203    EXCLUSIVE OR  TEMP      (exclusive or of two inputs)
204    OUTPUT        3         (output result)
205    JUMP          200       (repeat)
```

Note that TEMP is a temporary storage location in memory and that the EXCLUSIVE OR instruction is used to perform the necessary logical computation. The JUMP instruction causes the instructions to be repeated as rapidly as possible, and the logic function is continually performed until the program is halted. Operating speeds vary greatly, but each pass through the program would take less than 10 microseconds on many computers.

5.5

PROGRAM-DRIVEN INPUT/OUTPUT

Computers typically perform operations much faster than I/O devices, and hence there is a need to synchronize the computer with these devices and the processes to which they are connected. In many cases the computer must wait or be

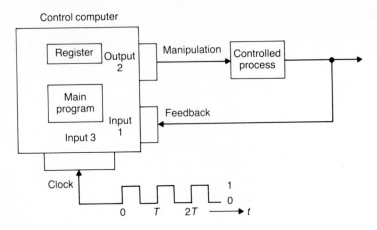

Figure 5.21 Control using program-driven I/O.

instructed to do other tasks between data transfers in order to allow I/O devices to finish their task and become ready. If these delays are program-controlled, the transfers are called program-driven I/O. Suppose that the control computer used to close the feedback loop in the system shown in Fig. 5.21 has one register and three I/O devices. Device 1 is the process feedback input device, device 2 is the process manipulation output device, and device 3 is a clock input device that is used to establish the sample period. The program to implement the proportional control tasks outlined in Section 4.2 then could be as follows:

```
R      EQUALS            100     (location of command in memory)
K      EQUALS            101     (location of gain in memory)
       ORIGIN            200
200    INPUT             3       (wait for clock = 0)
201    JUMP IF NOT ZERO  200
202    INPUT             3       (wait for clock ≠ 0)
203    JUMP IF ZERO      202
204    INPUT             1       (input feedback into register)
205    NEGATE                    (negate)
206    ADD               R       (add command to get error)
207    MULTIPLY          K       (multiply error by gain)
208    OUTPUT            2       (output manipulation from register)
209    JUMP              200     (repeat)
```

The first INPUT instruction transfers the value of the clock into the register. This value changes from zero to 1 at the beginning of each new sample period. The JUMP IF NOT ZERO instruction is used to repeat the input until the clock value is zero. The JUMP IF ZERO instruction then causes the clock input to be repeated until the clock value that is input into the register changes to 1. The third INPUT instruction transfers the value of the feedback into the register. The error is computed by negating the feedback value and adding the command value stored in the memory location associated with R. The manipulation is calculated by multiplying by the gain stored in the memory location associated with K. The OUTPUT instruction is then used to send out the manipulation calculated in the register. Next, the instruction sequence is repeated using the JUMP instruction. In this case, the computer is synchronized with the clock input so that control is performed only once per sample period. The synchronization is established using the INPUT and JUMP instructions at the beginning of the program, and hence the synchronization is explicitly programmed. It should be noted that this program is very much simplified and ignores many possible problems, including any checking of feedback device status that may be required prior to input consideration of word lengths, and overflow detection and correction in the control calculations. Also, the desired command and gain values must be placed in memory locations R and K prior to execution of this program.

5.6

INTERRUPT-DRIVEN INPUT/OUTPUT

When a computer is communicating with peripheral devices or controlled processes, a large amount of the computer's time can be spent waiting for these devices or processes; hence a more efficient use of the computer's time is desirable. The **interrupt** facility available on most computers allows utilization of computer time that might otherwise be spent waiting for relatively low-speed devices and process control operations.

As the name suggests, the interrupt philosophy is such that a device can interrupt the normal operation of the computer when it is ready to send or receive data. An analogy is a ringing telephone. The fact that it rings when attention is required means that it is not necessary to constantly check the telephone to determine whether anyone is on the line. A program interrupt is usually generated by activating a special interrupt input on the CPU. If an interrupt is to be used with an I/O device, the device status can be connected to the interrupt input. An interrupt can then be generated on the computer whenever the I/O device is ready to transfer data to or from the CPU.

The interrupt facility usually can be enabled and disabled under program control, and special instructions are used for this purpose. When the facility is enabled and a device causes an interrupt, the computer recognizes the interrupt, saves the state of the computer (including the address of the next instruction that would have been executed), and jumps to a specific location in memory where the user has placed an interrupt-handling procedure (subroutine). This procedure typically performs the desired I/O operations. After the interrupt has been serviced, a return instruction is executed to cause control to be returned to the interrupted program at the saved address. When interrupts are disabled, no interrupts can occur.

If there is more than one device that can cause an interrupt, it must first be determined which device generated the interrupt. Control can then be transferred to the proper interrupt procedure to service that device. In some computers there is more than one interrupt input, each of which has associated with it a specific location in memory to which the computer should automatically jump when an interrupt occurs on that input. In this way, the software does not have to be written to search for the device causing the interrupt. In some computers, all registers and the program counter are automatically saved when an interrupt occurs and are restored automatically at the end of the interrupt-servicing procedure. In other computers, this must be done via the interrupt procedure. Interrupt inputs must often be reset in the interrupt procedure to prevent a given interrupt event from erroneously generating another interrupt.

In control situations, a sample-period clock is often connected to an interrupt input, and the control tasks described in Section 4.2 are included in the interrupt-servicing procedure. The clock then causes an interrupt at the beginning of every sample period. Alarm and emergency situation indicators can also be connected to interrupt inputs so that the computer software can immediately respond to such contingencies.

The control program given in the previous section can readily be modified to use the interrupt-driven I/O approach rather than the program-driven I/O approach. In this case the clock signal is connected to the CPU as an interrupt (see Figure 5.22) rather than as an input device. Given that an interrupt is generated on the rising edge of the clock signal and that the contents of the register used in

Figure 5.22 Control using interrupt-driven I/O.

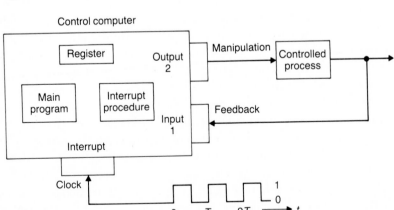

the interrupt procedure (and the main program) are not automatically saved, the new main program and interrrupt procedure could be as follows:

```
          ORIGIN      200      (main program)
200  CONNECT      300      (connect interrupt program to interrupt)
201  LOAD         A        (increment A)
202  INCREMENT
203  STORE        A
204  JUMP         201      (repeat)

          ORIGIN      300      (interrupt procedure)
300  STORE        SAVE     (save register contents)
301  INPUT        1        (get feedback)
302  NEGATE                (change sign)
303  ADD          R        (calculate error)
304  MULTIPLY     K        (calculate manipulation)
305  OUTPUT       2        (output manipulation)
306  LOAD         SAVE     (restore register contents)
307  RETURN                (return to main program where
                            interrupted)
```

The first instruction in the main program is a CONNECT instruction and is used to establish, in the CPU, the address of the procedure to be called when an interrupt occurs. The next four instructions are continually repeated (except when interrupted), and they perform the (unnecessary) function of continually incrementing the number stored in the memory location associated with A. When a clock interrupt occurs, execution of these instructions is halted and the next instruction to be performed is the STORE instruction in the interrupt procedure. When the control computations have been completed and the manipulation has been output, the RETURN instruction is used to transfer back to the instruction that was interrupted in the main program. Here the majority of the main program serves little useful purpose, but in practice, these instructions could be replaced by many instuctions performing useful functions while control interrupts are not being processed. The most important thing to note in this case is that synchronization between the computer and the clock is accomplished by the interrupt hardware in the CPU to which the clock is connected.

5.7

HIGH-LEVEL LANGUAGES FOR COMPUTER CONTROL

The ease or difficulty of implementing control algorithms on a control computer is highly dependent on the nature of the programming language used and on the presence of appropriate hardware and software for input, output, and timing.

Hardware interfaces for computer control are discribed in the next chapter. In this section, a set of procedures (subroutines) are defined that are callable from whatever language is being used, be it assembly language or a high-level language such as BASIC, FORTRAN, Pascal, Ada or "C." These procedures can be called in control programs and used to provide the input, output, and timing functions necessary to perform both open-loop and closed-loop control. This system is illustrated in Fig. 5.23 and the procedures are defined in the following paragraphs.

Figure 5.23 Control computer with analog I/O, logic I/O, and interrupt-handling procedures.

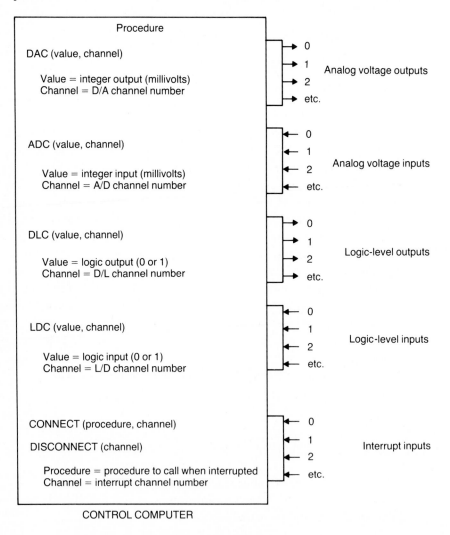

DAC (Digital-to-Analog Conversion)

The **DAC** procedure is used to output a binary integer to a digital-to-analog (D/A) converter device, producing a voltage proportional to the binary integer. Here the **D/A converter** is assumed to have several output channels numbered 0, 1, 2, ... The channel on which it is desired to set the output voltage is a parameter in the calling statement for the procedure, along with the integer value to be converted. For example, the D/A converter could produce a voltage in millivolts at its output that is equal to the binary integer. If it were desired to output 1.067 volts on D/A converter channel 3, the statement used would be

$$\text{DAC (1067,3)}$$

The D/A converter is typically used to output manipulating voltages to the process being controlled. The scaling of one bit per millivolt chosen here was arbitrary, as will be seen in the discussion of D/A converter hardware in Section 6.4.

ADC (Analog-to-Digital Conversion)

ADC is the procedure used to input the integer result of an analog-to-digital (A/D) conversion. The **A/D converter** can have a number of channels, and one must specify the channel number when calling the procedure. The procedure returns an integer proportional to the input voltage on the specified channel in millivolts. For example, if the gain of the A/D converter is one bit/millivolt and −2.5 volts are present on A/D converter input channel 2, then

$$\text{ADC (I,2)}$$

results in the integer variable *I* being set to −2500. A/D conversion and the scaling input voltages to binary data will be treated in greater detail in Section 6.5.

DLC (Logic Output)

The **DLC** procedure is used to set the state of a specified output to either logic 1 or logic 0 (true or false). Again, a number of output channels may be present on the logic output device. If it is desired to set output channel 5 to logic 1, the statement used is

$$\text{DLC (1,5)}$$

Hardware for logic-level output from a computer is described in Section 6.3.

LDC (Logic Input)

The **LDC** procedure is used to input the present logic value of a specified input channel. For example, if input channel 8 is at logic 1, then after

LDC (L,8)

the value of the logical variable L will be logic 1 (true). If input channel 8 is at logic 0, then the value of logical variable L will be logic 0 (false). Logic-level input hardware is described in Section 6.3.

CONNECT (Connect Control Procedure to Interrupt Input)

The **CONNECT** procedure is used to associate a given interrupt input with a specific control procedure provided by the programmer. It is assumed that when an interrupt signal is applied to an interrupt input before the CONNECT procedure is called, the computer ignores it and no interrupt occurs.

When CONNECT is called with the name of the procedure that is to be called when an interrupt on the specified interrupt input channel occurs, the appropriate hardware and software linkage is established in the computer to allow the specified interrupt procedure to be called. For example, if the programmer has written an interrupt procedure called CONTROL that is to perform control calculations, then

CONNECT (CONTROL,1)

associates procedure CONTROL with interrupt input channel 1. Whenever an interrupt signal is applied to interrupt input 1 (perhaps by connecting it to a clock signal with period T), whatever program is currently being processed by the computer is interrupted, its state is completely saved, and the CONTROL procedure is called. Interrupt interfacing hardware is described in Section 6.8.

DISCONNECT (Disconnect Control Procedure from Interrupt Input)

The **DISCONNECT** procedure is used when it is no longer desired to have an interrupt input cause interrupts. If the CONNECT procedure has been used to link a procedure CONTROL to interrupt input channel 1, then

DISCONNECT (1)

causes CONTROL no longer to be called when a signal is applied to interrupt input 1. An interrupt signal on a given input channel can be assumed to be ignored when no procedure is connected to that channel.

These procedures have been defined here so that they can be used in examples throughout the remainder of this text. They will be used to program both open-loop and closed-loop control, using both program-driven interrupt-driven I/O approaches. They are representative of those found on many computer systems equipped for real-time control. In the event that a given computer system is to be used for real-time control and does not possess standard procedures or subroutines such as these, they can be programmed either in assembly language or in a high-level language once the details of the interfaces are known.

5.8

CLOSED-LOOP COMPUTER CONTROL

A typical system block diagram for closed-loop control is shown in Figure 5.24. Algorithm 5.1 describes the control function to be performed by the computer. The command r_n is either generated by the computer or obtained from an external signal such as a keyboard input or sampled analog signal. Program 5.1 is an example of a control procedure written to perform approximate PID control using the control equation

$$m_n = m_{n-1} + k_0 e_n + k_1 e_{n-1} + k_2 e_{n-2} \tag{5.3}$$

If k_p, k_i, and k_d are the desired proportional, integral, and derivative gains and T is the sample period, then, as developed in Section 2.3,

$$k_0 = k_p + k_i T + \frac{k_d}{T} \tag{5.4}$$

$$k_1 = -k_p - \frac{2k_d}{T} \tag{5.5}$$

$$k_2 = \frac{k_d}{T} \tag{5.6}$$

Figure 5.24 Closed-loop process control.

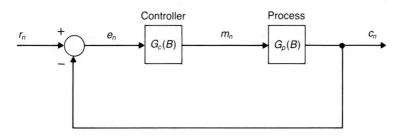

1. Sample c_n
2. Calculate $e_n \leftarrow r_n - c_n$
3. Calculate $m_n \leftarrow m_{n-1} + k_0 e_n + k_1 e_{n-1} + k_2 e_{n-2}$
4. Output m_n
5. Update m_{n-1}, e_{n-1}, and e_{n-2}
6. Done

Algorithm 5.1 Control algorithm.

The control procedure CONTROL given in Program 5.1 uses A/D channel 1 for process feedback and D/A channel 1 for manipulation output.

Programmed Synchronization to an External Clock Signal

In order to carry out control with a given sample period T, it is necessary to supply a clock with period T and to synchronize control software with this clock signal. As discussed in Sections 5.5 and 5.6, there are two major alternatives for synchronizing the execution of the control procedure to a logic-level clock signal: program-driven synchronization and interrupt-driven synchronization.

The first alternative involves continually testing the clock input signal and detecting the instant at which it changes from logic 0 to logic 1. Figure 5.25 illustrates a control system that uses a logic input signal to establish the sample period. Algorithm 5.2 describes how the computer synchronizes the calling of the control procedure with the logic-0-to-logic-1 transition of the clock on logical input channel 1. Program 5.2 is one possible implementation of Algorithm 5.2.

Note that when this algorithm is used, r_n is not changed once closed-loop control begins. This illustrates a major disadvantage of this approach. New values of r_n cannot be input easily from the keyboard while the process is under control, because the computer must be constantly checking the clock signal and activating the control procedure at the appropriate instants in time. It is usually not possible

Program 5.1 Procedure for PID control.

```
procedure CONTROL;
begin
    ADC(Cn, 1);                                {sample on A/D channel 1}
    En := Rn - Cn;                             {calculate error}
    Mn := Mn1 + K0 * En + K1 * En1 + K2 * En2; {PID control equation}
    DAC(Mn, 1);                                {output on D/A channel 1}
    En2 := En1;                                {update error history}
    En1 := En;
    Mn1 := Mn                                   {update manipulation history}
end;
```

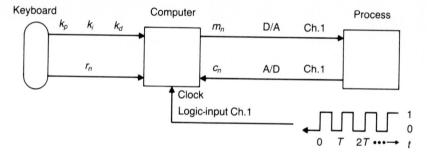

Figure 5.25 Control system with logic input for clock.

to "open the loop" while waiting for an operator to enter a new value of r_n on the keyboard, particularly when the sample period is shorter than the time required to key in a new value of r_n.

Synchronization with an External Clock Using Interrupts

Figure 5.26 illustrates a closed-loop computer control system synchronized to a clock connected as an interrupt input. In this case, Algorithm 5.3 describes the control system. Here the CONNECT routine is used to associate the interrupt procedure CONTROL with interrupt channel 1. An initial command r_n is read prior to activation of the controller, which occurs when the CONNECT procedure is executed. Program 5.3 is one possible implementation of Algorithm 5.3.

Because interrupts are used, the interrupt procedure CONTROL is automatically called whenever the interrupt input signal goes from 0 to 1 (false to true) regardless of what is occurring in the main program. Hence this portion of the program can be performing other operations and computations that are briefly interrupted by the control program. In this example, the command variable r_n is

Algorithm 5.2 PID control synchronization using logic input.

1. Input k_p, k_i, and k_d from keyboard
2. Calculate k_0, k_1, and k_2
3. Initialize m_{n-1}, e_{n-1}, and e_{n-2}
4. Input r_n from keyboard
5. Wait for clock = 0 (false)
6. Wait for clock = 1 (true)
7. Call CONTROL routine
8. Repeat steps 5 through 8

```
program Program_Driven_Control;
  const
    forever = false;                    {for endless repeat forever loop}
  var
    T, Kp, Ki, Kd : real;
    K0, K1, K2 : real;
    Rn : real;                          {command}
    Cn : real;                          {process output}
    En, En1, En2 : real;                {error history}
    Mn, Mn1 : real;                     {manipulation history}
    clock : boolean;
    procedure CONTROL;                  {this procedure is defined in Program 5.1}
    forward;
  begin
    readln(T, Kp, Ki, Kd);              {input sample period and PID gains}
    K0 := Kp + Ki * T + Kd / T;         {calculate discrete control equation gains}
    K1 := -Kp - 2.0 * Kd / T;
    K2 := Kd / T;
    En2 := 0.0;                         {initialize error and manipulation history}
    En1 := 0.0;
    Mn1 := 0.0;
    readln(Rn);                         {input command setting}
    repeat
      repeat
        LDC(clock, 1)                   {input clock on logic input 1}
      until not clock;                  {wait for clock to be false (logic 0)}
      repeat
        LDC(clock, 1)                   {input clock on logic input 1}
      until clock;                      {wait for clock to be true (logic 1)}
      CONTROL                           {call control procedure}
    until forever                       {endless loop}
  end.
```

Program 5.2 Program-driven PID control.

Figure 5.26 Control system with interrupt input for clock.

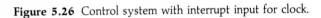

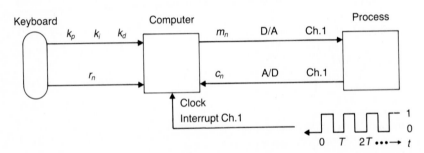

1. Input k_p, k_i, and k_d from keyboard
2. Calculate k_0, k_1, and k_2
3. Initialize m_{n-1}, e_{n-1}, and e_{n-2}
4. Input initial r_n from keyboard
5. Connect CONTROL routine to interrupt input
6. Change r_n when desired using keyboard
7. Repeat steps 6 and 7

Algorithm 5.3 PID control synchronization using interrupt input.

read from the keyboard whenever the operator keys in a new value, and the periodic execution of the control routine while a new r_n is being entered does not affect this input. This illustrates the greater flexibility of the interrupt-driven alternative. It allows valuable computer time to be spent on performing useful functions rather than continuously checking a clock signal. Here we were able to change r_n whenever we wanted without interfering with the control process.

Program 5.3 Interrupt-driven PID control.

```
program Interrupt_Driven_Control;
  const
    forever = false;                {for endless repeat forever loop}
  var
    T, Kp, Ki, Kd : real;
    K0, K1, K2 : real;
    Rn : real;                      {command}
    Cn : real;                      {process output}
    En, En1, En2 : real;            {error history}
    Mn, Mn1 : real;                 {manipulation history}
    clock : boolean;
  procedure CONTROL;
  forward;                          {this procedure is defined in Program 5.1}
begin
  readln(T, Kp, Ki, Kd);           {input sample period and PID gains}
  K0 := Kp + Ki * T + Kd / T;      {calculate discrete control equation gains}
  K1 := -Kp - 2.0 * Kd / T;
  K2 := Kd / T;
  En2 := 0.0;                      {initialize error and manipulation history}
  En1 := 0.0;
  Mn1 := 0.0;
  readln(Rn);                      {input initial command}
  connect(CONTROL, 1);             {connect control procedure to interrupt 1}
  repeat
    readln(Rn);                    {input new commands whenever desired}
  until forever                    {endless loop}
end.
```

Control Routine Timing

So far, we have not considered the time required by the control computer to perform the steps in Algorithm 5.1. The significant portions of this algorithm are sampling, computation, and manipulation output. A given computer requires a finite length of time to perform each of these steps, and the time required for each step can be defined as follows:

T_s = time required to sample the feedback

T_c = time required to compute the manipulation

T_o = time required to output the manipulation

These steps are performed sequentially by the control computer, as indicated by the timing diagram in Figure 5.27. Clearly, a basic requirement is

$$T_s + T_c + T_o < T \tag{5.7}$$

However, in analyzing closed-loop systems and designing controllers, we have assumed up to this point that there is no delay between the sampling of the feedback and the output of the manipulation. This implies in practice that

$$T_s + T_c + T_o \ll T \tag{5.8}$$

The effects of a nonzero delay in the controller depend on controller gains and on the relationship between the sample period T and the process time constants. Controller delays tend to make systems relatively less stable, an effect that can be minimized by ensuring that

$$T_s + T_c + T_o < \frac{T}{10} \tag{5.9}$$

This sometimes can be difficult to achieve in practice, particularly when complex control computations are required or it is necesary to use as high a sample rate as possible to obtain high performance. In this case, it is often necessary to approach the limit where

$$T_s + T_c + T_o \approx T \tag{5.10}$$

These computation delays can be significant, but they often cannot be conveniently modeled because they represent a fraction of a sample period and can be

Figure 5.27 Timing diagram for Control Algorithm 5.1.

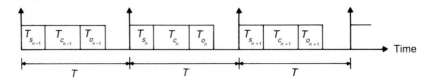

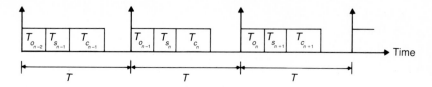

Figure 5.28 Timing diagram for Control Algorithm 5.4.

variable. This variability can be a result of the dependence of arithmetic and I/O operations on the magnitude and sign of variables at any given time and the possibility of programmed changes in control algorithms and logic to achieve nonlinear control objectives.

Control Algorithms with One Sample Delay

One approach that allows modeling of systems with significant or variable delays in their control algorithms is to reorganize the control algorithm by delaying the manipulation output by one sample period. The requirement of Eq. (5.7) still holds, but the timing diagram of Fig. 5.27 becomes that shown in Fig. 5.28. Algorithm 5.4 is then the new control algorithm. If the original controller transfer function was $G_c(B)$, the new controller transfer function is $BG_c(B)$. Note that the reason for introducing a delay of exactly one sample period is that the model $G_c(B)$ was not accurate because of existing delays. $BG_c(B)$ will be a better model as long as

$$T_o + T_s \ll T \tag{5.11}$$

This is often the case because control calculations are usually more time-consuming than sampling or output.

Program 5.4 implements the modified algorithm. Comparing Program 5.4 to Program 5.1 reveals that the manipulation output has been moved to the beginning of the program and that the control calculations now occur at the end. It would therefore be possible to replace the control algorithm given with a more complex algorithm without changing the time of the control I/O, as long as the

Algorithm 5.4 Modified control algorithm with delayed output

1. Output m_{n-1}
2. Sample c_n
3. Calculate $e_n \leftarrow r_n - c_n$
4. Calculate $m_n \leftarrow m_{n-1} + k_0 e_n + k_1 e_{n-1} + k_2 e_{n-2}$
5. Update m_{n-1}, e_{n-1}, and e_{n-2}
6. Done

```
procedure CONTROL;
begin
  DAC(Mn1, 1);                              {output on D/A channel 1}
  ADC(Cn, 1);                               {sample on A/D channel 1}
  En := Rn - Cn;                            {calculate error}
  Mn := Mn1 + K0 * En + K1 * En1 + K2 * En2;   {PID control equation}
  En2 := En1;                               {update history}
  En1 := En;
  Mn1 := Mn
end;
```

Program 5.4 Modified procedure for PID control with output delayed one sample period.

total time required for control I/O and calculations did not exceed the sample period.[2]

Control Algorithms with Precalculation

Another solution to the problem of excessive delay between sampling of the feedback and output of the manipulation is to break up the manipulation calculation into two parts: the portion that is only a function of variables available from the previous sample period and the portion that is a function of the variables that may have changed in the current sample period. For example, consider Eq. (5.3), in which only e_n changes in the current sample period. Hence Eq. (5.3) can be rewritten as

$$m_n = p_n + k_0 e_n \tag{5.12}$$

where

$$p_n = m_{n-1} + k_1 e_{n-1} + k_2 e_{n-2} \tag{5.13}$$

Note that it is not necessary to wait until sample period n to calculate p_n, because all of the variables in Eq. (5.13) relate to previous sample periods. It is therefore possible to calculate p_n at the end of the sample period $n - 1$. The resulting timing diagram is shown in Fig. 5.29, where T_c' and T_p are defined as follows:

$T_c' = $ time required to compute manipulation
using result of precalculation

$T_p = $ time required to precalculate part of
next manipulation

---

[2] It is a good idea to add the test for this error condition to all control programs, particularly in the development stage.

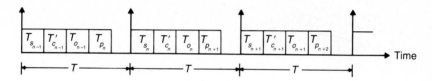

Figure 5.29 Timing diagram for Control Algorithm 5.5.

The new algorithm is given in Algorithm 5.5. A program example is given in Program 5.5.

Reduction of Computation Time

It is often desirable to reduce computation time as much as possible. Some of the benefits of reduced computation time are:

Higher sample rates can be used.
More feedback loops can be closed.
More functions can be performed in unused time between samples.

The available time between samples is

$$T_a = T - T_s - T_c - T_o \tag{5.14}$$

Clearly, if T_c is reduced, the computer time available for other functions will increase. Alternatively, the sample period T can be decreased as T_c is decreased, which may result in better performance.

Some of the many ways to reduce computation time are noted in the following paragraphs.

Using a faster computer

A computer that uses less time than another computer to execute instructions is able to perform control calculations more quickly.

Algorithm 5.5 Modified control algorithm with precalculation

1. Sample c_n
2. Calculate $e_n \leftarrow r_n - c_n$
3. Calculate $m_n \leftarrow p_n + k_o e_n$
4. Output m_n
5. Update m_{n-1}, e_{n-1}, and e_{n-2}
6. Precalculate $p_n \leftarrow m_{n-1} + k_1 e_{n-1} + k_2 e_{n-2}$
7. Done

```
procedure CONTROL;
begin
  ADC(Cn, 1);                          {sample on A/D channel 1}
  En := Rn - Cn;                       {calculate error}
  Mn := Pn + K0 * En;                  {PID control equation with precalculation}
  DAC(Mn, 1);                          {output on D/A channel 1}
  En2 := En1;                          {update history}
  En1 := En;
  Mn1 := Mn;
  Pn := Mn1 + K1 * En1 + K2 * En2      {precalculation for PID control equation}
end;
```

Program 5.5 Modified procedure for PID control with precalculation.

Using compiled languages rather than interpreted languages

Compiled high-level language programs are translated into machine language by a program called a **compiler**. The machine-language program then executes relatively quickly. Interpreted programs are continuously translated by a program called an **interpreter**, and they therefore tend to execute much more slowly. Some compilers generate more nearly optimal (reduced computation time) machine-language programs than others.

Reduction of the complexity of control algorithms

Comparison of the proportional control (P) equation

$$m_n = k_p e_n \tag{5.15}$$

with the proportional plus integal plus derivative (PID) control equation, Eq. (5.3), reveals that more arithmetic operations are required in the PID control than in the P control. It will take longer to perform the PID computation than the P computation.

Using integer or fractional arithmetic rather than real-number arithmetic

Integer and fractional arithmetic operations are faster than real-number operations, because the binary point is fixed for integers and fractions and there is no exponent. The floating-point storage formats normally used for real numbers require binary-point alignment for addition and subtraction and normalization after operations. These strictures together with the necessity of handling exponents, lead to longer computation times for real numbers.

Replacement of multiplication by shifts

Integer multiplication is generally slower than integer addition. The multiplication process can be speeded up by restricting gains (multipliers) in control algorithms to powers of 2 and using shifts instead. A binary number N with i bits can be represented as

$$N = b_{i-1}2^{i-1} + \cdots + b_1 2^1 + b_0 2^0 \qquad (5.16)$$

It can be observed that

$$2N = b_{i-1}2^i + \cdots + b_1 2^2 + b_0 2^1 \qquad (5.17)$$

and

$$\tfrac{1}{2}N = b_{i-1}2^{i-2} + \cdots + b_1 2^0 + b_0 2^{-1} \qquad (5.18)$$

If N is the 8-bit number 00101110_2, then $2N$ is 01011100_2 and $\tfrac{1}{2}N = 00010111_2$. These results can be obtained by shifting the bits in N one bit to the left or one bit to the right to obtain $2N$ or $N/2$, respectively. Shifting operations are provided on most computers and generally are faster than multiplication. These shifting operations can be generalized to obtain gains of $2^{\pm k}N$.

An example of closed-loop control using this approach is illustrated in Figs. 5.30 and 5.31. A gain of $2^{\pm k}$ is used in the control computer, and the D/A converter gain $k_{D/A}$ is used to adjust the output voltage of the D/A converter to obtain the desired controller gain k_c. If a controller gain k_c of 0.0018 volt per bit (error) is desired and the nominal D/A converter gain is 0.001 volt per bit, then $k = 1$ can be selected. A shift of one bit to the left can therefore be performed on the error e_n, and the gain of the D/A converter should be set to

$$k_{D/A} = \frac{0.0018}{2^1} = 0.0009 \text{ volt/bit} \qquad (5.19)$$

Figure 5.30 Proportional controller with computer control gain k_c.

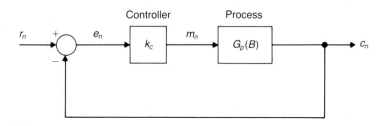

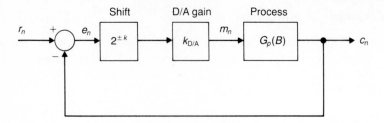

Figure 5.31 Separation of proportional gain into shift and external gain.

Reduction of word size in arithmetic operations

Word size and the number of bits of precision required in control computations can often be reduced by using an incremental control approach. Consider the alternative control configuration shown in Fig. 5.32. This control configuration is functionally equivalent to that of Fig. 5.30. However, the new variable Δc_n is the incremental change in the process output and is likely always to be a small number, even if c_n is large, because it is the difference:

$$\Delta c_n = c_n - c_{n-1} \tag{5.20}$$

If large step changes in r_n are not put into the system, then Δr_n is small because it is found from

$$\Delta r_n = r_n - r_{n-1} \tag{5.21}$$

Δe_n is found from

$$\Delta e_n = \Delta r_n - \Delta c_n \tag{5.22}$$

and it should be small. The error e_n, the sum of the incremental errors Δe_n, is found from

$$e_n = e_{n-1} + \Delta e_n \tag{5.23}$$

If the control system is functioning properly, e_n should also be small.

All of the variables in Fig. 5.32 are therefore likely to be small, with the exception of c_n and r_n, and it is possible to avoid high-precision operations. For

Figure 5.32 Incremental proportional controller.

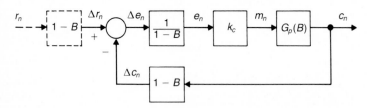

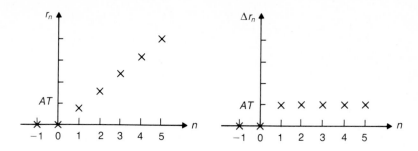

Figure 5.33 Generation of r_n and Δr_n for a ramp input.

processes such as machine drive systems, Δc_n is often generated by feedback devices and interfaces rather than c_n. This is because these drives may operate without being reversed for long periods of time, and c_n therefore can become very large, causing a problem of range and resolution.

Also, it may be more convenient to generate Δr_n rather than r_n. For example, a ramp input of slope A can be generated using

$$r_n = \frac{ATB}{(1 - B)^2} S_n \tag{5.24}$$

and r_n becomes large as n increases, as shown in Fig. 5.33. Δr_n is found from Eqs. (5.21) and (5.24) to be

$$\Delta r_n = (1 - B)r_n = \frac{ATB}{(1 - B)} S_n \tag{5.25}$$

This is merely a delayed step input and can be generated easily by setting Δr_n equal to the constant AT.

Step inputs in r_n can also be generated easily by using the incremental approach. Here

$$r_n = \frac{A}{1 - B} S_n \tag{5.26}$$

and

$$\Delta r_n = AS_n \tag{5.27}$$

As shown in Fig. 5.34, Δr_n in this case has a single nonzero sample of value A at $n = 0$.

It should be emphasized that the incremental controller is equivalent to the conventional controller and produces the same manipulation output. However,

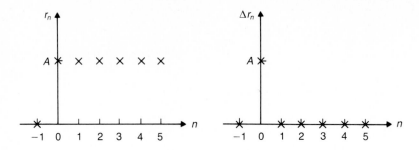

Figure 5.34 Generation of r_n and Δr_n for a step input.

the performance of the closed-loop system may be better because computation delays are reduced.

SUMMARY

The operation of process control computers has been discussed in this chapter. It is necessary to have a basic understanding of all the levels at which a computer system operates (binary logic, machine language, high-level language, and so on) in order to recognize and fully utilize all of the options available to the computer control designer. Our discussion has emphasized general characteristics rather than a particular computer architecture. Detailed explanations of specific computer architectures and instruction sets can be found in manufacturers' literature, and the principles described here can be readily applied in most cases.

Programming methods have been discussed for establishing closed-loop control on a digital computer. The main alternatives of program-driven and interrupt-driven synchronization with a clock determining the sample period have also been discussed, along with a number of methods for improving controller performance. It should be noted that there are significant trade-offs between using high-level languages and using assembly language. Software development and debugging can be easier with high-level languages, but they can be computationally slower and can prevent access to features of computer architectures, such as interrupts, that are extremely useful for control. It has been possible since the 1960s to purchase high-level languages for large computer systems that support the real-time control features described in this chapter. Unfortunately, these high-level language capabilities have not always been available for small computer systems, and users sometimes have been required either to implement the required features in assembly language or to find ways to adapt their computer systems to real-time control needs.

BIBLIOGRAPHY

Auslander, D. M., and P. Sayres. *Microprocessors for Measurement and Control*, New York, NY: McGraw-Hill, 1981.

Ball, R., and R. Pratt. *Engineering Applications of Microcomputers: Instrumentation and Control*. Englewood Cliffs, NJ: Prentice-Hall, 1987.

Bibbero, R. J. *Microprocessors in Instrumentation and Control*. New York, NY: Wiley, 1977.

Craine, J. F., and G. R. Martin. *Microcomputers in Engineering and Science*. Reading, MA: Addison-Wesley, 1985.

Dorf, R. C. *Modern Control Systems, 4/e*. Reading, MA: Addison-Wesley, 1986.

Eccles, W. J. *Microprocessor Systems: A 16-Bit Approach*. Reading, MA: Addison-Wesley, 1985.

Ferguson, J. *Microprocessor Systems Engineering*. Reading, MA: Addison-Wesley, 1985.

Fletcher, W. I. *An Engineering Approach to Digital Design*. Englewood Cliffs, NJ: Prentice-Hall, 1980.

Givone, D. D., and R. P. Roesser. *Microprocessors/Microcomputers*. New York, NY: McGraw-Hill, 1980.

Katz, P. *Digital Control Using Microprocessors*. Englewood Cliffs, NJ: Prentice-Hall, 1982.

Knuth, D. E., *The Art of Computer Programming, Vol. 1., 2/e*. Reading, MA: Addison-Wesley, 1972.

Lawrence, P. D., and K. Mauch. *Microcomputer System Design: An Introduction*. New York: McGraw-Hill, 1987.

Leventhal, L. A. *Introduction to Microprocessors: Software, Hardware, Programming*. Englewood Cliffs, NJ: Prentice-Hall, 1978.

Mano, M. *Computer Systems Architecture, 2/e*. Englewood Cliffs, NJ: Prentice-Hall, 1982.

Mano, M. *Digital Design*. Englewood Cliffs, NJ: Prentice-Hall, 1983.

Short, K. L. *Microprocessors and Programmed Logic*. Englewood Cliffs, NJ: Prentice-Hall, 1980.

Stone, H. S. *Microcomputer Interfacing*. Reading, MA: Addison-Wesley, 1982.

Wakerly, J. *Microcomputer Architecture and Programming*. New York, NY: Wiley, 1981.

PROBLEMS

5.1 Design a logic circuit that has inputs R and N and generates an output R' that is equal to R if N is logic 0 and to the logical complement of R if N is logic 1.

5.2 Repeat Problem 5.1, but consider the case where there are n inputs and outputs, R_i and R_i' respectively, for $i = 0, 1, 2, \ldots, n - 1$, all controlled by a single input N.

5.3 Suppose an interrupt on a particular computer is generated when logic input INT is logic 0. It is desired to generate an interrupt whenever PB, the logic-level output of a push-button switch, is logic 1. When an interrupt occurs, INT must be immediately set to logic 1 by pulse PR (logic-0-to-1-to-0 transition) generated by the interrupt procedure programmed on the computer in order to prevent the same interrupt from being generated again at the moment that the interrupt procedure is completed. Design a logic circuit to generate INT, given PB and PR. Draw a timing diagram illustrating the circuit's operation.

5.4 What are the outputs of the flip-flop shown in Fig. 5.10 when both inputs are nearly simultaneously changed from logic 0 to logic 1?

5.5 Design a logic circuit to implement the subtract and borrow functions in Table B.3 in Appendix B.

5.6 Design an accumulator register that will add an n-bit binary number A to the number in the register and then place the result back in the register when a pulse is received on input C.

5.7 Design a logic circuit to generate the negative (2's complement) of an n-bit binary number A.

5.8 Design a logic circuit that will perform the following operations on an n-bit register when a pulse is received on input C depending on the instruction code I1, I0.

I1	I0	Operation
0	0	set all bits to logic 0
0	1	arithmetic shift 1 bit to right
1	0	arithmetic shift 1 bit to left
1	1	set all bits to logic 1

5.9 Assume that the word size of a computer is only 4 bits. The computer has one 4-bit register R containing bits R3, R2, R1, and R0 that is used for arithmetic and logic operations and another 4-bit register I containing bits I3, I2, I1, and I0 that is to be loaded from memory with the next instruction to be executed. When instructions or data are loaded from memory, the bit pattern appears on signals M3, M2, M1, and M0. Design a logic circuit that loads register I from memory the first time timing signal C is pulsed. The second time C is pulsed, the circuit is to perform the operation on register R indicated in the following table and the result is to be placed back in register R. Show I1 and I0 as outputs that could be used by another circuit to select the memory address to be accessed by the operation.

I3	I2	Operation
0	0	load register R from memory
0	1	or register R and memory
1	0	and register R and memory
1	1	exclusive or register R and memory

5.10 Write an assembly language program that computes the following control equation:

$$m_n = m_{n-1} + k_o e_n + k_1 e_{n-1}$$

Assume that

m_n is stored in memory location 1000

m_{n-1} is stored in memory location 1001

k_o is stored in memory location 1002

e_n is stored in memory location 1003

k_1 is stored in memory location 1004

e_{n-1} is stored in memory location 1005

5.11 Write an assembly-language program that multiplies A times B by performing B additions of A to a register.

5.12 Write an assembly-language program that serially inputs the digits of a number, combines them, and outputs the result as indicated in the flowchart in Fig. 5.35.

Figure 5.35 Flowchart for Problem 5.12.

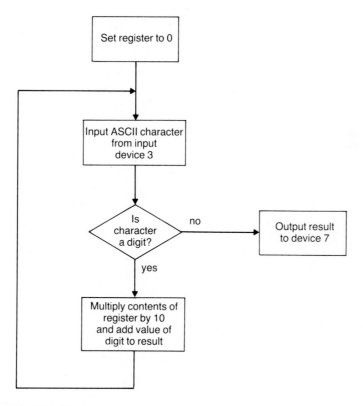

5.13 Write an assembly-language program to perform double-precision addition. (Double precision is described in Appendix B.)

5.14 Show how double-precision division by 2 can be accomplished by using the shift operations illustated in Figs. 5.19 and 5.20.

5.15 An incremental optical encoder is used to acquire shaft position feedback in a control system. Absolute position is accumulated in memory location P from position increments ΔP calculated from new and old encoder readings, P_{new} and P_{old}, at the beginning of each sample period using the flowchart shown in Fig. 5.36. Write an assembly-language program to implement this flowchart.

5.16 When the START button is pushed in the system shown in Fig. 5.37, the HALT lamp is turned off, the RUN lamp is turned on, and a proportional controller with gain K and command R is allowed to control the process. The sample period T is established by the time between 0 to 1 transitions in an external signal CLOCK. The control loop remains closed until limit switch LS1, limit switch LS2, or the STOP button is activated, at which point the proportional controller is deactivated, the process manipulation is set to zero, the RUN lamp is turned off, and the HALT

Figure 5.36 Flowchart for Problem 5.15.

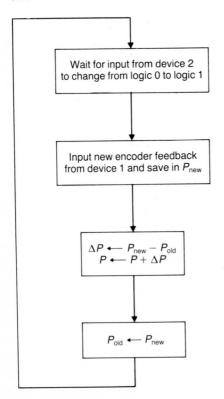

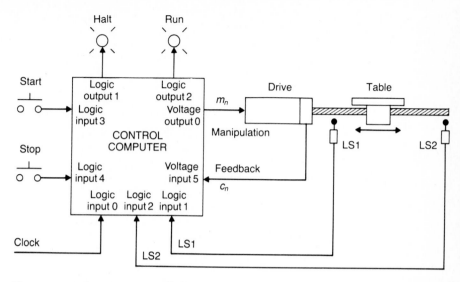

Figure 5.37 Computer-controlled system for Problem 5.16.

lamp is turned on. The system then waits for the START button to be pushed again.

a. Write an assembly-language program for the system, using the program-driven approach for synchronization with the external clock.

b. Write an assembly-language program for the system, using the interrupt-driven approach for synchronization with the external clock.

5.17 Repeat Problem 5.16, but use a high-level language and the procedures (subroutines) shown in Fig. 5.23.

5.18 The shaft of the stepping motor shown in Fig. 5.38 rotates 1/200 of a revolution whenever a pulse (logic-0-to-1-to-0 transition) is applied at the STEP input of its

Figure 5.38 Stepping-motor system for Problem 5.18.

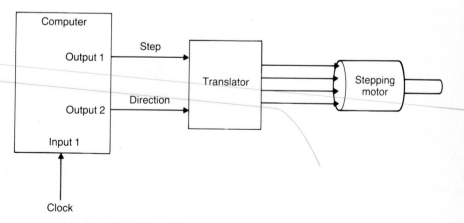

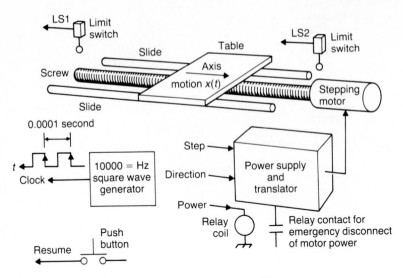

Figure 5.39 Hardware for control system in Problem 5.19.

translator. The direction of rotation is determined by the DIRECTION input (clockwise = logic 1). Write a program that uses the high-level language procedures in Figure 5.23 to repeatedly turn the motor one revolution clockwise and then one revolution counterclockwise. The speed of rotation is to be 360 rpm. The frequency of the clock signal is 6000 Hz.

5.19 Figure 5.39 shows a positioning table driven by a stepping motor. The system is interfaced to a control computer in the following manner:

The channel of the **STEP** logic output is 16.

The channel of the **DIRECTION** logic output is 8.

The channel of the **POWER** logic output is 4.

The channel of the **CLOCK** logic input is 2.

The channel of the **RESUME** logic input is 1.

The channel of the **LS1** interrupt is 1.

The channel of the **LS2** interrupt is 2.

Show how Program 5.6, which uses the procedures in Figure 5.23, should be modified to turn off the **POWER** relay when an interrupt occurs and to return to normal operation when the **RESUME** button is pushed.

5.20 A high-performance control system is being designed that requires a sample rate of 2000 Hz (sample period = 0.0005 s). The system is to be marketed in large quantities, and a particular microprocessor has been selected that is available in a number of models. The only difference between the models is in the maximum-frequency crystal with which they can be operated. The number of instructions that can be executed per second is proportional to crystal frequency. As indicated in the following table, a number of different crystal frequencies are available, but unfortunately, the cost of the processor and the rest of the components in the

```
program Stepping_Motor_Control;
  var
    P, V : real;
    N, I : integer;
    C : boolean;
begin
  P := 0.0;
  V := 0.03333;
  N := 1020;
  DLC(1, 8);
  DLC(0, 16);
  DLC(1, 4);
  for I := 1 to N do
    begin
      repeat
        LDC(C, 2)
      until not C;
      repeat
        LDC(C, 2)
      until C;
      P := P + V;
      if P > 1.0 then
        begin
          P := P - 1.0;
          DLC(1, 16);
          DLC(0, 16)
        end
    end
end.
```

Program 5.6 Stepping motor control
program for Problem 5.19.

computer control system rises rapidly with crystal frequency. If the controller delay
is 1.5 millisecond on the lowest-frequency crystal model, which (if any) processor
model should be selected for the system?

Model	Cost	Crystal
1	1	2 MHz
2	2.5	4 MHz
3	7.5	8 MHz
4	12	16 MHz

5.21 A computer is required to control the two processes shown in Fig. 5.40, each with
its own proportional control gain. Write an interrupt-driven program using the
procedures (subroutines) given in Figure 5.23 to close the loop for process 1 and
process 2 during alternate sample periods.

5.22 A precision relationship is to be maintained between the outputs of the processes
in Fig. 5.40, with the output of process 1 being used as the command input to

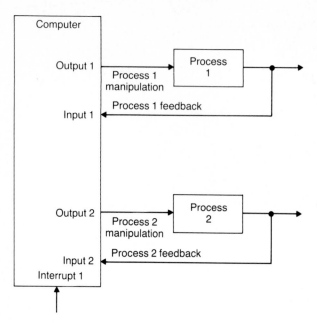

Figure 5.40 Two processes controlled by a single computer for Problems 5.21 and 5.22.

process 2. Write an interrupt-driven program to control the processes, using the procedures in Fig. 5.23.

5.23 A proportional controller with a sample period of 1 millisecond has been designed for an integrating process with a gain of 6.35 to achieve dead-beat response with zero error after one sample period for changes in the command input. Unfortunately, it is later discovered that non-linearities in the process feedback must be compensated for, resulting in a 300- to a 500-μs controller delay depending on the magnitude of the feedback. Assume that most of the delay is due to computations.
  a. Plot the response of the system to a unit step input, with a controller delay of first 300 μs and then 500 μs. Why doesn't the system respond as designed?
  b. Redesign the controller so that the response of the system is independent of variations in computation time.

5.24 The sample period of a high-performance computer control that has been designed for a process is 0.004 s. It takes 70 μs to sample the feedback from the process and 20 μs to output the manipulation to the process. The control calculations take 0.0012 s using floating-point arithmetic (real numbers). Given the following information, assess the timing of this control software and determine what action, if any, should be taken to redesign the system.

Calculations using integer arithmetic takes 1/10 as much time as calculations using floating-point arithmetic.

Fifty percent of the control algorithm can be precalculated.

Chapter 6
Computer Interfacing

INTRODUCTION

To be effective at implementing real-time computer control, it is necessary to understand how the control computer can be linked to external devices. For example, it is necessary to be able to sample process feedback data from sensors that provide either analog or digital information and to be able to manipulate process actuators on the basis of control computations. The interfaces between physical devices and the internal computer architecture can take many forms. Actual implementation is sometimes complicated by basic differences among the internal architectures of various control computers and by the variety of devices available.

Following the philosophy of the previous chapter, the discussion that will be presented here draws on a simplified computer architecture that is closely related to the architecture of many control computers. This structure is used to describe how devices such as push-buttons, limit switches, lamps, numeric displays, alpha-numeric displays, and keyboards can be connected to the computer. These interfaces establish machine–computer and operator–computer interactions. Interfaces for digital-to-analog conversion, interfaces for analog-to-digital conversion, and digital interfaces to sensors allow the computer to perform closed-loop control of processes and process condition monitoring.

221

6.2
COMPUTER ARCHITECTURE

As computer technology has progressed, there has been a trend toward establishing "bus" structures for communication between the central processing unit (CPU) of the computer and memory and input/output devices. The term **bus** refers to a group of logic-level inputs or outputs that are closely related to each other, operate in parallel, and together serve a common function. The simplified computer architecture considered here is shown in Fig. 6.1. The important features are the memory address bus, the memory data bus, two memory control signals generated by the CPU called READ and WRITE, the I/O address bus, the I/O data bus, two I/O control signals generated by the CPU called IN and OUT, and an input to the CPU for interrupt generation called INT.

Memory Address Bus (MAB)

The **memory address bus (MAB)** is used to specify which memory location is to be accessed by the CPU. It typically consists of 16 to 20 separate signals, each specifying one bit in the address of the memory location to be read from or written into.

Memory Data Bus (MDB)

The **memory data bus (MDB)** is used to transfer binary data to or from the memory location specified by the memory address bus. In the case of a memory

Figure 6.1 Simplified computer architecture.

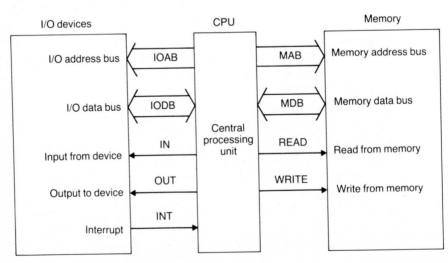

read operation, such as a LOAD instruction, the memory hardware places the binary contents of the selected word in memory on the memory data bus so that it can be read by the CPU. The number of signals in the memory data bus is therefore usually equal to the word size of the computer, which is commonly 8, 16, 24, 32, or more bits.

READ and WRITE Control Signals

The **READ control signal** and the **WRITE control signal** are generated by the CPU and specify, respectively, when data is to be transferred from memory to the CPU and when data is to be transferred from the CPU to memory. They also indicate when the binary information on the memory address bus and the memory data bus is valid so that transfers can take place when these signals are correct and not in a state of transition. The functions of these logic-level signals are as follows:

READ	WRITE	Function
0	0	No memory read/write operation is required.
0	1	Memory address and data are valid for write operation.
1	0	Memory address is valid for read operation.
1	1	Not used.

Input/Output Address Bus (IOAB)

The function of the **input/output address bus (IOAB)** is to specify which input/output device is to communicate with the computer at a given point in time. The I/O address bus consists of a number of separate signals, each specifying one bit in the address of the I/O device to be accessed by the CPU. If there are m of these signals in the bus, then 2^m different I/O addresses can be selected and, hence, a large number of devices accessed. However, more than one address may often be associated with a given I/O device, particularly if the device is complex in nature. The IOAB may be physically shared with other buses, such as the memory address bus, but this is not always the case.

Input/Output Data Bus (IODB)

The **input/output data bus (IODB)** is responsible for transferring data words to and from I/O devices. Once an I/O device has been selected by the computer using the I/O address bus, that device can access the I/O data bus. In the case of input to the CPU, the device can set the logic-level signals on the individual lines in the bus so that the CPU can read them. In the case of output from the CPU, the logic levels on the I/O data bus are set by the CPU and can be sampled by the device. The I/O data bus is therefore bi-directional; the direction is determined by whether an input or an output operation is being performed by the CPU.

The number of lines in the I/O data bus is usually the same as the computer word size. This may be 8 to 16 bits in a small computer and more in larger computers. In some computers, the I/O data bus is physically shared with the memory data bus.

IN and OUT Control Signals

The **IN control signal** and the **OUT control signal** generated by the CPU serve two purposes: They specify whether an input or an output operation is to be performed, and they indicate when the addresses and data specified by the I/O address bus and the I/O data bus are valid (as opposed to being in a state of transition). The functions of these logic-level signals are as follows:

IN	OUT	Function
0	0	No I/O operation is to be performed.
0	1	I/O address and data are valid for output operation.
1	0	I/O address is valid for input operation.
1	1	Not used

A transition to logic 1 of the OUT control signal should cause the I/O device selected by the I/O address bus to accept data output from the CPU on the I/O data bus. The I/O address bus and the I/O data bus are assumed to be held constant by the CPU when OUT is in the logic-1 state. OUT is returned to logic 0 by the CPU after a specified period of time.

A transition to logic 1 of the IN control signal should cause the I/O device selected by the I/O address bus to set the I/O data bus to the appropriate data to be input to the CPU. The I/O address bus is assumed to be constant while IN is in a logic-1 state, and the I/O device must hold the input data on the I/O data bus constant until IN is returned to logic 0 by the CPU after a specified period of time.

CPU Interrupt Signal (INT)

The **CPU interrupt signal (INT)** is an input to the CPU that causes it to interrupt the program being executed and to begin execution of an interrupt procedure. Some computers have more than one interrupt input, which makes possible easy assignment of priority to interrupt-generating devices and simplified identification of which device has generated the interrupt.

The functions of the interrupt signal are as follows:

INT	Function
0	No interrupt requested.
1	Interrupt requested.

Example 6.1 Output instruction   An output instruction for a simplified computer with only one register from which input/output can be performed could have the form

OUTPUT 1

This instruction indicates that the CPU is to output a binary word from the register to the I/O device with address 1.

Of course, different computer architectures vary with respect in their specific synchronization requirements, but a typical circuit is shown in Fig. 6.2. This output device interface uses the OUT signal described earlier to indicate to the output device that it is time to receive a word of data from the CPU.

Figure 6.2 Output device interface (address $= 1$).

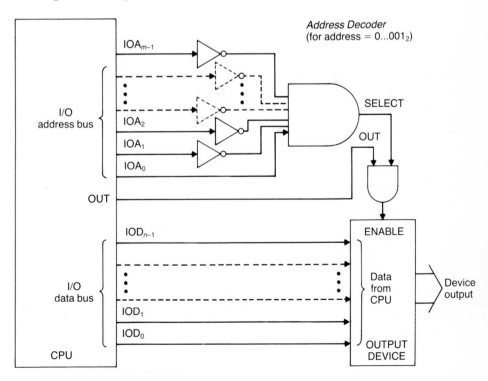

Communication between the CPU and the output device takes place according to the following steps:

1. The OUT signal is at logic 0.
2. The CPU fetches an "OUTPUT 1" instruction from memory.
3. The I/O address bus is set by the CPU to the binary address of the output device ($00\ldots001_2$ in this example), causing SELECT to be set to logic 1.
4. The I/O data bus is set by the CPU to the binary word in the accumulator.
5. The OUT signal is set by the CPU to logic 1.
6. The output device is enabled because the OUT and SELECT signals are at logic 1.
7. The output device samples the binary word on the I/O data bus and holds it for further use.
8. The CPU resets the OUT signal to logic 0.
9. The CPU fetches the next instruction from memory (I/O address bus and I/O data buses change).

A timing diagram for the output device interface is shown in Fig. 6.3. Note that the I/O address bus and the I/O Data Bus are valid (set to specific values) only for specific periods of time. The OUT signal is activated only when the buses are

Figure 6.3 Timing diagram for output from CPU to I/O device.

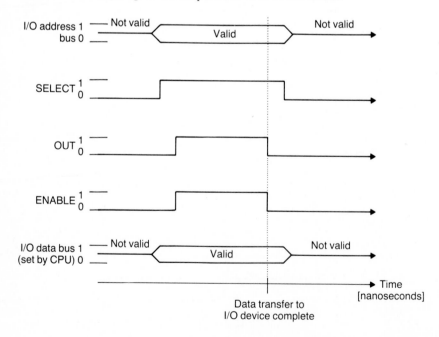

valid, indicating that it is time to perform the output data transfer. The data transfer must be completed when OUT returns to logic 0.

At the end of this sequence, the output device has acquired the binary data from the accumulator and is free to do whatever is necessary with it. The interpretation of this data is determined by the hardware of the output device. A lamp interface may simply turn on lamps that correspond to which bits are set in the binary word. If the output device is a digital-to-analog converter, further processing by the device is necessary to produce an analog voltage proportional to the binary word. However, what is important at the interface between the CPU and the I/O device is the synchronization of the device with the CPU and the transfer of binary information. □

Example 6.2 Input instruction   An input instruction for a simplified computer with only one register from which input/output can be performed could have the form

INPUT 2

This instruction specifies that a binary word will be input to the register in the CPU from the device with address 2. Fig. 6.4 shows a circuit for an input device

Figure 6.4 Input device interface (address $= 2$).

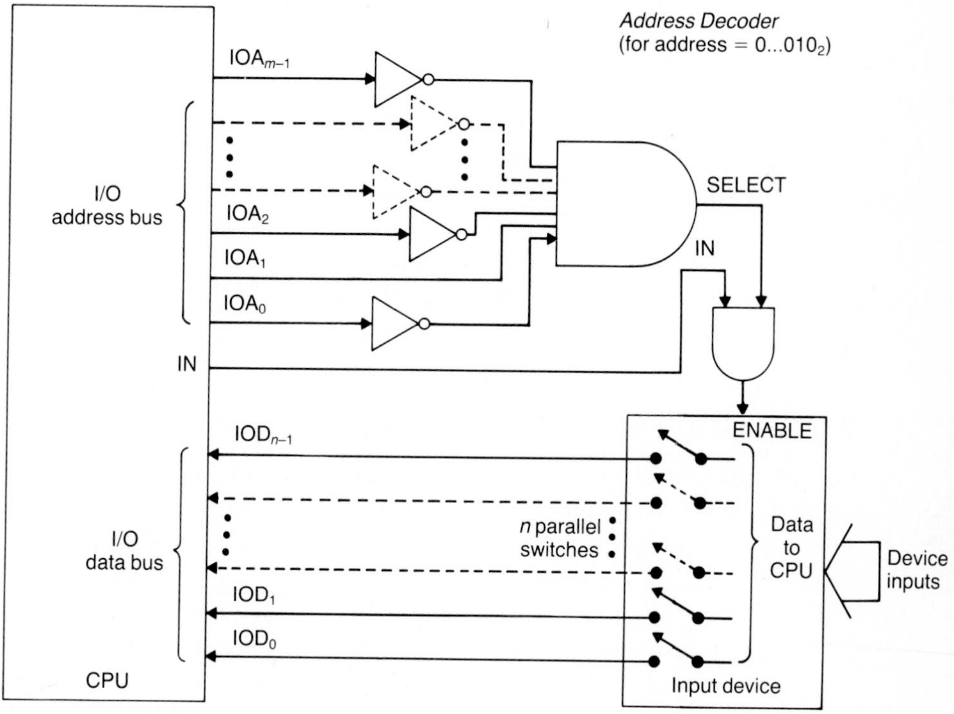

interface that performs this function. Comparing Fig. 6.4 to Fig. 6.2 reveals that a new device address is selected ($00 \ldots 010_2$ rather than $00 \ldots 001_2$) and that the bi-directional I/O data bus is operated in the opposite direction (input rather than output). Because only a single device can control the I/O data bus at any given time, switches are shown in Fig. 6.4 to connect the input device to the I/O data bus when the device is selected by the CPU. These switches are closed only when the IN signal is set by the CPU and, in addition, when the address on the I/O address bus is the address of this device (which enables SELECT).

The following sequence of events takes place during an input operation:

1. The IN signal is at logic 0.
2. The CPU fetches an "INPUT 2" instruction from memory.
3. The CPU sets the I/O address bus to the binary address of the input device ($00 \ldots 010_2$ in this example), thus setting SELECT to logic 1.
4. The CPU sets the IN signal to logic 1.
5. The input device is enabled, because both the IN signal and the SELECT signals are at logic 1.
6. The input device connects itself to the I/O data bus.
7. The IO data bus is set by the input device to the binary value to be input to the CPU.
8. The CPU transfers the binary value on the I/O data bus to the register.
9. The CPU resets the IN signal to logic 0.
10. The input device disconnects itself from the I/O data bus.
11. The CPU fetches the next instruction from memory (I/O address bus changes).

A timing diagram for the input device interface is shown in Fig. 6.5. Note that the I/O device must connect itself to the I/O data bus and establish valid data on the bus as soon as possible after the IN signal becomes logic 1. The data transfer to the CPU is completed when the IN signal is changed back to logic 0. □

Programmed-Driven vs. Interrupt-Driven I/O

In the previous chapter it was mentioned that input and output operations can be performed either in a program-driven mode or in an interrupt-driven mode. I/O devices are usually assigned at least two addresses: one to transfer data and one to allow the CPU to determine the status of the device. In the program-driven mode, the CPU is able to continually input and test the status of the devices from one address and to input or output data using another address. The status is treated as data by the CPU but interpreted in an appropriate manner by program

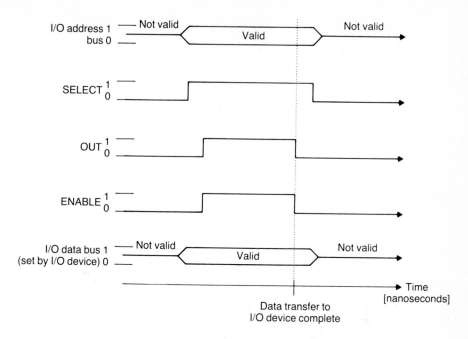

Figure 6.5 Timing diagram for input to CPU from I/O device.

software. As an example, the binary word that is input from the status address of the keyboard input interface shown in Fig. 6.6(a) might be

Status	Meaning
0	No key has been pressed.
1	A key has been pressed on the keyboard, and data is ready indicating which key it is.

When 0 is input from the status address, there is no data for the CPU to read. When a 1 is input from the status address, however, the computer should proceed to input the binary word defining which key was pushed from the data address. Obviously, the status should then be returned to zero until another key is pressed. This is usually done automatically by the device. Note that for this mode of operation, the CPU must continually test the status of the device, and this must be done often enough so that no information (keystrokes in this example) will be missed.

For interrupt-driven input/output, an interrupt is generated by the I/O device when it is ready to input or output data. This is done by having the device set the interrupt signal INT to logic 1. This forces the CPU to execute a software subroutine specifically written to perform the required I/O operation. In this case,

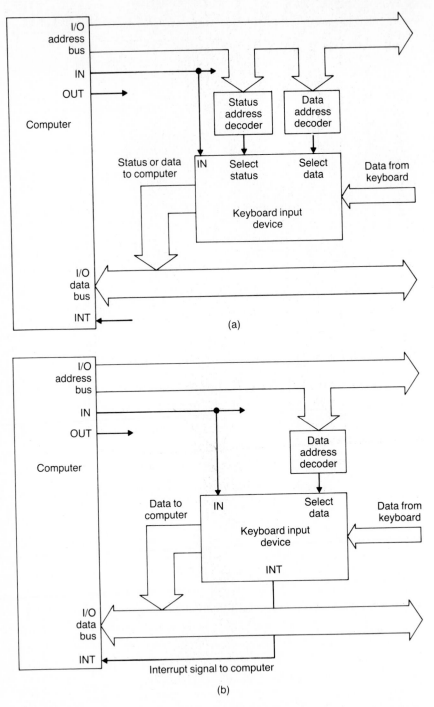

Figure 6.6 Device for (a) program-driven keyboard input and (b) interrupt-driven keyboard input.

it is unnecessary to continually test the status of the device to see whether an I/O operation is to be performed; the occurrence of the interrupt tells the CPU that such an operation is required. In the example of the keyboard input device shown in Fig. 6.6(b), if the device interrupts the CPU whenever a key is pressed, then the CPU can immediately input data indicating which key has been pressed, and no computing time is consumed in checking the status of the device.

6.3
LOGIC-LEVEL INPUT/OUTPUT

As described in the previous section, data communication between the CPU and I/O devices take place on the I/O data bus, which consists of a set of parallel logic-level signals. These can be used to interface the CPU directly to many devices (including counters, lamps, and switches) as illustrated in the following examples. Throughout these examples, it will be assumed that a logic signal has been derived from the I/O address bus to indicate that a particular I/O device address has been selected. This signal, which has been called SELECT, is shown in both Fig. 6.2 and Fig. 6.4. Its functions are as follows:

Select	Function
0	I/O address does not specify this device.
1	I/O address does specify this device.

Example 6.3 Output to single LED   Consider an interface to control the state (on or off) of a light-emitting diode (LED). Figure 6.7 shows a simple circuit

Figure 6.7 LED interface.

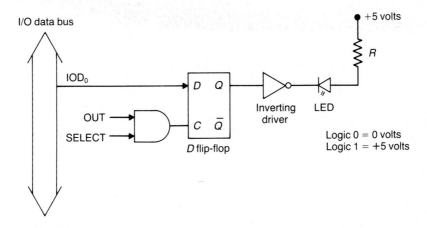

in which the lowest bit of the I/O data bus (IOD_0) is used to convey to a D flip-flop the desired state of the lamp. To control the lamp, 0 or 1 is first loaded into the lowest bit of a CPU register. An output instruction is then executed to place this data on the I/O data bus and the address of the LED interface on the I/O address bus. The CPU then sets the OUT signal to logic 1, indicating that all is ready for the LED interface to act. The SELECT signal detects the LED interface address, and the presence of both this signal and the OUT signal causes the D flip-flop to sample the level (0 or 1) of IOD_0, which is then held at the output Q. If this output is logic 0, the output of the inverting driver is logic 1, the current through the LED is zero, and the LED is off. On the other hand, if the output of the D flip-flop is logic 1, the output of the inverting driver is logic 0 and the LED is on with an intensity determined by the resistance R. □

Example 6.4 Input from a switch Suppose it is desired to sample a switch periodically to determine whether it is open or closed. A 0 is to be input into the lowest bit of a CPU register if the switch is open. A 1 is to be input if the switch is closed.

A push-button switch interface circuit is shown in Fig. 6.8. The bus driver functions as a switch and allows the interface to be disconnected from the I/O data bus. When both the SELECT signal and the IN signal are at logic 1, the interface is connected to the lowest bit of the I/O data bus (IOD_0). If the switch is open, then logic 1 is sensed and inverted, and logic 0 is applied to IOD_0. A 0 is then input into the lowest bit of the register. On the other hand, if the switch is closed, the bus driver applies logic 1 to IOD_0. A 1 is therefore input into the lowest bit of the register. The bus driver immediately disconnects the interface from the I/O data bus when the IN signal is reset to logic 0 by the CPU. □

Figure 6.8 Push-button switch interface.

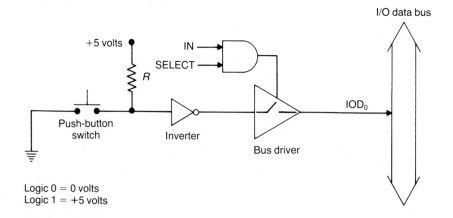

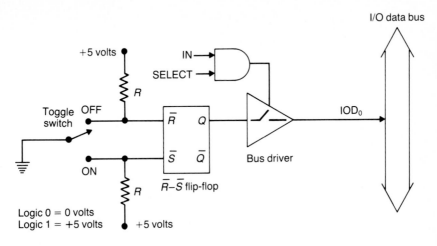

Figure 6.9 Toggle switch interface with debouncing.

Example 6.5 Input from a toggle switch with debouncing   Input from a toggle switch can be accomplished via the circuit shown in Fig. 6.9. The switch has two positions: ON and OFF. It is desired to input a 0 into the lowest bit of a CPU register if the switch is off and a 1 if the switch is on. Although this could be accomplished without using the $\bar{R}$-$\bar{S}$ flip-flop, it is desirable to eliminate the effects of contact bounce. (Contact bounce is the rapid, repeated opening and closing of switch contacts while the switch position is changing.) The flip-flop is set and outputs logic 1 when the switch first contacts the ON terminal. Subsequent contacts with the terminal without intervening contact with the OFF terminal do not affect the output of the flip-flop. In a similar manner, the flip-flop is reset and its output is logic 0 upon the first contact with the OFF terminal. The use of the bus driver to connect the circuit to the I/O data bus is identical to that discussed in the previous example. This interface device illustrates the type of "positive" thinking that leads to reliable computer control systems. For example, a double-contact limit switch used to detect the end of travel of a machine axis may be more expensive than a single-contact limit switch, but the former can result in a more reliable system.   □

Example 6.6 Parallel input from a counter   In the previous examples, it was shown how to input the state of a single switch to the CPU. It is also useful to be able to input a number of parallel signals at the same time, taking full advantage of the fact that the I/O data bus can transfer an entire word of binary information to or from the CPU.

Consider a binary counter that counts the number of pulses that have been generated by some external source. If the number of bits in the counter is less than or equal to the number of bits in the I/O data bus, the entire count of the

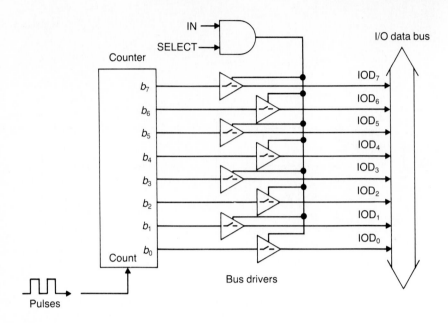

Figure 6.10 Counter interface for 8-bit parallel input.

number of pulses received can be obtained with one input operation. In this example, it is assumed that an 8-bit counter is used and that there are at least 8 bits in the I/O data bus.

Figure 6.10 shows the interface for input to the CPU from the counter. When the IN and SELECT signals are present, the eight bus drivers connect the individual counter output to the I/O data bus.[1] The CPU can then load the contents of the counter into a register and process this information in whatever way is desired. ☐

6.4

DIGITAL-TO-ANALOG CONVERSION

The **digital-to-analog (D/A) converter** uses a binary-number output by the CPU to produce an analog voltage proportional to that number. It is commonly used when there are requirements for a computer control system to supply time-varying voltage outputs to various types of actuators.

---

[1] Other aspects of logic circuit design such as synchronization, latching, and multiplexing have to be considered in implementing the circuit in Fig. 6.10. This is particularly true if the number of bits in the counter is greater than the number of bits in the I/O data bus so that all bits in the counter cannot be transferred at once.

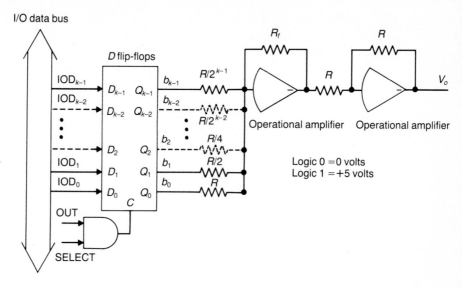

Figure 6.11 Digital-to-analog converter.

The basic D/A converter circuit shown in Fig. 6.11 uses operational amplifiers to produce an output proportional to a k-bit unsigned binary word. When the OUT signal is set to logic 1 by the CPU and the address of the D/A converter is present on the I/O address bus (as indicated by the SELECT signal), the binary word on the I/O data bus is latched by D flip-flops (the number needed is k). The outputs of these flip-flops are fed through a set of input resistors to an operational amplifier. When the OUT signal returns to logic 0; the outputs of these flip-flops remain unchanged, providing a constant input to the remainder of the circuit.

The values of the input resistors on the operational amplifier are chosen so that its output voltage is

$$V_o = V_h\left(\frac{R_f}{R/2^{k-1}}b_{k-1} + \frac{R_f}{R/2^{k-2}}b_{k-2} + \cdots + \frac{R_f}{R/2}b_1 + \frac{R_f}{R}b_0\right) \quad (6.1)$$

where V_h is the logic-1 output voltage of the D flip-flops (nominally $+5$ volts) and $b_{k-1}, b_{k-2}, \ldots, b_1, b_0$ are the binary outputs of the flip-flops (0 or 1). Equation 6.1 can be rewritten as

$$V_o = V_h\frac{R_f}{R}N \quad (6.2)$$

where

$$N = b_{k-1}2^{k-1} + b_{k-2}2^{k-2} + b_1 2^1 + b_0 2^0 \quad (6.3)$$

It is evident that N in Eq. (6.3) is the binary-number output to the D/A converter by the CPU and that the output of the D/A converter is therefore proportional to N. The gain of the D/A converter can be adjusted as desired by changing R or R_f.

Range

The maximum output voltage is produced when all bits are 1. In this case, $N = 2^k - 1$ and

$$V_{max} = V_h \frac{R_f}{R} (2^k - 1) \tag{6.4}$$

The minimum output voltage is produced when all bits are zero. In this case, $N = 0$ and

$$V_{min} = 0 \tag{6.5}$$

Signed output can be obtained from the D/A converter for k-bit (including sign) 2's-complement numbers by inverting the sign bit and adding a bias voltage of $-V_h(R_f/R)(2^{k-1})$ to the output V_o. In this case, the 2's-complement number zero produces the expected $V_o = 0$. Also,

$$V_{max} = V_h \frac{R_f}{R} (2^{k-1} - 1) \tag{6.6}$$

and

$$V_{min} = -V_h \frac{R_f}{R} (2^{k-1}) \tag{6.7}$$

The range of the D/A converter is then from V_{min} to V_{max}.

Resolution

The minimum nonzero voltage that can be produced is obtained with $N = 1$. Then from Eq. (6.2),

$$V_q = V_h \frac{R_f}{R} \tag{6.8}$$

where V_q represents the resolution of the D/A converter. If both the resolution and the range are specified, the minimum number of bits required in the D/A converter can be found from

$$k = 1 \frac{\log\left(\dfrac{V_{max} - V_{min}}{V_q} + 1\right)}{\log 2} \tag{6.9}$$

Because k must be an integer, any fractional part in the result of Eq. (6.9) should be rounded up to the next highest integer.

Accuracy

In the circuit shown in Fig. 6.11, the accuracy of the conversion and its linearity depends very much on the precision of the input resistors on the operational amplifier. Because the range of these values (R to $R/2^{k-1}$) may be large, high accuracy is difficult to achieve with this circuit. For this reason, other hardware implementations with improved accuracy and thermal stability are normally used. Complete units with high accuracy and low cost are commercially available in sizes generally ranging from 8 to 16 bits.

Example 6.7 Digital-to-analog converter   The unsigned decimal number 261 is output to a 10-bit D/A converter constructed as shown in Fig. 6.11. The 10-bit binary equivalent of 261_{10} is 0100000101_2; hence, from Eq. (6.1), the D/A converter output is

$$V_o = V_h \left[\frac{R_f}{R/512} (0) + \frac{R_f}{R/128} (1) + \frac{R_f}{R/64} (0) \right.$$

$$\left. + \cdots + \frac{R_f}{R/8} (0) + \frac{R_f}{R/4} (1) + \frac{R_f}{R/2} (0) + \frac{R_f}{R} (1) \right]$$

or, from Eq. (6.2),

$$V_o = V_h \frac{R_f}{R} (261)$$

If

$$V_h = 10 \text{ volts}$$

and

$$R_f = R/1024$$

then

$$V_o = 10 \left(\frac{1}{1024} \right) 261 = 2.55 \text{ volts}$$

In this case, the maximum voltage that can be output from the D/A converter is found from Eq. (6.4) as

$$V_{max} = 10 \left(\frac{1}{1024} \right) (2^{10} - 1) = 9.99 \text{ volts}$$

The range of output voltages is therefore 0 to 9.99 volts, and resolution can be found from Eq. (6.8) as

$$V_q = 10\left(\frac{1}{1024}\right) = 0.00977 \text{ volt}$$

If an improved resolution of 0.0025 volt is required over a range of 0 to 10 volts, the number of bits required in the D/A converter can be found from Eq. (6.9) as

$$k = \frac{\log\left(\dfrac{10 - 0}{0.0025} + 1\right)}{\log 2} = 11.97 \text{ bits}$$

A 12-bit D/A converter would therefore be needed to satisfy this range and resolution requirement. ☐

6.5
ANALOG-TO-DIGITAL CONVERSION

The **analog-to-digital (A/D) converter** is a device that samples an analog input voltage and encodes that voltage as a binary number. It allows the control computer to sample the output of feedback transducers that produce an analog voltage proportional to some feedback variable. The necessity of converting analog signals to digital representation makes the A/D converter a common device in control applications.

Figure 6.12 shows one possible A/D converter circuit in which a D/A converter is used along with computer software to perform the analog-to-digital conversion. In the circuit, the analog voltage output of the D/A converter, V_o, is compared to the input voltage V_i, which is to be converted to a digital value. The output of the comparator is a logic-level signal indicating whether $V_o < V_i$ or $V_o > V_i$. This output is input to the computer and the result is used to successively modify V_o until, eventually, $V_o \approx V_i$.

One strategy for accomplishing this conversion is to write the A/D conversion software so that the output of the D/A converter is swept from its minimum value toward its maximum value. As long as V_i is within the range of the D/A converter, the output of the comparator initially indicates $V_o < V_i$, but it immediately changes state when $V_o > V_i$. The last binary-number output to the D/A converter then can be used as the result of the A/D conversion.

The **conversion time** can be defined as the time delay between the start of the A/D conversion and attainment of the digital representation of V_i. In the worst case, all possible binary values are successively tried before the correct

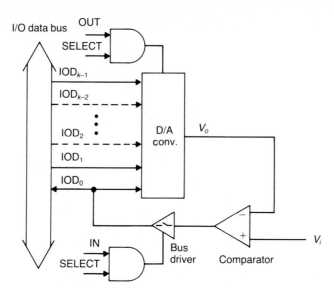

Figure 6.12 Hardware/software analog-to-digital converter.

value is found. The maximum conversion time for this strategy is proportional to 2^k and therefore can be excessively long. Conversion time can be improved by using a modified strategy in which all bits are initially set to zero. The proper value of the most significant bit (b_{k-1}) is determined first by observing whether the state of the comparator changes when the bit is changed to 1. The bit is left at 1 if no change occurred and is set back to zero if a change occurred. The same procedure is followed for the next most significant bit (b_{k-2}), and so on down to the least significant bit (b_0). The conversion time in this case is proportional to k rather than 2^k; a significant improvement. In either case, the A/D conversion time can be considerable because software is extensively involved in the conversion.

It is faster to implement the entire A/D conversion process in hardware. Complete units are commercially available and generally incorporate from 8 to 16 bits in the conversion. Conversion times can be as fast as 0.1 μs. Figure 6.13 illustrates an A/D converter in which a logic circuit is used to generate digital values for the D/A converter on the basis of feedback from the comparator. A start command is required and can be generated by an output instruction (no data need actually be transferred to the A/D converter from the I/O data bus). When the conversion is complete, the digital value being supplied to the D/A converter can be input to the CPU, representing the best encoded value for the analog input voltage. Depending on the speed of conversion, a software delay may have to be implemented to wait until the conversion is complete. This can be facilitated if a DONE logic signal is generated by the A/D converter and interfaced to the CPU.

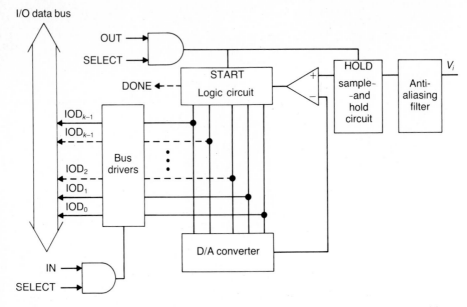

Figure 6.13 Hardware analog-to-digital converter with filter and sample and hold.

Filtering

The rate of data sampling using an A/D converter should be greater than twice the frequency present in the analog input voltage. **Aliasing** and associated control difficulties can result when this principle is not observed. Aliasing is a phenomenon whereby sampling a signal of frequency f at a frequency of f_s yields the same data as sampling a signal of frequency $f + f_s, f - f_s, f + 2f_s, f - 2f_s$ or, in general, $f \pm nf_s$, where n is an integer. The most significant manifestation of this phenomenon from a control point of view is that sampled high-frequency signals can appear to be lower-frequency signals because of aliasing. When a signal with a frequency of greater than $\frac{1}{2}f_s$ is sampled, the apparent lower frequencies due to aliasing are

$$f_a = |f - nf_s| \tag{6.10}$$

for integer n and $f_a < f$. The lowest frequency can be found by choosing n such that

$$f_a \le \frac{1}{2}f_s \tag{6.11}$$

Figure 6.14 shows an example wherein $f > f_s/2$. The figure shows that the apparent frequency is much lower than the actual signal frequency. If the possibility of aliasing is to be eliminated, then only frequencies less than $f_s/2$ can be permitted to reach the A/D converter. Because it may not be convenient or

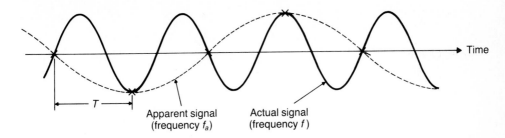

Sample frequency: $f_s = 1/T$

Figure 6.14 Apparent low-frequency signal as a result of aliasing due to $f \geq \frac{1}{2}f_s$.

practical to eliminate the source of high frequencies in a signal to be sampled, a low-pass analog **filter** is usually placed at the input to the A/D converter as shown in Fig. 6.13. The function of this filter is to remove (ideally) all frequencies that are not less than half the sample frequency.

Example 6.8 Aliasing Suppose that the power amplifier for a DC motor uses full-wave rectification of 60-Hz AC line voltage. Close examination of motor velocity reveals a 120-Hz ripple caused by the amplifier. Suppose further that motor velocity is to be controlled by a computer as shown in Fig. 6.15(a) via feedback from a tachometer, the output of which is sampled by an A/D converter. The sample period is to be $T = 0.008$ s.

As shown in Fig. 6.15(b), there is a 120-Hz ripple in the tachometer voltage. If no anti-aliasing filter is used, then with the sample frequency

$$f_s = \frac{1}{0.008 \text{ s}} = 125 \text{ Hz}$$

the apparent frequency of the ripple after sampling is obtained from Eq. (6.10) as

$$f_a = |120 - 125| = 5 \text{ Hz}$$

It therefore appears to the control computer that the motor velocity is oscillating at a frequency of 5 Hz. This is a relatively low frequency, and the computer corrects the perceived variation, resulting in a 5-Hz variation in actual motor velocity as shown in Fig. 6.15(c).

This aliasing problem can be prevented by filtering the feedback voltage from the tachometer to remove the ripple before it is sampled by the A/D converter. The cut-off frequency for the filter must be less than $fs/2$. That is,

$$f_c < 62.5 \text{ Hz} \quad \square$$

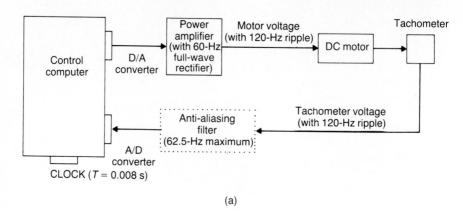

(a)

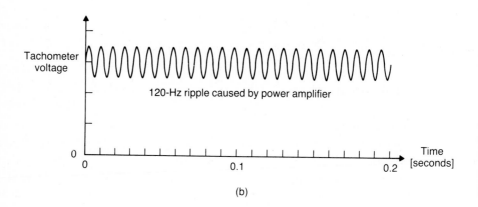

(b)

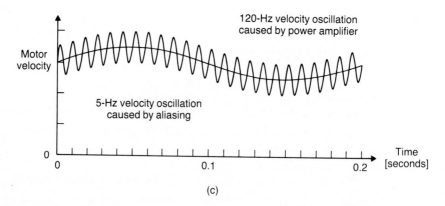

(c)

Figure 6.15 (a) Computer-controlled motor with 120-Hz ripple in amplifier voltage, (b) tachometer voltage with constant D/A output, and (c) motor velocity resulting from computer control when an anti-aliasing filter is not used.

Resolution

The **resolution** of the A/D converter is a direct function of the number of bits used in the conversion. If there are k bits and only positive input voltages are to be converted, the resolution of the A/D converter is

$$V_q = \frac{(V_{max} - V_{min})}{2^k - 1} \tag{6.12}$$

where V_{max} is the maximum input voltage, V_{min} is the minimum input voltage, and $V_{max} - V_{min}$ is the range of the A/D converter. V_q can also be referred to as the quantization level. The effects of quantization are illustrated in Fig. 6.16(a), where the digital result of the A/D conversion, N, is plotted against the input voltage V_i. The error between the digitized voltage and V_i can be found from

$$V_e = V_i - NV_q \tag{6.13}$$

The error V_e is plotted as a function of V_i in Fig. 6.16(b).

Figure 6.16(b) shows that the maximum error due to quantization in converting the input voltage V_i is V_q. If the allowable quantization error and the maximum input voltage are known for a given application, the minimum number of bits required in the A/D converter can be found from Eq. (6.9). Again, k must be an integer and any fractional part should be rounded up to the next highest integer.

Example 6.9 Analog-to-digital resolution   The output voltage of an aircraft altimeter is to be sampled using an A/D converter. The sensor outputs 0 volts at 0 meters altitude and outputs 10 volts at 10,000 meters altitude. If the allowable error in sensing is 10 meters, then from Eq. (6.9) the minimum of bits required is

$$k = \frac{\log\left(\dfrac{10000}{10} + 1\right)}{\log 2} = 9.97$$

and a 10-bit A/D converter can be used to sample the altimeter output. □

Figure 6.16 Quantization effects: (a) digitized result N vs. input voltage V_i and (b) quantization error V_e vs. input voltage V_i.

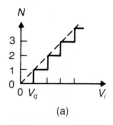
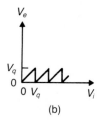

(a)   (b)

Amplitude Uncertainty

When input signals are constant or are varying slowly, the quantization error is the major accuracy consideration in A/D conversion. However, if input voltages are changing rapidly, the effect of the rate of change of these signals on the accuracy of the conversion process must be considered. As Fig. 6.17 shows, changes in the input voltage during the conversion process can result in deviation of the converted value from the desired value that would have been obtained had the conversion been instantaneous. If the rate of change of the input voltage is dV_i/dt, an error can occur in the converted value of

$$\Delta V_i = \frac{dV_i}{dt} T_c \qquad (6.14)$$

where T_c is the conversion time. This error or **amplitude uncertainty** can be significant, particularly when conversion times are relatively long.

Example 6.10 Analog-to-digital conversion error   Consider the case where a sine wave of frequency ω and amplitude A is to be sampled with an A/D converter. Here,

$$V_i = A \sin \omega t$$

$$\frac{dV_i}{dt} = A\omega \cos \omega t$$

and

$$\left. \frac{dV_i}{dt} \right|_{max} = A\omega$$

Figure 6.17 Error in A/D conversion due to changing input signal.

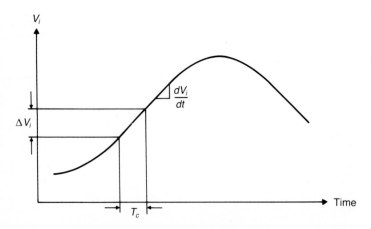

From Eq. (6.14), the maximum change in the sinusoidal input during the conversion time is

$$\Delta V_{i_{max}} = A\omega T_c$$

If the frequency of the sinusoidal signal is 600 rad/s and the conversion time T_c is 50 μs, the maximum error due to the change in the input signal during the conversion process is

$$\Delta V_{i_{max}} = 600 \ (50 \times 10^{-6})A = 0.03A$$

Thus there can be up to a 3 % error in the converted value of the input signal in this example. □

Sample-and-Hold

The magnitude of errors due to nonconstant input voltages can obviously be decreased by decreasing the conversion time T_c in Eq. (6.14). This approach can be costly, however, and an alternative method is to add a **sample-and-hold circuit** to the input to the A/D converter as shown in Fig. 6.13. The purpose of this circuit is to hold the input signal constant during the A/D conversion. The sample-and-hold circuit shown in Fig. 6.18 has as its main element a "hold" capacitor C_{hold}. Two operational amplifiers are used. The first provides a high-impedance input for the sample-and-hold circuit and also can be used to filter the input signal. The second amplifier provides a high-impedance output for the "hold" capacitor.

When an A/D conversion is not in progress, the "hold" capacitor is connected to the input amplifier and the voltage across the capacitor "tracks" the output of the amplifier. When an A/D conversion is desired, the capacitor is disconnected from the input amplifier. The input voltage to the output amplifier is then effectively constant, because the rate of discharge of the capacitor is low as a result of the high-impedance input of this amplifier. Once the A/D conversion is complete, the "hold" capacitor is again connected to the input amplifier to allow it to "track" the input voltage. The use of the sample-and-hold circuit ensures that the converted value accurately represents the value of the

Figure 6.18 Sample-and-hold circuit for A/D converter.

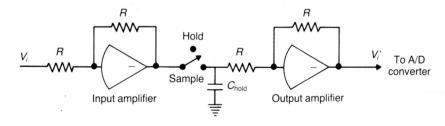

input signal at the time the conversion began, regardless of the conversion time of the A/D converter.

Multiplexing

Often, when sampling several input signals, it is desirable to use a single A/D converter to reduce hardware costs. The input to the A/D converter can then be switched from one input signal to another by means of a **multiplexer** circuit such as that shown in Fig. 6.19. This implies that the input signals are sampled sequentially rather than in parallel. If no sample-and-hold circuits are used and n inputs are sampled, then Eq. (6.14) becomes

$$\Delta V_{i_n} = \frac{dV_{i_n}}{dt} nT_c \tag{6.15}$$

where ΔV_{i_n} is the error resulting from the change in the nth input signal, V_{i_n}, between the beginning of conversion of the first input to the completion of conversion of the last input, V_{i_n}. Hence, if several signals are to be converted, the converted value of the last input may have varied considerably from its value at the instant that conversion of the first input began. This **slewing error** may be unacceptable when samples of all inputs are required at precisely the same instant in time for purposes of comparison or accurate control calculations.

This problem can be eliminated either by using separate A/D converters or by putting a sample-and-hold circuit on each input as shown in Fig. 6.19. All inputs are then sampled at the same time, and the output of the individual sample-and-hold circuits can be converted at will without inaccuracy due to slewing effects.

Figure 6.19 Multiple-input A/D converter with multiplexer.

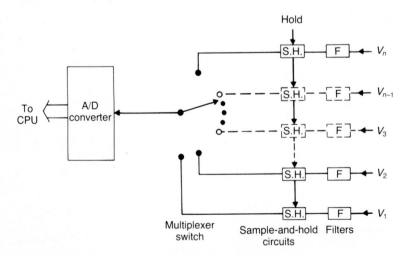

Example 6.11 Simultaneous sampling Instantaneous power is to be measured for the motor shown in Fig. 6.20(a) by sampling motor voltage V_m and current shunt voltage V_s via an A/D converter. The power is then computed using

$$P = (V_m - V_s) \frac{V_s}{R}$$

Figure 6.20 A/D conversion system for motor power measurement.

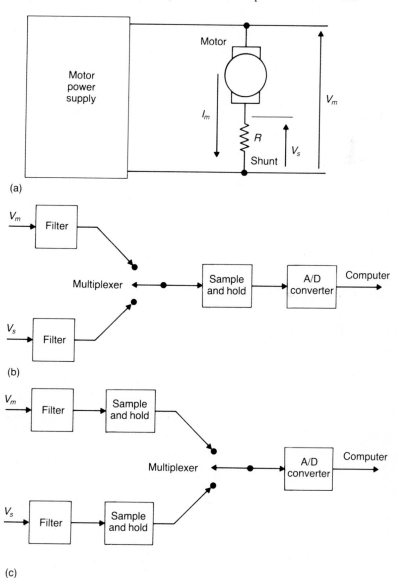

(a)

(b)

(c)

where R is the current shunt resistance. If the circuit in Fig. 6.20(b) is used and V_m is sampled and converted before V_s is sampled and converted, then the rate of change of these variables and the delay between the two samples due to the conversion time can decrease the accuracy of the power measurement. On the other hand, if the circuit in Fig. 6.20(c) is used, both variables are sampled at the same instant in time. There is no delay between samples to affect accuracy, and first V_m and then V_s can be converted and used in the power computation. □

Conversion Time and Stability

The conversion time is an important consideration for control systems in terms of its effect on closed-loop system stability. Conversion time has been defined earlier as the time delay between the start of the A/D conversion by the CPU and acquisition of the digital value corresponding to the analog voltage input. Like other time delays in control systems, long conversion times have a detrimental effect on closed-loop stability. Also, in the control algorithm and analysis presented in Chapter 4, it was assumed that essentially no time is required to sample the feedback, compute the manipulation, and output it. This may not be the case for sampling using an A/D converter unless T_c is relatively small and the sample period is relatively large. As a result, the validity of control analysis and controller designs can be adversely affected unless steps such as those suggested in Section 5.8 are taken.

6.6

ALPHA-NUMERIC INPUT/OUTPUT

There is often a need in control systems for communication of alphabetic or numeric information with an operator. The computer must therefore be interfaced to devices such as numeric displays, thumbwheel switches, keyboards, printers, and video displays. Although the computer operates only on binary numbers, it is inconvenient to require an operator to enter numbers and other information in binary form and to look at binary displays. It is therefore necessary to communicate with the operator using decimal numbers and his or her own "natural" language. If alpha-numeric characters must be input or output, then these characters (A to Z, 0 to 9, ., ?, !, and so on) must be represented by binary numbers so that they can be manipulated by the computer.

A number of character coding schemes are in existence, but standardization has made the 7-bit ASCII coding scheme one of the most widely used. ASCII character codes are provided for all letters, digits, and punctuation marks, along with a number of codes that allow device control functions such as "space," "line feed," and "carriage return" to be performed. Table 6.1 shows the ASCII

Table 6.1

Selected ASCII Character Codes

	ASCII		
CHARACTER	**Octal**	**Hexadecimal**	**BINARY**
"A"	101_8	41_{16}	1000001
"B"	102_8	42_{16}	1000010
"5"	065_8	35_{16}	0110101
"6"	066_8	36_{16}	0110110
"?"	077_8	$3F_{16}$	0111111
"carriage return"	015_8	$0D_{16}$	0001101

character codes for a small set of alphabetic, numeric, punctuation, and control characters. The code for each character is given in octal and hexadecimal representation and consists of seven bits of binary information. Appendix C contains a complete set of ASCII character codes.

Output to Alpha-Numeric Displays

Alpha-numeric display devices are used in control systems to allow a variety of information to be displayed for the operator of the system. Video displays and printers are examples of these devices. The main advantage of the video display is that information can be displayed at high speeds and in an easily readable format. Its main disadvantages are that only a limited amount of information can be displayed at one time and that no permanent copy of the displayed information is available, as it would be with a printer. Video displays therefore tend to be used to display information of short-term importance. Keyboard/ video-display combinations are sometimes packaged as one unit and form a convenient means for communication with a computer.

The display interface shown in Fig. 6.21 allows the CPU to output 7-bit ASCII character codes to a video display or printer. The display interface is often a commercially available unit and may include a keyboard interface as well. A READY signal is usually present and can be input by the CPU to indicate whether the display interface is ready to accept another character. If the device is ready, the 7-bit ASCII character code for the desired character is loaded by the CPU into a register and then output to the display interface via the I/O data bus. Character codes received by the display interface are output to the display device in either a serial or a parallel format. For serial transmission, all 7 bits of ASCII codes are sequentially multiplexed into a single signal, together with control information that allows the signal to be decoded and the character code

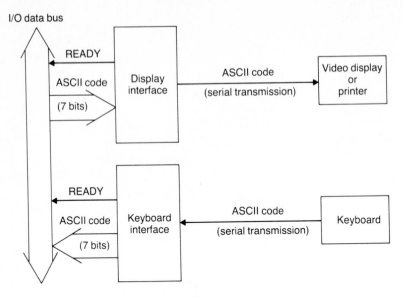

Figure 6.21 Keyboard/display interface.

reconstructed by the video display or printer. Serial transmission has the obvious advantage of minimizing wiring costs, particularly when transmission over considerable distances is required. Hence many keyboard/video-display units and printers use serial rather than parallel communication with the computer interface.

If the character "A" is to be displayed, a register is loaded with 41_{16}, which is the code for "A" shown in Table 6.1. Control codes such as carriage return ($0D_{16}$) are available by which to manipulate where characters will next appear on the display. The display of decimal numeric information derived from binary numbers requires converting each binary number into the corresponding decimal digits and then outputting the ASCII character code for each digit. If one wants to display in decimal form the binary number 1000001 (65_{10}), it is necessary to output the characters "6" and "5" to the display, using the corresponding character codes 36_{16} and 35_{16}. Note that if 1000001 (41_{16}) were to be output to the display without conversion to decimal form, it would be interpreted as a character code, and the corresponding character "A" would be displayed. A technique for converting binary numbers into decimal digits is described in Section 6.7.

Keyboard Input

In control applications, a **keyboard** can be useful in allowing the operator to enter a variety of commands. Alpha-numeric keyboards are usually capable of generating all of the ASCII characters shown in Table 6.1 and Appendix C, plus many control codes. When a key is pressed on the keyboard, the appropriate 7-bit ASCII character code is generated by the keyboard and transferred to the keyboard interface. These seven bits of information can then be input to the CPU via the I/O data bus and presented to the computer by an appropriate interface.

As shown in Fig. 6.21, the keyboard interface is responsible for receiving ASCII character codes from the keyboard. Codes received in a serial format are reconstructed to form the binary ASCII code. When the keyboard interface has received a character, it sets its READY signal, indicating to the CPU that a character has been received. An input operation can then be performed by the CPU to load the ASCII character code into a register. The register then contains the ASCII code for the key pressed on the keyboard. For example, the "A" key results in 41_{16} being loaded into the register, and the "carriage return" key results in $0D_{16}$ being loaded into the register.

Once the character code has been input by the CPU, it is up to the program to determine how it should be interpreted or further processed. Single characters may indicate that specific functions are to be performed by a control system. Strings of characters may be input and stored for later use. Strings of decimal digits to be interpreted as numbers must be converted numerically to produce the corresponding binary number. A technique for performing this conversion is described in the following section.

6.7
NUMERIC I/O

Numeric input and output to devices such as keyboards, video displays, and printers normally require conversion of binary numbers to and from character codes. A 16-bit unsigned binary number, for example, has a decimal value between 0 and 65535 and hence may require input or output of up to five decimal digits. Table 6.2 gives three commonly used representations of the characters "0" through "9." Four binary bits are required to represent the decimal digit 9, and it is convenient to allocate 4 bits to represent the decimal digits 0 through 9. This code is called the binary-code-decimal (BCD) representation of the digit. It can be seen in Table 6.2 that the BCD code for a decimal digit is just the binary number that corresponds to the digit. Possible BCD codes above 1001_2 are not used.

Some devices use BCD codes for input/output, but it is more common to require conversion to character codes first. It is worth noting that one can obtain the ASCII code from the BCD code, and vice versa, by adding or subtracting 30_{16}. Because the input and output of numeric information plays an important part in communicating with a machine or process operator, techniques for performing these conversions will now be discussed.

Table 6.2

Codes for Decimal Digits

DECIMAL DIGIT	BCD CODE	ASCII CODE	7-SEGMENT CODE*
0	0000	30_{16}	$3F_{16}$
1	0001	31_{16}	06_{16}
2	0010	32_{16}	$5B_{16}$
3	0011	33_{16}	$4F_{16}$
4	0100	34_{16}	66_{16}
5	0101	35_{16}	$6D_{16}$
6	0110	36_{16}	$7D_{16}$
7	0111	37_{16}	07_{16}
8	1000	38_{16}	$7F_{16}$
9	1001	39_{16}	$6F_{16}$

* For the segment sequence g, f, e, d, c, b, a shown in Fig. 6.24.

Binary-to-Decimal Conversion

A basic algorithm is presented here for converting an unsigned binary integer to BCD representation. It also is shown how to obtain the ASCII codes for display purposes; this is a simple matter once the BCD codes have been found. For 16-bit integers, 5 decimal digits must be obtained but the method is generalized. Decimal digits will be obtained in BCD form, converted to ASCII, and displayed.

As indicated in the flowchart shown in Fig. 6.22, the BCD representation of the most significant decimal digit is obtained by taking the integer result of dividing the binary number by the appropriate power of 10. The integer quotient, the result of the DIV operation, is taken as the BCD digit. The integer remainder, the result of the MOD operation, replaces the binary number in the next iteration and is used to find the next lower significant digit. The BCD representation of each digit is converted to ASCII character code by adding 30_{16} to the BCD code as it is produced. The ASCII digit can be displayed immediately as described in Section 6.6. When the least significant digit has been produced, the algorithm is terminated. Often the computer can generate characters faster than they can be displayed, and a delay may need to be implemented in order to wait until the display device is ready before the next character can be output.

Example 6.12 Conversion of binary to ASCII To illustrate binary-to-decimal conversion and the subsequent generation of an ASCII character string representing a binary number in decimal form, consider conversion of the unsigned, 8-bit binary integer

$$BIN = 01000001_2$$

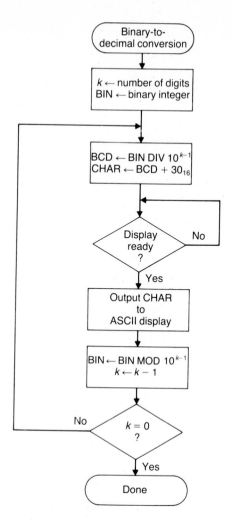

Figure 6.22 Binary-to-decimal
conversion with ASCII output.

Unsigned 8-bit binary integers are less than 256_{10} and hence require at most
three digits in the decimal-integer form.

Referring to the flowchart in Fig. 6.22, we can trace the binary-to-decimal
conversion as it proceeds as follows:

k	**BIN**	10^{k-1}	**BCD**	**CHAR**
3	01000001_2	$100_{10} = 01100100_2$	$0000_2 = 0_{16}$	$30_{16} = $ "0"
2	01000001_2	$10_{10} = 00001010_2$	$0110_2 = 6_{16}$	$36_{16} = $ "6"
1	00000101_2	$1_{10} = 00000001_2$	$0101_2 = 5_{16}$	$35_{16} = $ "5"

The result is the character string "065," which can be output as the decimal equivalent of the original binary number

$$065_{10} = 01000001_2 \quad \square$$

Decimal-to-Binary Conversion

When a decimal number is input from a device such as a keyboard, the data received by the CPU will be the character code representing the decimal digits in the number. To operate on these numbers and utilize them in further computations, it is necessary first to convert them into binary representations.

The flowchart shown in Fig. 6.23 illustrates one technique for performing this conversion for unsigned binary integers. The result, BIN, is initially set to zero before any input occurs. The first ASCII code received will be the most significant decimal digit. This character is input by the CPU as described in Section 6.6 and converted to BCD by subtracting 30_{16}. BIN is then multiplied by 10_{10}, and the newly received BCD code is added to it. The multiplication by 10_{10} "shifts"

Figure 6.23 Decimal-to-binary conversion with ASCII input.

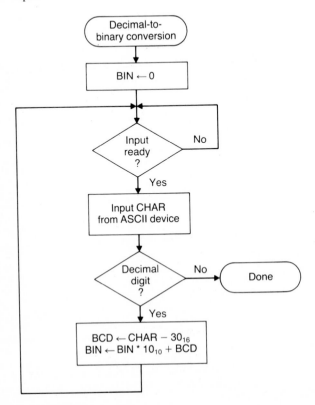

precious binary results by the equivalent one decimal place prior to addition of the new digit. This process is repeated as successively less significant digits are input. The algorithm can be conveniently terminated when a character is received that is not a decimal digit.

Example 6.13 Conversion of ASCII to binary To illustrate decimal-to-binary conversion, consider the ASCII character string "65" followed by a carriage return. In this case, the values of the variables in the flowchart in Fig. 6.23 are as follows:

CHAR	BCD	BIN
"6" = 36_{16}	6_{16}	$00000000_2 {}^* 1010_2 + 00000110_2 = 00000110_2$
"5" = 35_{16}	5_{16}	$00000110_2 {}^* 1010_2 + 00000101_2 = 01000001_2$
carriage return = $0D_{16}$	none	01000001_2

The result, as expected, is

$$01000001_2 = 65_{10} \quad \square$$

Seven-Segment Display Output

It is often unnecessary to incorporate a video display into a control system when only numeric information needs to be displayed. **Seven-segment displays** are one inexpensive and widely used means of displaying numeric information. Figure 6.24 shows the layout of a typical 7-segment display, several of which can be located side-by-side to display multiple digits.

Figure 6.24 Seven-segment display with BCD input.

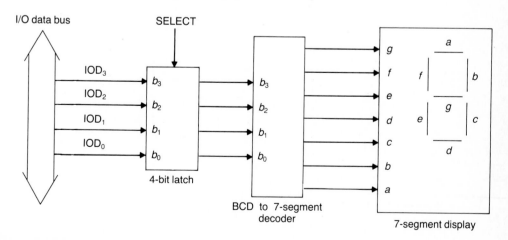

Table 6.2 gives the 7-segment codes for display of the decimal digits "0" through "9"; the illumination of each of the segments in the display can be controlled independently by using a circuit such as that shown in Fig. 6.2. An alternative method is to use a BCD–to–7-segment decoder as shown in Fig. 6.24. If it is desired to display a particular digit, the corresponding BCD code is first computed by the CPU and then output to the 7-segment display interface. For multiple-digit displays, the BCD code for each digit is obtained and these BCD codes can be output to side-by-side 7-segment displays.

Thumbwheel Switch Input

Just as video displays may be unnecessary in many control systems, so may a keyboard be replaced by more simple devices such as push-buttons and thumbwheel switches. Figure 6.25 schematically shows a 10-position **thumbwheel switch** accompanied by a BCD encoder. Although individual thumbwheel switch contacts can be interfaced to the CPU using a circuit such as that shown in Fig. 6.4, it is often more convenient to use hardware that will directly produce BCD codes for input to the CPU so that decimal-to-binary conversion can be carried out easily. In the circuit shown in Fig. 6.25, the 4 bits output by the encoder indicate the position of the switch in BCD code. These are interfaced to the I/O data bus, and an input operation can be performed by the CPU to load them into an accumulator, allowing the CPU to sense the switch position.

If more than one decimal digit must be input, a set of side-by-side thumbwheel switches can be used. The BCD inputs from each switch can then be

Figure 6.25 Ten-position thumbwheel switch with BCD encoder.

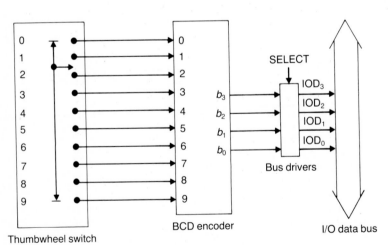

combined, via decimal-to-binary conversion techniques, to obtain the binary representation of the multidigit decimal number.

6.8

INTERRUPT INTERFACING

The handling of interrupts and the hardware required to generate interrupts vary greatly among computer systems. Small computers may have a single interrupt input that is manipulated by external devices to cause an interrupt on the CPU. If there is more than one such device, the CPU must use input operations to test the status of each device and determine which device caused the interrupt. More sophisticated systems provide multiple interrupt inputs to the CPU and a priority structure that allows a more important interrupt to override the servicing of a less important interrupt. In these systems, the interrupt input to which the device is connected can be automatically associated with appropriate interrupt-handling software.

Example 6.14 Interrupt generation by means of a push-button As a simple example, consider a CPU with a single interrupt input INT that causes an interrupt to occur whenever the input is at a logic-1 level. Figure 6.26 shows a circuit for generating such an interrupt with a push-button. When the push-button switch is closed, the output of the inverter changes from logic 0 to logic 1. This is connected to the INT input of the CPU and causes the program being

Figure 6.26 Interrupt generation by means of a push-button.

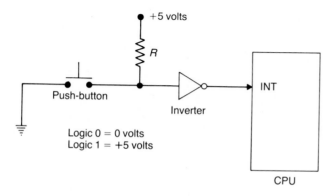

executed by the CPU to be interrupted. Program control is transferred to specific-interrupt handling software that performs whatever function is to be initiated by pressing the push-button. When this function is complete, control is returned to the interrupted program at the point where processing stopped when the interrupt occurred. The use of the interrupt feature removes any necessity for the CPU to continually test the state of the push-button to determine whether it has been pressed. As long as the interrupt system on the CPU is enabled and the button is pushed, the CPU will respond regardless of its current activities.

Care must be taken when using interrupts to ensure that they do not occur either when they are unexpected or when they cannot be serviced. This is particularly true on start-up when the entire software and hardware system is being initialized. For this reason, a program using interrupts should not enable the interrupt system until initialization is complete and interrupts can be safely allowed. A convenient means of enabling and disabling the interrupt system is usually provided. Also, it is generally necessary to disable interrupts internally in the CPU during the servicing of an interrupt to prevent the possibility of the interrupt interrupting itself. This function is performed automatically by most CPUs. □

Example 6.15 Real-time clock As another example, consider a circuit to generate interrupts at specific time intervals. Such a device is called a real-time clock and is used in control applications to establish the sample period as T seconds for control algorithms. The circuit shown in Fig. 6.27 is supplied with a square wave with a period equal to the desired time interval between interrupts. Additional logic is used in the circuit to overcome the potentially serious problem that occurs when the interrupt input to the CPU has not returned to logic 0 when processing of the interrupt is complete. This causes another interrupt to occur immediately rather than after the required interval of time.

Figure 6.27 Interrupt interface for a real-time clock.

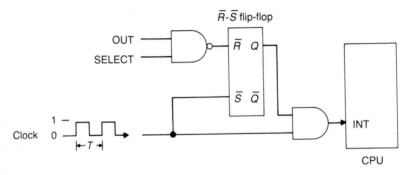

In Fig. 6.27, output Q of the $\bar{R}$–$\bar{S}$ flip-flop is logic 1 before the rising edge of the square-wave clock because input $\bar{S}$ is logic 0. Flip-flop output Q remains at logic 1 when the clock makes a transition to logic 1. The INT input of the CPU is then logic 1 and an interrupt is generated. This interrupt must be recognized immediately, and can be used to begin the processing of a control procedure. The first step of this procedure should be to reset the INT interrupt input of the CPU to logic 0 so that an interrupt cannot occur again until the next rising edge of the square-wave clock. This is done by using an OUTPUT instruction to momentarily set $\bar{R}$ to logic 0, which resets flip-flop output Q to logic 0, and also changes INT to logic 0. Output Q returns to logic 1 when the clock returns to logic 0, but INT remains at logic 0. The control procedure will then be processed each time the clock signal changes to logic 1, and the resulting sample period will be the period of the clock. □

6.9

COMPUTER CONTROL EXAMPLE

A complete computer control system is likely to require a number of input/ output interfaces. Some allow the computer to manipulate the process to be controlled, some supply monitoring information for human operators, and some allow the control computer to communicate with other computers in larger manufacturing or process control systems. As an example, consider the interfaces necessary to allow a computer to control the liquid-level process shown in Fig. 6.28(a). The level command is established by a human operator using a set of four thumbwheel switches, and a level readout is provided on four 7-segment LEDs for monitoring purposes. The computer obtains feedback from a liquid-level sensor via an A/D converter and manipulates a flow-control valve using a D/A converter. A 20-Hz clock is used to generate interrupts so that PI control can be implemented in computer software. Status information can be sent to, and received from, a system-monitoring computer via a communication network.

Figure 6.28(b) shows the interfaces required for this control application. All are assigned different I/O addresses and include appropriate address-decoding and device-selection circuits. Data is transferred to and from all devices via the I/O data bus. Control software can be written in assembly language, in a high-level language, or in a combination of both. The relatively low sample rate (20 Hz) may allow all software to be written in a high-level language. If a higher sample rate were used (250 Hz, for example), it might be necessary to write the interrupt procedure for closed-loop control in assembly language to ensure that all necessary control computations could be performed rapidly enough.

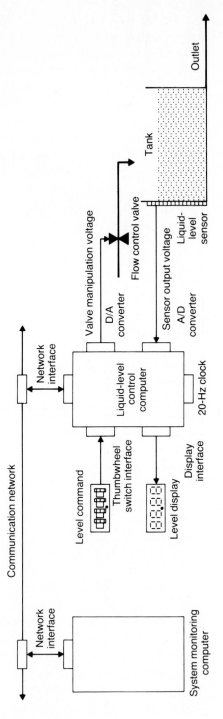

(a)

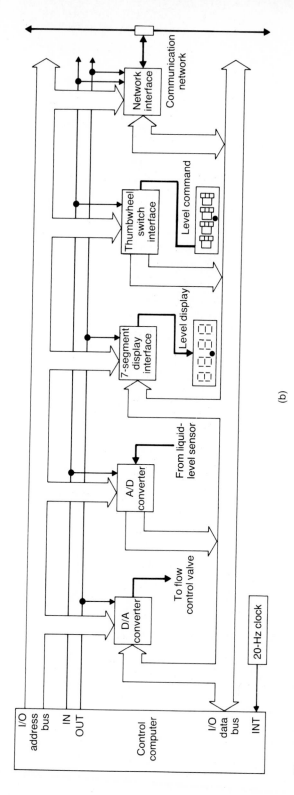

Figure 6.28 (a) System with monitoring computer and liquid-level process control computer, and (b) control computer interfaces for liquid-level process.

(b)

261

6.10
SUMMARY

The primary objective of this chapter has been to present the basic concepts of interfacing a control computer to various input/output devices. In light of the wide variety of control computers available, we considered examples of typical devices and how they can be interfaced to a simple but representative computer architecture. Through a review of specific hardware requirements, the principles given here can be related to a great many interfacing tasks.

It is necessary, of course, to have a sound understanding of digital logic and computer hardware to be able to produce reliable and efficient interfaces for control computers. Interfaces of proven reliability, an indispensable attribute for control systems, are available from many computer manufacturers and suppliers of peripherals. It often proves more cost-effective to buy "off-the-shelf" interface hardware than to develop hardware "in house," particularly when a system is in the development stage. But designing one's own hardware will always be necessary when no appropriate interface is commercially available.

BIBLIOGRAPHY

Artwick, B. A. *Microcomputer Interfacing.* Englewood Cliffs, NJ: Prentice-Hall, 1980.

Craine, J. F., and G. R. Martin. *Microcomputers in Engineering and Science.* Reading, MA: Addison-Wesley, 1985.

Eccles, W. J. *Microprocessor Systems: A 16-Bit Approach.* Reading, MA: Addison-Wesley, 1985.

Ferguson, J. *Microprocessor Systems Engineering.* Reading, MA: Addison-Wesley, 1985.

Fletcher, W. I. *An Engineering Approach to Digital Design.* Englewood Cliffs, NJ: Prentice-Hall, 1980.

Garrett, P. H. *Computer Interface Engineering with Model-Based Analysis.* Englewood Cliffs, NJ: Prentice-Hall, 1987.

Givone, D. D., and R. P. Roesser. *Microprocessors/Microcomputers.* New York, NY: McGraw-Hill, 1980.

Lawrence, P. D., and K. Mauch. *Microcomputer System Design: An Introduction.* New York: McGraw-Hill, 1987.

Leventhal, L. A. *Introduction to Microprocessors: Software, Hardware, Programming.* Englewood Cliffs, NJ: Prentice-Hall, 1978.

Mano, M. *Computer Systems Architecture, 2/e.* Englewood Cliffs, NJ: Prentice-Hall, 1982.

Mano, M. *Digital Design.* Englewood Cliffs, NJ: Prentice-Hall, 1983.

Short, K. L. *Microprocessors and Programmed Logic.* Englewood Cliffs, NJ: Prentice-Hall, 1980.

Stone, H. S. *Microcomputer Interfacing.* Reading, MA: Addison-Wesley, 1982.

PROBLEMS

6.1 Design a logic circuit that will generate a SELECT signal for an interface that will respond to the 8-bit I/O address $3A_{16}$.

6.2 Design an address-decoding and device-selection circuit that generates four I/O device-enabling signals according to the following table:

ADDRESS (8 bits)	IN	OUT	SIGNAL
20_{16}	1	0	STATUS
20_{16}	0	1	MODE
21_{16}	1	0	DATA-IN
21_{16}	0	1	DATA-OUT

6.3 Design an interface to use output IOD_0 to disable and enable an interrupt input CLOCK using an $\bar{R}\text{–}\bar{S}$ flip-flop with output Q and any necessary logic gates. Assume that SELECT and OUT signals have already been generated and are available as defined in Section 6.2. An interrupt input INT to the CPU is to be generated according to the following table:

Q	CLOCK	INT
0	0	0
0	1	0
1	0	0
1	1	1

6.4 It is desired to illuminate an LED when switches SW3, SW2, SW1, and SW0 are set to C_{16} (SW0 is the least significant bit). The LED and switch elements are shown in Fig. 6.29.

a. Design a logic circuit to perform the required function.

b. Design a computer interface and the required software to perform the required function. Compare this solution with that of part (a).

6.5 Two 3-bit binary numbers A and B are to be input from a set of switches, added, and the result displayed on LEDs as indicated in Fig. 6.30. This action is to be performed each time the push-button labeled UPDATE is pushed. The result consists of a 3-bit sum S and a carry bit labeled CARRY. Design the computer interfaces and software required to accomplish this task.

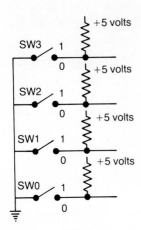

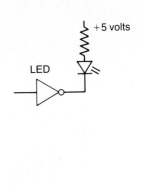

Figure 6.29 Switches and LED for Problem 6.4.

Figure 6.30 Switch-selectable binary numbers to be added and displayed in Problem 6.5.

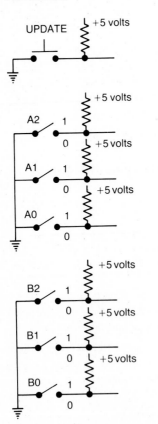

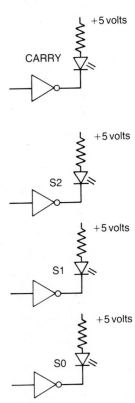

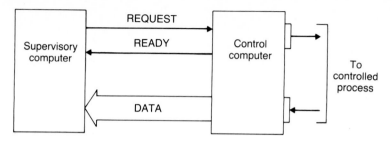

Figure 6.31 "Handshaking" signals and data transfer between computers for Problem 6.6.

6.6 A control computer used in a manufacturing system is required to send status data to a supervisory computer as indicated in Fig. 6.31. First, a REQUEST signal is set to logic 1 by the supervisory computer, indicating to the control computer that current status data is to be output to a buffer. After the control computer has output the necessary data to the buffer, a READY signal is set to logic 1, indicating to the supervisory computer that it is time to input the status data from the buffer. After the supervisory computer has input the data, it resets the REQUEST signal to logic 0, and when the control computer senses that the REQUEST signal is logic 0, it resets the READY signal to logic 0. The supervisory computer must ensure that the READY signal is logic 0 before it begins another status information transfer cycle. Design for both computers the hardware and software that implement this data transfer and "handshaking" sequence. Assume that the control computer is concurrently performing closed-loop control using interrupts and that this function should have higher priority than communication of status data.

6.7 A 10,000-Hz clock has been connected to IOD_0 of the I/O data bus as an input device with address 2.
   a. Write assembly-language software to output a 100-Hz square wave from a D flip-flop connected to IOD_0 as an output device with address 4.
   b. Repeat part (a), but connect the 10,000-Hz clock to the interrupt input.

6.8 The shaft of a stepping motor shown in Fig. 6.32 rotates one angular increment in position whenever a pulse is received at its STEP input. The direction of rotation is determined by its DIR input. If the motor rotates in the positive direction when DIR is logic 0 and the resolution of the motor is 250 steps per revolution, then setting DIR to logic 1 and delivering 50 pulses (logic 0-1-0) to the motor on the STEP input causes the motor to turn one-quarter of a revolution in the negative direction.
   a. Design an interface that will allow a control computer with the architecture shown in Fig. 6.1 to output signals STEP and DIR to the motor. Interrupts are to be generated by using a 10,000-Hz clock so that STEP pulses can be output from the computer at a maximum rate of 10,000 per second.
   b. Design interfaces for two limit switches, LS1 and LS2, that are used to reverse the direction of motion of a table positioned by a screw directly coupled to the

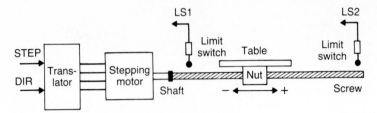

Figure 6.32 Table driven by stepping motor and control computer for Problem 6.8.

motor. Assume that 10 cm of table motion are provided between the limit switches and that the pitch of the screw is 4 revolutions per centimeter.

c. Design assembly-language software to continually move the table from one limit switch to another at a constant velocity of 0.5 cm/s.

6.9 A process is to be manipulated using a D/A converter with a range of ± 10 volts and a resolution 0.05 volts. How many bits must be present in the D/A converter?

6.10 A D/A converter is to be used to deliver velocity commands to a motor. The maximum velocity is to be 3000 rpm, and the minimum nonzero velocity is to be 2 rpm. How many bits must be present in the D/A converter for bi-directional operation?

6.11 Design a D/A converter to output ± 5 volts from an 8-bit, signed binary-integer output by the CPU. The integer zero should produce zero D/A output voltage. Can -5 volts be output? Can $+5$ volts be output?

6.12 The voltage range of feedback from a process is ± 5 volts, and a resolution of 0.01% of the voltage range is required. How many bits must be in the A/D converter used to sense the process feedback?

6.13 Eight channels of analog information are being used by a computer to close eight control loops. Assume that all analog signals have identical frequency content and are multiplexed into a single A/D converter. The A/D converter requires 100 microseconds per conversion. The closed-loop control software requires 400 microseconds of computation and output time for four of the loops, and the other four require 200 microseconds. What is the maximum frequency content that the analog signals can have and still be consistent with good practice?

6.14 A rotary potentiometer is to be used as a remote rotational displacement sensor. The maximum angular displacement to be measured is $180°$, and the potentiometer is rated for 10 volts and $270°$ of rotation.

a. What voltage increment must be resolved by an A/D converter in order to resolve an angular displacement of 0.75 degrees? How many bits would be required in the A/D converter for full-range detection?

b. The A/D converter requires a 10-volt input voltage for full-scale binary output. If an amplifier is placed between the potentiometer and the A/D converter, what amplifier gain should be used in order to take advantage of the full range of the A/D converter?

6.15 The rotational velocity of a motor is controlled with a proportional digital controller using the output voltage of a tachometer sampled at a rate of 120 Hz with an A/D converter. Motor velocity manipulation voltage is output using a D/A converter. The motor operates over a speed range of 2000–3000 rpm in one direction only.

    a. It is necessary to achieve a velocity resolution of 1 rpm. How many bits are required in the A/D converter?

    b. It is observed that the motor velocity oscillates at a frequency of about 1 Hz during operation at 2220 rpm. The tachometer is known to have 13 commutator bars. What is likely to be the source of the oscillation, and how can it be corrected?

6.16 The three-phase, half-wave SCR power amplifier used to drive a DC motor produces the motor voltage waveform shown in Fig. 6.33 when load torques are low. Motor speed is to be sampled at a frequency of 200 Hz using an A/D converter and the output voltage of a tachometer. Design an anti-aliasing filter to be used at the input of the A/D converter.

6.17 The x-y-z position-sensing probe shown in Fig. 6.34 is to be interfaced to a computer and its outputs sampled at a rate of 100 Hz via an A/D converter. The sampled data will be used to find $x = f(z)$ and $y = g(z)$. The maximum x, y, and z velocities are 0.02 μs m/s, the A/D converter has one analog input, and its conversion time is 70 μs. Complete the A/D conversion system shown in Fig. 6.34, and explain why each functional component that you added is required.

6.18 A 50-Hz signal is sampled once every 1.5 s. What is the lowest apparent frequency in the sampled data that arises from aliasing? What range of sample frequencies ensure that aliasing does not occur when a 50-Hz signal is being sampled?

6.19 A torque sensor has been mounted on a power take-off (PTO) unit driven by a farm tractor engine. The voltage produced by the torque sensor is to be sampled by an A/D converter. The rotational speed of the PTO shaft is 600 rpm. Significant frequency content is present in the PTO torque at twice the shaft rotation frequency because of speed fluctuations caused by the universal joint connecting the PTO to the engine. Also, approximately 12 power pulsations are generated by the piston engine during each PTO shaft revolution. What is the minimum sampling period that can be used that ensures that aliasing will not occur?

Figure 6.33 Motor voltage waveform for Problem 6.16.

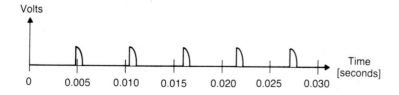

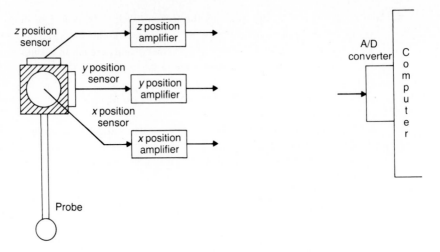

Figure 6.34 x-y-z probe and A/D converter for Problem 6.17.

6.20 Show how each of the following unsigned binary numbers would be converted to an ASCII representation of the equivalent decimal number via the binary-to-decimal conversion algorithm shown in the flowchart in Fig. 6.22. Give the values of the variables at the end of each pass through the flowchart, using 8-bit unsigned binary numbers.

a. 01000110_2
b. 10011111_2

6.21 Show how the decimal number "138" represented by ASCII characters would be converted to binary form, using the decimal-to-binary conversion algorithm shown in the flowchart in Fig. 6.23. Give the values of the variables at the end of each pass through the flowchart, using 8-bit unsigned binary numbers.

6.22 Modify the decimal-to-binary conversion algorithm shown in Fig. 6.23 to allow the character "+" or the character "−" to be optionally present as the first character of a numeric character string.

6.23 Modify the binary-to-decimal conversion algorithm shown in Fig. 6.22 to output spaces rather than leading zeros.

6.24 Modify the binary-to-decimal conversion algorithm shown in Fig. 6.22 to work with 2's-complement signed binary numbers and to output "−" as the first character if the binary number is negative.

6.25 Design an interface for the computer and liquid-level process shown in Fig. 6.28 that allows both the 20-Hz clock and a pushbutton labeled "EMERGENCY FLOW CUTOFF" to generate interrupts via the INT input. The CPU must be able to determine the source of each interrupt generated so that it can take the necessary action.

6.26 Design a circuit that will interface two interrupts, INT0 and INT1, to the INT input of a CPU such that INT0 has the highest priority and INT1 the lowest. A signal on INT0 is to be able to interrupt the CPU even when the CPU is currently handling an interrupt generated by INT1—but not vice versa. In other words, INT0 can interrupt INT1, but INT1 cannot interrupt INT0. Also design any software required to operate the interface and handle the interrupts.

Chapter 7
Sensors for
Computer Control

INTRODUCTION

A **sensor** can be defined as a device that outputs a signal for purposes of detection or measurement of a physical property. Examples include temperature sensors, which have outputs that change as a function of temperature, and position sensors, which output a measurement of displacement. Developments in electronic instrumentation have led to the availability of a variety of types of sensors that can be readily interfaced to a control computer. Sensors are available for the measurement of position, velocity, acceleration, force, torque, temperature, and so on, allowing feedback loops to be established for the control of these variables in machines and processes. Sensors are often referred to as **transducers** because they convert one physical quantity into another. An example is a tachometer's conversion of rotational velocity into voltage.

Sensors can be roughly classified as analog or digital. **Digital sensors** have logic-level outputs that can be directly interfaced to a control computer. Some pre-processing is usually involved in the sensor electronics to obtain a digital output. **Analog sensors** typically output a voltage proportional to the measured variable. An analog-to-digital converter must usually be employed to interface analog sensors to control computers.

The choice of sensors to be used in a given control application depends on (among other things) the nature of the variables to be measured, the cost of the sensor, and the levels of precision and performance required. Factors to be considered include those described in the following paragraphs.

Accuracy

The **accuracy** of a sensor can be defined as the agreement between the actual value of the measured variable and the measurement as it is output by the sensor. It is frequently expressed in terms of the error that can be expected between the measurement and the measured variable. If the distribution of errors has a known standard deviation σ, then the accuracy is frequently expressed as $\pm 2\sigma$ or $\pm 3\sigma$. For some sensors it is more common to express accuracy as a percentage of the measured value.

Resolution

Resolution refers to the change in the measured variable to which the sensor will respond with a like change in the measurement.

Repeatability

Repeatability expresses the variation over repeated measurements of a given value of the measured variable. For example, it is the variation in the output of a position sensor that is observed when a machine axis being measured is returned to exactly the same position several times.

Range

The **range** of measurement of a sensor specifies the upper and lower limits of the measured variable for which measurements can be made. For example, an angular position sensor may be limited to one revolution, or a range of $0°$ to $360°$.

Dynamic Response

Dynamic response refers to the maximum sinusoidal frequency of change in a measured variable for which agreement can be maintained between the measurement and the measured variable. Dynamic response is usually limited by the electrical and mechanical characteristics of the sensor. It may be expressed in terms of the 3-db attenuation bandwidth of the sensor's frequency response.

In this chapter, a number of types of sensors are discussed. Those sensors that are most commonly found in machine and process control systems receive the most attention. These include sensors of linear and angular position, velocity sensors, and temperature sensors. Sensors for other physical quantities (such as acceleration, strain, and pressure) are also discussed.

7.2

POSITION SENSORS

Position sensors are by far the most common type of sensor in machine control systems because of the obvious need to control the position of various moving parts of the machine. Position sensors are also commonly used for measuring other physical quantities that can be converted into positional changes. (An example is a pressure sensor constructed with a diaphragm or bellows.) In the following sections, various types of sensors of linear and angular position are discussed.

Potentiometer

Potentiometers are available in linear and rotary form, as illustrated in Fig. 7.1. A voltage V is applied across the resistive element of the potentiometer, and an output voltage V_o is produced that is proportional to the position of a brush that slides along the resistive element. The brush is connected mechanically to the mechanical component whose translational or rotational position is to be measured. The resistive element is normally either wire-wound or coated with a conductive film. The resistance per unit length is usually constant along the element. The output can then be expressed as

$$V_o = kVx \qquad \text{(linear)} \qquad\qquad (7.1)$$

or

$$V_o = kV\theta \qquad \text{(rotary)} \qquad\qquad (7.2)$$

Special potentiometers can be obtained with variations in resistance that result in a nonlinear relationship between output voltage and brush position. Examples include sinusoidal and logarithmic windings on rotary potentiometers.

Figure 7.1 (a) Linear potentiometer; (b) rotary potentiometer.

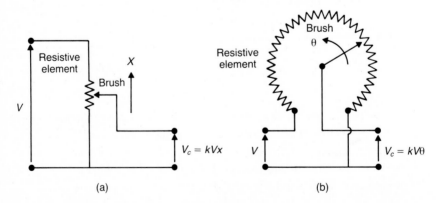

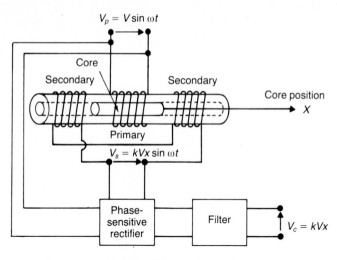

Figure 7.2 Linear variable differential transformer (LVDT).

Wear can be a problem when potentiometers are used, particularly if the brush hovers near a single point. Accuracies can be better than 0.1% of measurement range. Range is limited because the brush must eventually reach its end of travel, but cost is relatively low. Potentiometers are analog devices, and an analog-to-digital converter is required at the interface between a potentiometer and a control computer to convert the output voltage V_o into digital form.

Linear Variable Differential Transformer

The **linear variable differential transformer (LVDT)** is an inductive position-sensing device typically consisting of one primary winding and two secondary windings on a hollow cylinder. As shown in Fig. 7.2, the central winding is designated as the primary winding and is excited by applying a sinusoidal voltage V_p with a typical frequency of between 1000 and 5000 Hz. The outer windings, designated as secondary windings, have equal numbers of turns. A ferro-magnetic core moves axially within the hollow cylinder, and movement of the core varies the magnetic flux linking the primary winding to each of the secondary windings. This flux variation can be detected and forms the basis for position measurement via the LVDT.

The secondary windings of the LVDT are usually connected in series opposition so that the difference in voltage between the two windings can be detected. When the core is centered, the magnetic flux and induced voltages in both secondary windings are equal, and the voltage difference is zero. As the core moves toward one secondary winding, the induced voltage in that winding increases and the induced voltage in the other secondary winding decreases, causing a voltage difference to appear.

The windings can be constructed so that the relationship between core position and output voltage is linear. If

$$V_p = V \sin \omega t \tag{7.3}$$

then the voltage difference as a function of core position x can be expressed as

$$V_s = kVx \sin \omega t \tag{7.4}$$

where k is a constant. Plots of V_s for several core positions are shown in Fig. 7.3.

Figure 7.3 Differential voltage as a function of various core displacements. (a) Primary voltage, V_p; (b) V_s for $X > 0$; (c) V_s for $X = 0$; (d) V_s for $X < 0$.

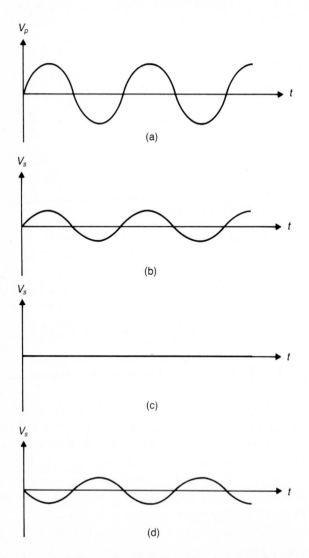

A phase-sensitive rectifier can be used to convert this signal into a non-sinusoidal output voltage that is proportional to core position. The magnitude of the output voltage is equal to the amplitude of the differential voltage V_s, and its sign is determined by whether the differential voltage is in phase or out of phase with the signal applied to the primary winding. The output voltage after filtering is then

$$V_o = kVx \qquad (7.5)$$

LVDTs can be obtained with ranges of motion from about 1 to 50 mm and resolutions of 1 μm or better. To a large extent, resolution is a function of the design of the LVDT output electronics, and it tends to decrease with range. An analog-to-digital converter is required to convert the output voltage into computer-readable form.

Rotary Resolver

Resolvers are inductive angular transducers that work on the same principles as transformers. Alternating current in stationary (stator) coils is used to induce an alternating current in separate rotating (rotor) coils, or vice versa. These devices therefore function as rotary transformers, and the coupling between coils is varied by changing the angle between them. Figure 7.4 illustrates the change in induced voltage in a rotating coil as a function of the angular position of a rotating winding when a voltage V_s is applied across the stationary winding.

$$V_s = V \sin \omega t \qquad (7.6)$$

If it is assumed that the coupling coefficient between coils is unity, then the rotor voltage V_r is related to the stator voltage as follows:

$$V_r = V_s \cos \theta \qquad (7.7)$$

or

$$V_r = V \cos \theta \sin \omega t \qquad (7.8)$$

where θ is the angular position of the rotating coil with respect to the stationary coil. Note that there is a 180° phase shift in the induced voltage after 180° of rotation. Furthermore, after 360° of rotation, the output is the same as at 0°. In fact, the rotor can be rotated indefinitely and the relationship between rotor and stator voltages repeats every revolution. The range of these devices is therefore one revolution, but some external means of counting resolver revolutions can usually be provided if more than one revolution is required. Accuracies of better than ± 10 minutes of arc can be achieved with precision resolvers.

A two-phase rotary resolver is shown in Fig. 7.5. There are two stator windings oriented at right angles to each other. If sinusoidal voltages are applied across the stator windings, the induced rotor voltage is the sum of the

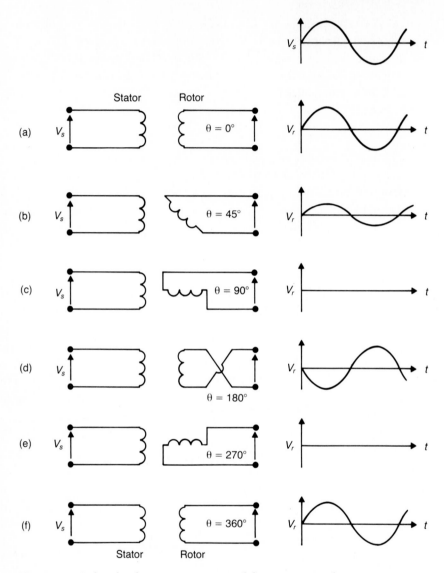

Figure 7.4 Induced voltage in a rotating coil for various angular positions.

contributions from these windings as determined by the rotor angle. Alternatively, a sinusoidal voltage can be applied to the rotor. In this case, the voltage induced in one stator winding is a function of the sine of the rotor angle, and the voltage induced in the other stator winding is a function of the cosine of the rotor angle. Two techniques for using these relationships to measure rotor rotation will now be discussed.

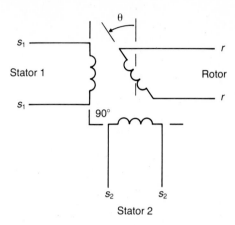

Figure 7.5 Two-phase rotary resolver.

Phase Modulation

Figure 7.6 shows a system for detecting the angular position of the rotor of a two-phase resolver by observing the phase shift in rotor voltage with respect to stator voltage. Sinusoidal voltages V_{s_1} and V_{s_2} are applied to the stator windings $90°$ out of phase, such that

$$V_{s_1} = V \sin \omega t \tag{7.9}$$

$$V_{s_2} = -V \cos \omega t \tag{7.10}$$

The voltage induced in the rotor V_r is a function of the rotor angle θ, and, given a unity coupling coefficient, the rotor voltage is

$$V_r = V_{s_1} \cos \theta - V_{s_2} \sin \theta \tag{7.11}$$

Figure 7.6 Angular position measurement with a two-phase resolver using phase modulation.

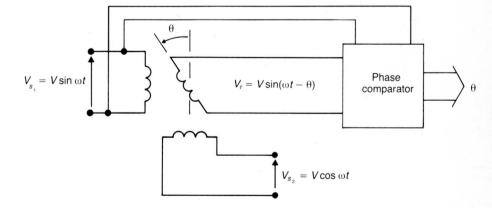

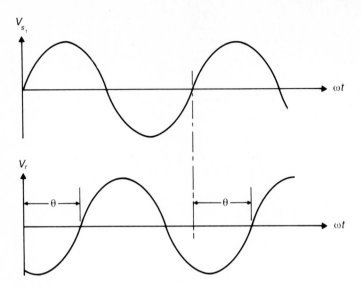

Figure 7.7 Phase shift in rotor voltage with respect to stator voltage.

Substituting Eqs. (7.9) and (7.10) for V_{s_1} and V_{s_2} yields

$$V_r = V \cos \theta \sin \omega t - V \sin \theta \cos \omega t \tag{7.12}$$

or

$$V_r = V \sin(\omega t - \theta) \tag{7.13}$$

A phase comparator can now be used to obtain the phase shift of V_r with respect to V_{s_1} shown in Fig. 7.7. This phase shift is equal to the rotor angle θ. One method of phase detection is timing the zero positive-going crossings of these two signals. This technique is relatively insensitive to voltage differences in the signals, and output can be provided directly in digital form through the use of binary counters in the zero-crossing timing circuitry.

Amplitude Modulation

Resolvers can be used to construct a system for sensing the difference between the angular positions of two resolver rotors. Such a system is shown in Fig. 7.8. One resolver can be mounted in a remote location and the second resolver rotor connected to a drive motor shaft, the angular position of which is to be controlled. The remote resolver serves as a command input, and the objective is to produce an error signal proportional to the difference between the positions of the resolver rotors.

As shown in Fig. 7.8, a sinusoidal reference voltage V_r is applied to the rotor of the remote resolver.

$$V_r = V \sin \omega t \tag{7.14}$$

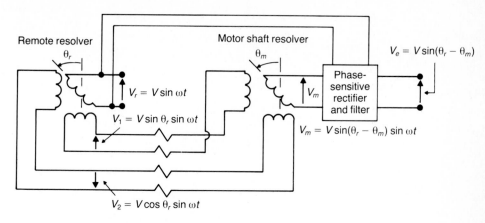

Figure 7.8 Angular error detection using two 2-phase resolvers.

If θ_r is the angular position of the remote resolver rotor, then given unity coupling, voltages induced in the two stator windings of the remote resolver are

$$V_1 = V_r \sin \theta_r = V \sin \theta_r \sin \omega t \qquad (7.15)$$

$$V_2 = V_r \cos \theta_r = V \cos \theta_r \sin \omega t \qquad (7.16)$$

If θ_m is the angular position of the motor shaft resolver rotor, the voltage V_m induced in this rotor is

$$V_m = V_1 \cos \theta_m - V_2 \sin \theta_m \qquad (7.17)$$

Substituting Eqs. (7.15) and (7.16) yields

$$V_m = V(\sin \theta_r \cos \theta_m - \cos \theta_r \sin \theta_m) \sin \omega t \qquad (7.18)$$

or

$$V_m = V \sin (\theta_r - \theta_m) \sin \omega t \qquad (7.19)$$

As Eq. (7.19) shows, the voltage at the motor shaft resolver rotor is a function of the angle error $\theta_r - \theta_m$. A phase-sensitive rectifier is used to convert the amplitude-modulated voltage V_m into an output voltage V_e. The sign of V_e is determined by whether V_m is in phase or out of phase with the reference input V_r. V_e can then be expressed as

$$V_e = V \sin(\theta_r - \theta_m) \qquad (7.20)$$

If the angle error is small, then

$$V_e \approx V(\theta_r - \theta_m) \qquad (7.21)$$

V_e is therefore proportional to $\theta_r - \theta_m$ for small angle errors. Because the output of the phase-sensitive rectifier is an analog signal, an analog-to-digital converter is required to input this error signal into a computer control system.

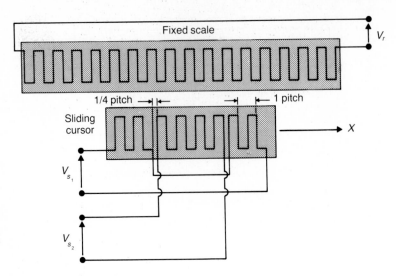

Figure 7.9 Linear resolver.

Linear Resolvers

Linear resolvers function in a manner similar to that of rotary resolvers. Their windings are on linear scales, however, so they are able to sense linear motion directly. In the design shown in Fig. 7.9, the equivalent of the rotary resolver rotor winding is printed on a stationary scale that is as long as the measurement range required. One "pitch" of this scale is about 2 mm (0.08 in.) in length and corresponds to one revolution of a rotary resolver. The equivalent of the two stators of the rotary resolvers are printed on a sliding cursor. These have the same pitch as the scale but are offset with respect to each other by one-quarter of a pitch. Nearly perfect sine-wave coupling can be achieved between the scale and the cursor with carefully designed conductor widths and spacings.

The phase-modulation method of position sensing described in the previous section for the rotary resolver can be used with the linear resolver as well. The sine and cosine voltages described by Eqs. (7.9) and (7.10) are applied to the cursor. The voltage induced in the scale has the same form as Eq. (7.14) except that its amplitude is greatly attenuated. Again, a phase comparator can be used to obtain the position measurement in computer-readable form.

Measurement accuracies of 10 μm (0.0004 in.) can readily be obtained with linear resolvers. The range of one pitch can be extended to any reasonable length by providing an external means of counting pitches.

Optical Angle Encoders

In an **optical angle encoder**, light is transmitted through a slotted or ruled disk on a rotating shaft onto photo-detectors that convert transmitted light into

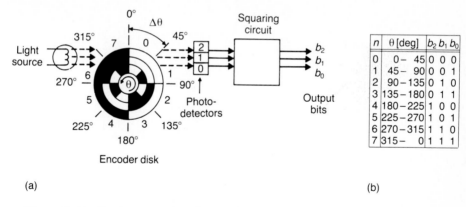

Figure 7.10 Absolute optical angle encoder with three output bits.

electrical signals. Opaque and transparent areas of the disk alternately occlude and reveal the light source as the disk rotates. Variations in photo-detector outputs are then processed electronically and interpreted as changes in shaft position. Absolute or incremental angular position encoding can be achieved, depending on the number of photo-detectors employed and on the arrangement of opaque and transparent areas on the disk.

Absolute Optical Angle Encoders

An **absolute optical angle encoder** produces a set of logic-level signals representing a binary word proportional to angular position. An absolute optical angle encoder with three bits of binary position output is shown in Fig. 7.10(a). Each bit is generated by one photo-detector according to the arrangement of opaque areas on the disk. Collectively, these bits generate the binary code for angular position shown in Fig. 7.10(b). Rotation of the disk by an angle $\Delta\theta$ is required to change the binary output by the value of one bit.

The disk arrangement shown in Fig. 7.10(a) is not normally used in practice, because more than one bit changes simultaneously in the binary output at boundaries between some regions on the disk. This can result in large momentary errors in position measurement. For example, as the disk rotates from 3 to 4, the binary output changes from 011_2 to 100_2. The synchronism of these changes usually cannot be absolutely guaranteed, and intermediate combinations such as 111_2 can occur at the boundary.

A number of alternative codes exist wherein only one bit changes at any given boundary. An absolute encoder disk that generates such a code is shown in Figure 7.11(a). The code, called a progressive or Gray code, is shown in Fig. 7.11(b), and it can be verified that only one bit changes at a time. The code can be converted readily into true binary code by either hardware or computer software.

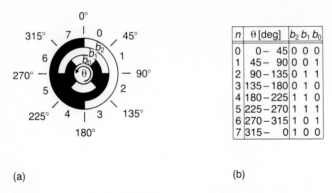

n	θ [deg]	b_2 b_1 b_0
0	0 – 45	0 0 0
1	45 – 90	0 0 1
2	90 – 135	0 1 1
3	135 – 180	0 1 0
4	180 – 225	1 1 0
5	225 – 270	1 1 1
6	270 – 315	1 0 1
7	315 – 0	1 0 0

(a)   (b)

Figure 7.11 Absolute optical angle encoder with progressive code (Gray code) output.

Absolute angle encoders are available with as many as 16 output bits. The code generated repeats after one revolution, and several geared encoders can be used to increase the resolution or range of angular position sensing. However, absolute angle encoders become impractical when range and resolution requirements are such that a large number of output bits are required. For this reason, incremental optical angle encoders are used for many position-sensing applications.

Incremental Optical Angle Encoders

An incremental indication of disk rotation can be obtained by using two bits of binary output. This has the advantage of simplifying the encoder because only two photo-detectors are required, but external counter hardware must be added to develop an absolute position measurement from the output of the incremental encoder. The arrangement of opaque areas on the disk of the incremental optical angle encoder is shown in Fig. 7.12. The angular spacing of the opaque areas is $\Delta\theta$. Track B is offset from track A by $\Delta\theta/4$, and the output of photo-detectors A

Figure 7.12 Incremental encoder with computer interface.

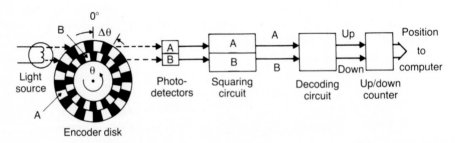

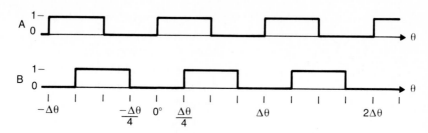

Figure 7.13 Incremental encoder output vs. rotation.

and B after processing by squaring circuits is the function of the rotation angle θ shown in Fig. 7.13. This quadrature relationship can also be achieved by using a single track and placing photo-detector B in a position offset from photo-detector A by $\Delta\theta/4$.

Although it is possible to monitor changes in angular position by directly sampling outputs A and B with a control computer, this is a task that can usually be performed more economically with counter hardware. Figure 7.14(a) shows

Figure 7.14 Incremental encoder and counter waveforms. (a) Positive rotational velocity $(+d\theta/dt)$; (b) negative rotational velocity $(-d\theta/dt)$.

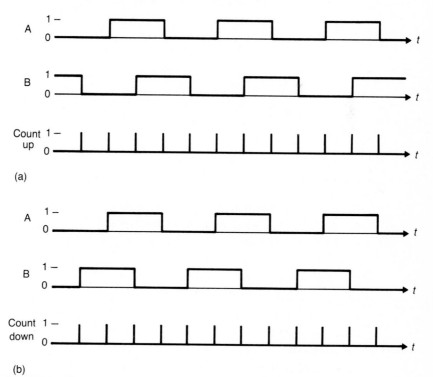

output A and B as a function of time for a positive rotational velocity. At each transition in A and B, a pulse can be generated and can be input to the counter. The binary output of the counter can be as many bits as are needed to represent the entire range of motion required (it is not limited to one revolution of the disk). Each pulse received during positive rotation causes the binary counter output to increase by one (count up). Provided that the counter is properly initialized so that its output is zero at the desired angular position, the counter gives an absolute reading of angular position.

Figure 7.14(b) shows outputs A and B as a function of time for a negative rotational velocity. Comparison with Figure 7.14(a) reveals that output B is now shifted $180°$ with respect to output A. The transitions in A and B in this case can therefore be distinguished from the case of positive rotation. A separate set of pulses can be generated for negative rotation and can be input to the counter, each pulse causing the counter output to be decremented by one (count down). In this way, a bi-directional sensor of angular position is formed. The additional sensor hardware is included in the decoding circuit in Fig. 7.12. The output of the counter changes by 4 for an angular rotation of $\Delta\theta$, so the effective resolution is $\Delta\theta/4$. The output of the counter can be directly interfaced to a control computer.

Incremental optical angle encoders are available with 100,000 or more opaque areas, or "lines," on the disk. However, the dynamic response of the photo-detectors limits the rotational velocity of the sensor. If the maximum frequency at which the photo-detectors can function is known, the maximum rotational velocity of the encoder can be found from

$$\omega_{max} = \frac{f_{max}\Delta\theta}{60} \quad \text{rpm} \tag{7.22}$$

where f_{max} = photo-detector cut-off frequency, in Hz
        $\Delta\theta$ = incremental angle, in deg

Capacitive Sensors

Noncontacting **capacitive sensors** use the change in capacitance with distance between two plates as a basis for position measurement. The capacitance of two plates, as illustrated in Figure 7.15(a) is found from

$$C = k\frac{A}{d} \tag{7.23}$$

where c = capacitance
        k = the dielectric constant
        A = area of the plates
        d = distance between plates

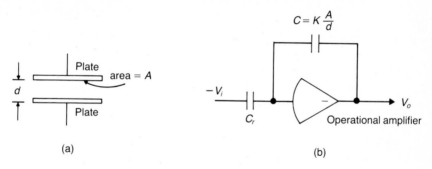

Figure 7.15 Capacitive sensor. (a) capacitor; (b) distance sensing circuit.

In practice, care must be taken in the construction of the sensor to ensure that the relationship in Eq. (7.23) holds.

The variable capacitance of the sensor is placed in the feedback of an operational amplifier as shown in Fig. 7.15(b), and a reference capacitance C_r is placed at the input. As long as the open-loop amplifier gain is very high, the output voltage V_o is related to the input voltage $-V_i$ by

$$V_o = V_i \frac{C_r}{C} \tag{7.24}$$

Substituting Eq. 7.23 into this equation gives

$$V_o = V_i \frac{C_r}{kA} d \tag{7.25}$$

If V_i is a constant-amplitude, sinusoidal voltage, the amplitude of the output voltage V_o is proportional to the plate distance d. A typical operating frequency is 16 kHz. The output must be demodulated and converted to digital form for input to a control computer. Resolutions of better than 10 μm and ranges of 5 mm or more can be obtained with capacitive position sensors. One serious drawback in the use of capacitive sensors, however, is their sensitivity to changing plate geometry and dirty environments.

Other Position Sensors

Of the many types of position sensors available, only a few have been described in this section. For example, no mention has been made of laser interferometers and Moire fringe optical gratings, which are capable of very high resolutions and are utilized where high-precision measurements are required.

7.3

VELOCITY SENSORS

Velocity sensors are used in machinery control systems for velocity control and for position loop stabilization when cascaded feedback control techniques are applied. In the latter case, the velocity feedback loop may be closed externally to the control computer. However, it may still be advantageous in some cases to input velocity measurements to the control computer.

d.c. Tachometers

Most velocity sensors use electromagnetic induction to produce a voltage proportional to the velocity of a conductor moving in a magnetic field. A **d.c. tachometer** (direct-current tachometer) consists of a wound armature and a permanent magnet stator with a set of commutator brushes. Rotation of the armature produces an induced voltage in the armature windings as they move through the field created by the permanent magnets. This output voltage can be as high as 20 volts/rpm, but more typically it is around 10 mv/rpm. As shown in Figure 7.16(a), the d.c. output voltage is directly proportional to the rotational speed of the armature. If ω is the armature speed and k_t is the tachometer gain, the d.c. output voltage V_o is

$$V_o = k_t \omega \qquad (7.26)$$

The commutator and armature cause an a.c. ripple (noise) to be superimposed on the d.c. output of the tachometer as shown in Figure 7.16(b). The amplitude of this ripple can be as much as 5% of the d.c. voltage, and its fundamental frequency is directly proportional to rotational velocity. The effect of this ripple on closed-loop performance should be carefully considered. For low speed and positioning servo systems, a tachometer with a low ripple amplitude should be used because the ripple frequency will be low and within the bandwidth of the closed-loop system. For closed-loop systems designed to operate in a given speed range, a tachometer can be selected with a ripple frequency that is well above the closed-loop bandwidth at design speed.

Linear Velocity Sensors

A **linear velocity sensor** can be constructed from a moving magnet and two stationary coils. As illustrated in Fig. 7.17, the coils are connected in series opposition so that the voltages induced by the two poles of the magnet add rather than subtract at the center position of the magnet. The displacement range of a moving-magnet velocity transducer is typically 50 mm. The magnet can be directly mounted without electrical connections on a moving mechanical

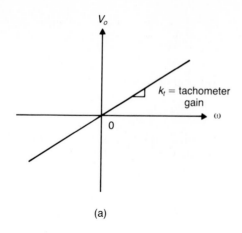

(a)

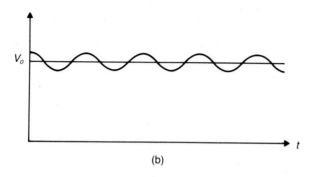

(b)

Figure 7.16 A d.c. tachometer. (a) Output voltage vs. rotational speed; (b) an a.c. ripple at constant rotational speed.

Figure 7.17 Moving-magnet velocity sensor.

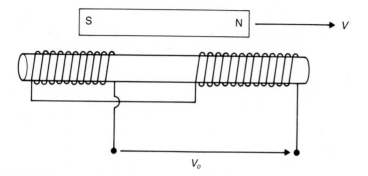

element. The output voltage V_o must be converted from analog to digital form prior to input to a digital computer.

Differentiation of Position Feedback

When a control system is equipped with a position sensor, it is possible to differentiate the output of the position sensor to obtain a velocity measurement. This technique can be used to eliminate the cost of additional velocity sensors in some control systems, particularly those with low performance requirements. However, for high-performance control systems the detrimental effects of differentiation tend to reduce the desirability of this technique.

Analog Differentiation

Output voltages from position sensors and their associated electronics can be differentiated using a circuit such as that shown in Fig. 7.18 prior to conversion and input to a control computer. The circuit produces an output voltage V_o that is proportional to the rate of change of the input voltage V_i as expressed by

$$V_o = -RC\frac{dV_i}{dt} \tag{7.27}$$

It is well known that differentiation of analog signals tends to increase the level of unwanted noise. This is because for an input voltage of the form

$$V_i = V \sin \omega t \tag{7.28}$$

the output voltage from the differentiator, as given by Eq. (7.27), is

$$V_o = -\omega RCV \cos \omega t \tag{7.29}$$

It can be seen from Eq. (7.29) that higher-frequency noise (greater ω) is preferentially amplified, resulting in reduced signal-to-noise ratios. Filtering techniques can be applied but, in general, filtering introduces phase lag in the signal, reducing its value for control feedback.

Figure 7.18 Analog differentiation circuit.

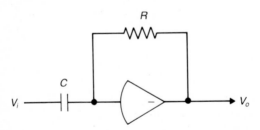

Digital Differentiation

Feedback derived from digital position sensors can be differentiated digitally with computer software. **Digital differentiation** is attractive because external hardware can be eliminated and velocity measurements can be synthesized within the control computer from position measurements. This attractiveness is reduced, in practice, by the same sensitivity to high frequency noise exhibited by analog differentiation.

It is possible to estimate velocity from position measurements by using difference formulas,[1] the simplest of which is

$$V_i = \frac{P_i - P_{i-1}}{T} \tag{7.30}$$

where
V_i = current estimate of velocity
P_i = current position
P_{i-1} = previously sampled position
T = time interval between samples P_{i-1} and P_i

If R_p is the position resolution of the sampled output of a position sensor, the corresponding velocity resolution R_v can be found from

$$R_v = \frac{R_p}{T} \tag{7.31}$$

Velocity resolution therefore depends on the sample interval T as well as on the position resolution. At a sample instant, the measured quantity is the average velocity over the sample interval T rather than the instantaneous velocity that would be obtained from a velocity sensor. It is important to note that velocity resolution improves (its magnitude decreases) as the sample interval becomes longer.

Example 7.1 Velocity calculation from position encoder output   As an example, consider a system in which there is a need for velocity feedback. An incremental optical angle encoder with a digital counter interface of the type we have described in Section 7.2 is present in the system. If the encoder has 800 "lines" per revolution; the position resolution of the sensor is

$$R_p = \frac{1}{4(800)} = 3.125 \times 10^{-4} \text{ rev}$$

---

[1] This concept is not limited to differentiation of position feedback. It can be used whenever the rate of change in a discrete variable needs to be obtained.

If the counter is to be sampled at a rate of 120 Hz, then

$$T = \frac{1}{120 \text{ s}}$$

From Eq. 7.31, the velocity resolution that results from the use of Eq. 7.30 is

$$R_v = \frac{120}{4(800)} = 0.0375 \text{ rev/s}$$

or

$$R_v = 2.25 \text{ rpm}$$

Increasing the sample rate (decreasing the sample period) only makes the velocity resolution worse, whereas decreasing the sample rate (increasing the sample period) may have adverse effects on system performance. Another option is to increase the position resolution of the optical encoder. However, as indicated by Eq. (7.20), this may limit the maximum velocity that can be measured and can significantly increase the cost of the encoder.

Although the position resolution in this example may be acceptable, the velocity resolution of 2.25 rpm is large compared to what would normally be obtained by sampling the output of a tachometer. For example, if the output of a tachometer with 10 mv/rpm gain is sampled via an A/D converter with a 2.5-mv resolution, then the resulting velocity resolution is

$$R_v = 2.5 \text{ mv}\left(\frac{1 \text{ rpm}}{10 \text{ mv}}\right) = 0.25 \text{ rpm}$$

which is an order of magnitude better than that obtained above by using digital differentiation. The tradeoff, therefore, is between the cost of a tachometer and A/D converter in a system that already contains a position sensor, and the reduced velocity resolution that results from the use of digital differentiation. □

7.4

ACCELERATION SENSORS

Accelerometers for sensing vibration can be constructed by using piezoelectric materials such as quartz or ceramic crystals, as illustrated in Fig. 7.19(a). The mass M is accelerated by the force F applied to the base of the crystal, which is usually rigidly attached to the vibrating mechanical body. The stress in the crystal produced by the force F over the cross-sectional area A of the crystal results in a minute deformation of the crystal lattice. This deformation causes an electrical

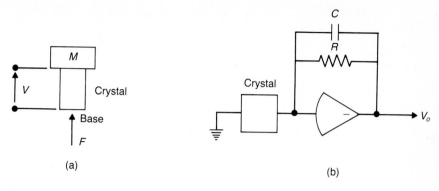

Figure 7.19 (a) Piezoelectric accelerometer; (b) charge amplifier.

polarization in the crystal, and a voltage is produced across the crystal. This voltage V can be expressed as

$$V = k\frac{F}{A}$$ (7.32)

where k is the piezoelectric voltage constant of the crystal.

Application of Newton's Law leads to the following expression for a, the acceleration of the mass, as a function of the voltage produced across the crystal.

$$a = V\frac{A}{kM}$$ (7.33)

A typical value of the piezoelectric voltage constant is 0.5 mv/Pa (3.4 volts/psi). Large voltages can be generated using small crystal masses, and in practice, these masses are made as small as possible to increase the dynamic response of the sensor.

The voltage produced by the piezoelectric accelerometer is detected by a charge amplifier, which can be constructed using an operational amplifier with capacitive feedback, as shown in Fig. 7.19(b). The charge amplifier output voltage is proportional to acceleration.

7.5

FORCE SENSORS

Force sensors (or load sensors) are often constructed using position transducers. Figure 7.20 illustrates two force sensors that use an electric member for converting force to displacement and an LVDT for measuring displacement. In Fig. 7.20(a), the force or load is applied to the core of the LVDT. The core is

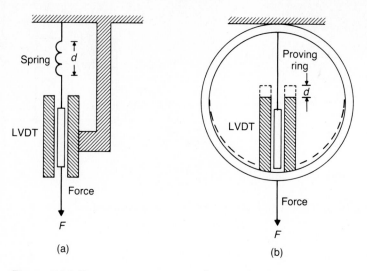

Figure 7.20 Force sensors constructed using LVDTs.

supported by a spring that extends as force is applied to the core. Displacement of the LVDT core is measured electronically as described in Section 7.2 and is directly interpreted as a force measurement.

The force sensor shown in Fig. 7.20(b) is a somewhat more robust design. In this case the force is applied to a proving ring that distorts as a result. The amount of distortion is again detected by an LVDT and interpreted as a force measurement. The spring shown in Fig. 7.20(a) and the ring shown in Fig. 7.20(b) can be sized to give the desired compliance so that core displacement stays within reasonable limits for the forces applied. In either case, the applied force F can then be found from

$$F = kd \tag{7.34}$$

where k = compliance of the spring or ring
        d = LVDT core displacement

7.6

TEMPERATURE SENSORS

Control of temperature is important in chemical processes and in other industrial processes such as the heat treatment of materials. Sensors used for the measurement of temperature include resistance thermometers, thermocouples; and thermistors. These devices are described in the following sections.

Resistance Thermometer

The **resistance thermometer** consists of a metallic conductor, the resistance of which increases with temperature. The metals used include copper, nickel, and platinum. The metal is either wound in a coil or deposited in a film on a suitable substrate. Resistance thermometers are available for temperature ranges between $-200°C$ and $800°C$, and design details depend on the temperature range selected. Generally, the thermometer is operated between two temperature extremes T_{min} and T_{max}. If the resistance of the device is measured at these two temperatures (R_{min} and R_{max}, respectively), then the resistance R can be expressed as a linear function of temperature T as follows:

$$R(T) = R_{min} + (T - T_{min})\left(\frac{R_{max} - R_{min}}{T_{max} - T_{min}}\right) \tag{7.35}$$

Temperatures commonly selected for this calibration are $0°C$ and $100°C$ (the melting point of ice and the boiling point of water), because they are convenient temperature references to maintain. If $T_{min} = 0°C$ and $T_{max} = 100°C$, then Eq. (7.35) can be simplified to

$$R(T) = R(0°) + kT \tag{7.36}$$

where

$$k = \frac{R(100°) - R(0°)}{100°} \quad \text{ohms/°C} \tag{7.37}$$

A Wheatstone bridge is generally used to detect the change in resistance over the operating temperature range. This change in resistance may be on the order of 40 ohms/100°C, with a resistance at 0°C of approximately 100 ohms. The resistance thermometer is placed as one of the arms of the Wheatstone bridge, as shown in Fig. 7.21. The voltage across the bridge V_o is then a function of temperature as measured by the resistance thermometer.

Figure 7.21 Wheatstone bridge with resistance thermometer.

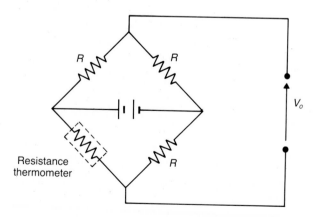

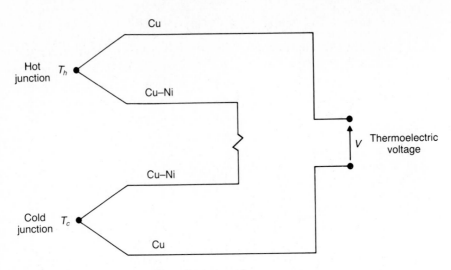

Figure 7.22 Copper/constantan thermocouples.

Thermocouples

Two dissimilar metals in contact at a point generate a thermoelectric voltage that can be used to measure temperature. As illustrated in Fig. 7.22, a **thermocouple** consists of pairs of wires that are made of different metals or metal alloys and are joined together at points called thermojunctions. The voltage developed by the circuit in Fig. 7.22 is a function of the choice of metals and junction temperatures. An empirical law for the voltage developed as a function of junction temperature is

$$V = k_1(T_h - T_c) + k_2(T_h^2 - T_c^2) \tag{7.38}$$

where T_h = absolute temperature of hot junction
            T_c = absolute temperature of cold junction
       k_1, k_2 = constants depending on the metals used

It can be seen from Eq. (7.38) that a temperature difference is required to produce the thermoelectric voltage V. Common pairings of metals for use in thermocouples are listed in Table 7.1. Also given is the temperature range over which the thermocouple is normally used. Thermocouples with lower maximum temperatures tend to be less expensive and to have greater sensitivity than those with higher maximum temperatures, making them ideal for low-temperature applications. Table 7.2 lists the boiling and melting points of various substances as an indication of the range of temperatures that may be encountered in industrial processes.

For temperature measurement, one junction is usually held at a constant reference temperature so that voltage will be a function solely of the temperature

Table 7.1

Thermocouple Temperature Ranges

METALS		TEMPERATURE RANGE	
Copper/constantan	(Cu/Cu-Ni)	−250 to 400°C	−400 to 750°F
Iron/constantan	(Fe/Cu-Ni)	−200 to 850°C	−350 to 1600°F
Cromel/constantan	(Ni-Cr/Cu-Ni)	−200 to 850°C	−350 to 1600°F
Cromel/alumel	(Ni-Cr/Ni-Al)	−200 to 1100°C	−350 to 2000°F
Platinum/platinum–rhodium	(Pt/Pt-Rh)	0 to 1400°C	50 to 2600°F

of the other junction. The cold junction can, for example, be immersed in a mixture of ice and water so that the melting ice creates a constant reference temperature of 0°C.

A temperature reference can also be provided by instrumentation. As illustrated in Fig. 7.23, an accurately regulated temperature environment is maintained within the instrument. The cold junction is situated in this environment at a constant reference temperature that is often somewhat above room temperature. The voltage developed as a function of hot-junction temperature can then be sampled by using an analog-to-digital converter. A table can be stored in software or hardware that relates hot-junction temperature T_h to output voltage V using Eq. (7.38). This function is fixed for a given pair of thermocouple materials and a given reference temperature. The measured temperature can be output in digital form, and measurement accuracies can be better than ±0.05°C (±0.1°F).

Table 7.2

Melting and Boiling Points of Various Substances

METALS	TEMPERATURE	
Absolute zero	−273°C	−460°F
Boiling point of oxygen	−183°C	−297°F
Melting point of water	0°C	32°F
Boiling point of water	100°C	212°F
Melting point of zinc	419°C	787°F
Melting point of silver	961°C	1761°F
Melting point of gold	1063°C	1956°F
Melting point of iron	1535°C	2795°F

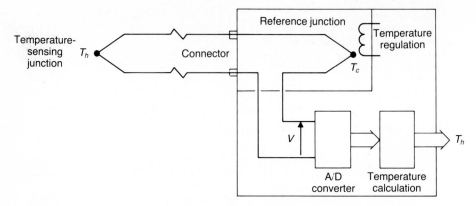

Figure 7.23 Instrumentation for temperature sensing using a thermocouple.

Thermistors

A **thermistor** (thermal resistor) is a semiconductor device that exhibits a change in resistance with temperature. In this respect thermistors are similar to resistance thermometers, but their rate of change in resistance with temperature is greater, and hence their sensitivity is much improved. Figure 7.24(a) shows the resistance characteristics of a typical thermistor. Note that for this thermistor, the resistance decreases exponentially as temperature increases. This function can approximately be expressed as

$$R = ae^{-bT} \qquad (7.39)$$

where R = thermistor resistance
T = temperature of the thermistor, in kelvins (K)
a, b = constants depending on the thermistor used

A circuit for detecting temperature via a thermistor is shown in Fig. 7.24(b). The output voltage of the circuit is a linear function of thermistor resistance and can be sampled using an analog-to-digital converter. The output voltage V_o is

$$V_o = \frac{V_i}{R_i} R \qquad (7.40)$$

where V_i = constant input voltage
R_i = input resistance

Temperature in K can be calculated from Eq. (7.39) as follows:

$$T = \frac{-\ln(R/a)}{b} \qquad (7.41)$$

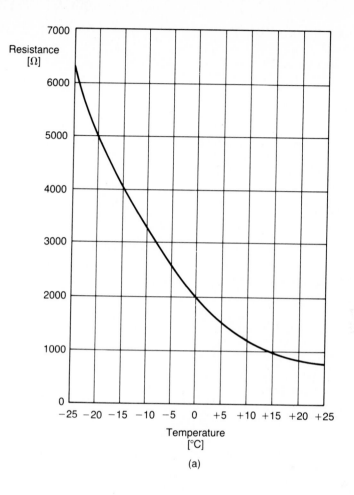

(a)

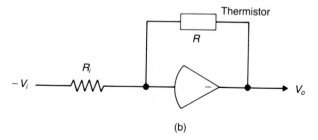

(b)

Figure 7.24 (a) Resistance vs. temperature for a typical thermistor; (b) Temperature sensing circuit.

Or, applying Eq. (7.40), we find that

$$T = c - \ln(V_o)/b \qquad (7.42)$$

where

$$c = -\ln(R_i/aV_i)/b \qquad (7.43)$$

7.7

STRAIN SENSORS

A conductor under tension or compression undergoes a change in length that results in a change in resistance of the conductor. The relationship between change in resistance and change in length is

$$\frac{\Delta R}{R} = S\frac{\Delta L}{L} \qquad (7.44)$$

where R = resistance of the conductor
        L = length of the conductor
        S = sensitivity or gauge factor of the conductor

The factors $\Delta R/R$ and $\Delta L/L$ are the resistance strain and the mechanical strain, respectively. The sensitivity of most strain gauge materials is about 2.0, but is in excess of 100 for some semiconductors.

    Strain gauges are typically constructed either from a fine resistive wire suspended under tension between two supports on a mechanical structure or from a resistive foil that can be bonded directly to the mechanical structure. These are illustrated in Fig. 7.25. Strain gauges are used to measure mechanical strain in the bodies to which they are attached, but when attached to a suitably designed mechanical structure, they can also be used to measure other physical quantities such as pressure, force, and deflection.

Strain Measurement

A Wheatstone bridge can be used to measure the small changes in resistance that result in most strain gauge applications. Because the resistance of strain gauges changes with temperature, temperature-compensated-foil strain gauges are available that minimize these effects. Another technique of temperature compensation is to use two strain gauges, one in tension and one in compression, as shown in Fig. 7.26. Both strain gauges experience equal changes in resistance that are due to temperature, so these changes do not affect the bridge output.

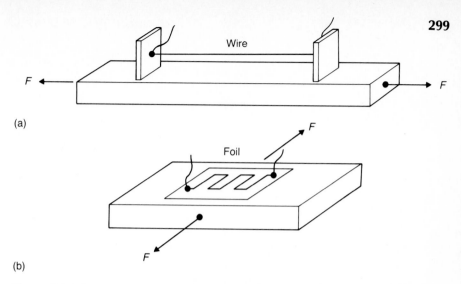

(a)

(b)

Figure 7.25 Strain gauges mounted on deformable body under stress. (a) Wire strain gauge; (b) foil strain gauge.

Figure 7.26 Strain measurement using gauges in tension and compression. (a) Strain gauges in tension and compression; (b) wheatstone bridge circuit.

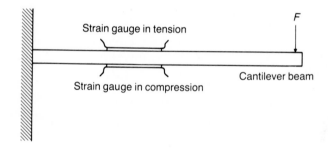

(a)

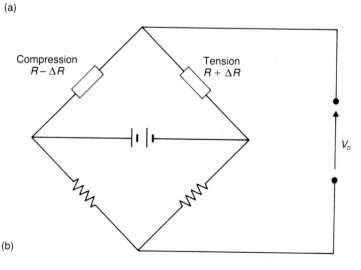

(b)

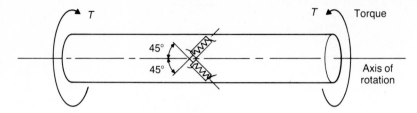

Figure 7.27 Torque measurement using strain gauges.

Torque Measurement

The principal axes of strain in a shaft under torsional stress are oriented at angles of 45° with respect to the axis of rotation of the shaft. Torque is usually measured by placing a pair of strain gauges on the shaft oriented at angles of $\pm 45°$ with respect to the axis of rotation, as shown in Fig. 7.27. When torque is applied to the shaft, one strain gauge is then in tension and one strain gauge is in compression. This situation is similar to that shown in Fig. 7.26, and an identical Wheatstone bridge circuit can be used.

7.8

OPTICAL SENSORS

The use of optical sensors is an attractive idea for control feedback and monitoring in many manufacturing processes. Optical encoders fall into this class of device and have already been described in Section 7.2. Vision systems also fall into this category. Unfortunately, the high bandwidth of video signals that are output by television cameras and the necessity for a great deal of post-processing of these signals in order to obtain usable control information have limited their application in many control problems. However, the development of solid-state optical sensor arrays has reduced processing requirements as well as costs and has opened up new possibilities for the use of these sensors.

Four-Quadrant Photodiode

A **four-quadrant photodiode** consists of a two-by-two array of silicon photodiodes as shown in Fig. 7.28(a). One application of the device is two-dimensional measurement of the displacement of a light beam, typically generated by a laser source. The array dimensions are often on the order of

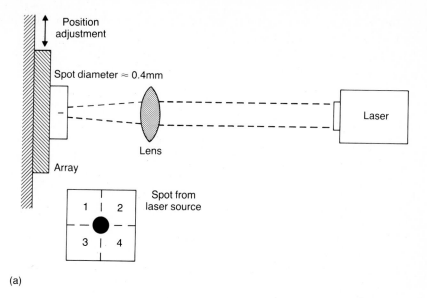

(a)

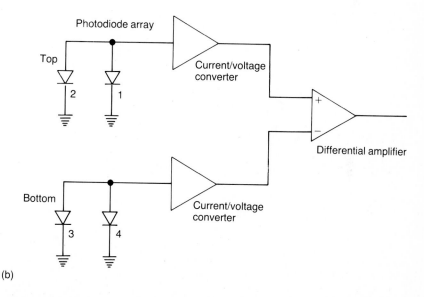

(b)

Figure 7.28 (a) Measurement of laser beam position using a four-quadrant photodiode, and (b) vertical position measurement circuit for a four-quadrant photodiode.

1/8 in. (3 mm) with an accuracy of 0.0001 in. (2.5 μm) for measurement of the displacement of a beam 0.04 in. (1 mm) in diameter.

The circuit shown in Fig. 7.28(b) is for a somewhat more simplified application wherein displacement in only one dimension is to be measured. In this case, the outputs of photodiodes 1 and 2 are summed and the outputs of photodiodes 3 and 4 are summed. The current generated by each of these photodiode pairs is proportional to the intensity of the light falling on that pair. The difference between these currents is proportional to the relative vertical position of the beam on the device. A voltage proportional to displacement of the beam can be generated by taking the difference between the outputs of two current-to-voltage converters. This output can then be sampled by a control computer via an analog-to-digital converter, and the result can be used as laser beam position feedback.

Linear Photodiode Arrays

A **linear photodiode array** consists of a row of silicon photodiodes as shown in Fig. 7.29. A storage capacitor is supplied with each photodiode to integrate its

Figure 7.29 An n-element linear photodiode array.

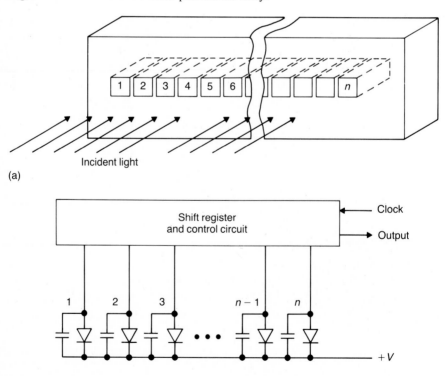

Incident light

(a)

(b)

photocurrent. Together, the photodiode and the capacitor form a cell, and there can be more than 2000 cells in an array. Cell dimensions vary, but typical dimensions for a 1024-cell array are 15 μm × 25 μm (0.0006 in. × 0.001 in.), with a center-to-center distance of 25 μm (0.001 in.). Incident light that falls between cells is collected by the nearest cell, so the center-to-center distance represents the resolution of the array. Because the linear array is a discrete device, only the average light intensity on each cell can be measured.

A simplified description of the operation of the photodiode array is as follows. Any initial charge is first removed from the capacitors and, when a start signal is received, the photodiodes are allowed to gradually generate a charge on their capacitors. The amount of time allowed for this charge to build up is called the **integration time**. The amount of charge is a function of light intensity and can be expressed as

$$V_i = S_d I_i \, \Delta t_\ell \qquad (7.45)$$

where V_i = charge on cell i
          S_d = photodiode sensitivity
          I_i = light intensity on cell i
          Δt_ℓ = integration time

At the end of the integration time, the charges on all cells are transferred to an analog shift register so that they can be sequentially output from the array to external interfacing circuits. Individual cell charges are shifted out of the register at intervals of Δt_s, and another set of charges can be integrating in the cells while the shift register is operating on the previously integrated charges. The dynamic response of the array is therefore limited by the rate at which the charge from each cell can be output from the array. The maximum frequency at which complete images can be obtained is therefore

$$f_{\max} = \frac{1}{n \, \Delta t_s} \qquad (7.46)$$

An example of the output of a photodiode array is shown in Fig. 7.30. The measure of the light intensity at the first cell appears first, and it is followed sequentially, at intervals of Δt_s, by the measure of the light intensity at the other cells in the array. If this output is to be sampled by an analog-to-digital converter on a control computer, the conversion of each segment of the output must begin at the rising edge of the shift register lock. The conversion must be complete before the next rising edge, so the conversion time of the analog-to-digital converter must be less than Δt_s. This is a critical consideration, because conversion time limits Δt_s and hence the obtainable dynamic response, as indicated by Eq. (7.46).

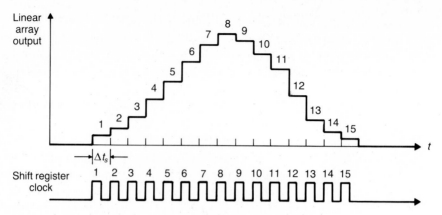

Figure 7.30 Example of the output of a linear photodiode array.

Two-Dimensional Photodiode Arrays

Two-dimensional images can be detected by using **two-dimensional photo-diode arrays**. As an example, the image of an object being viewed is projected on a two-dimensional array as shown in Fig. 7.31(a). The discrete image detected by the array is shown in Fig. 7.31(b). Again, only the average intensity of light incident on each element can be resolved.

The output of the cells in a two-dimensional array appears in a serial manner, as is the case with the linear array. There can be a large number of cells in a two-dimensional array. (For example, a 256×256 array has 65,536 individual elements). The dynamic response may therefore be extremely low, as indicated by Eq. (7.46).

Along with the problem of low dynamic response, problems often arise in interpreting the output of a two-dimensional array in order to obtain useful feedback information for contol purposes. A simple application such as sensing

Figure 7.31 Example of the output of a two-dimensional photodiode array.

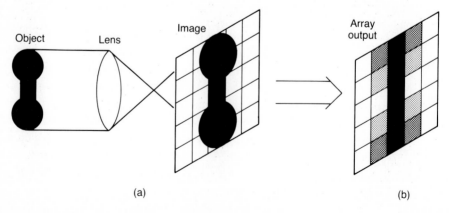

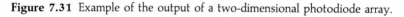

the position of a small-diameter moving spot can easily be performed by finding the cell with maximum output. However, if position information is to be obtained regarding the location and identity of objects with complex shapes, much further processing may be required. These image-processing and pattern-recognition functions tend to be mathematically complex and to consume large amounts of computer processing time. They therefore tend to be expensive and to impose a further limit on dynamic response.

7.9

SUMMARY

The technology of sensors is constantly undergoing development. New types of sensors are continuing to emerge that are smaller, less expensive, more accurate, and easier to use than their predecessors. The future is bright for semiconductor microsensors. These include accelerometers, strain gauges, and temperature sensors fabricated via integrated-circuit technology. Measurement of other phenomena, such as gas flow pressure and chemical composition, now can be performed with silicon integrated-circuit devices. The compatibility of this technology with microprocessor technology means that sensors can incorporate digital calibration and linearization as an integral part of their design, a concept that will have far-reaching effects on the future development of sensors.

Besides the development of microsensors, other technologies (such as laser, video, sound, and microwave sensors) have assumed new importance in control system design. This trend is bound to continue, given the increasing need for machinery and process control systems with ever-more-complex sensing and control requirements.

The subject of sensor design and application is far too broad to be treated comprehensively here. The sensor designs described in this chapter are, in most cases, only some of many that are suitable for the measurement of physical phenomena. The design of sensors and their associated electronics is an art in its own right, and the reader is referred to other texts on sensor and transducer technology.

BIBLIOGRAPHY

Beckwith, T., N. Buck, and R. Marangoni, *Mechanical Measurements, 3/e.* Reading, MA: Addison-Wesley, 1981.

Dally, J. W., W. F. Riley, and K. G. McConnell. *Instrumentation for Engineering Measurements.* New York, NY: Wiley, 1984.

Doebelin, E. O. *Measurement Systems: Application and Design*. New York, NY: McGraw-Hill, 1983.

Hunter, R. P. *Automated Process Control Systems, 2/e*. Englewood Cliffs, NJ: Prentice-Hall, 1987.

Johnson, C. *Process Control Instrumentation Technology, 2/e*. New York, NY: Wiley, 1982.

Lawrence and Mauch. *Real-Time Microcomputer System Design: An Intoduction*. New York, NY: McGraw-Hill, 1987.

PROBLEMS

7.1 The repeatability of a touch-trigger probe to be used in coordinate measurement applications has been tested as shown in Fig. 7.32(a) by repeatedly moving the probe tip in and out of contact with a gauge block. This motion repeatedly generates the contact-indicating signal TRIGGER, and the relationship between TRIGGER and probe position x is shown in Fig. 7.32(b). Given the data in the following table, find the $\pm 2\sigma$ variation in probe triggering position (σ is the standard deviation).

n	x [mm]
1	21.00045
2	21.00014
3	21.00008
4	21.00002
5	20.99984
6	20.99971
7	20.99984
8	20.99987
9	21.00035
10	20.99988
11	20.99999
12	20.99969
13	21.00008
14	21.00041
15	20.99970
16	21.00018
17	20.99996
18	21.00002
19	21.00022
20	20.99966

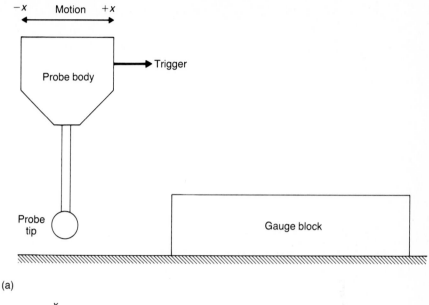

(a)

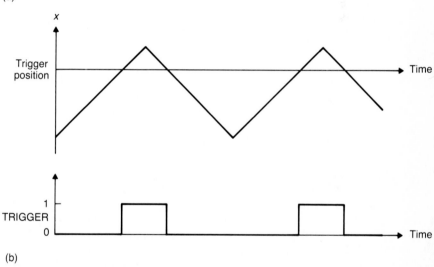

(b)

Figure 7.32 (a) Touch-trigger probe for Problem 7.1, and (b) trigger signal generated by probe in Problem 7.1.

7.2 In order to generate the TRIGGER signal, the probe shown in Fig. 7.32(a) requires a small amount of pre-travel between the position where contact is made and the position where TRIGGER is generated. When the probe is used for coordinate measurements in the x-y plane, its pretravel characteristics are roughly triangular, as shown in Fig. 7.33(a), and are caused by the internal mechanism of the probe. The data in the following table were taken by recording the x-y position of the probe body at the time the probe signal is generated when a probe tip of 1-mm radius approaches a precision cylinder of 1-cm radius from various directions in the x-y plane, as shown in Fig. 7.33(b).

θ [deg]	x [mm]	y [mm]
22.5	10.16875	4.21203
45.0	7.78190	7.78190
67.5	4.21147	10.16738
90.0	0.0	11.00578
112.5	−4.21262	10.17017
135.0	−7.78305	7.78305
157.5	−10.16759	4.21155
180.0	−11.00492	0.0
202.5	−10.16787	−4.21167
225.0	−7.78312	−7.78312
247.5	−4.21266	−10.17026
270.0	0.0	−11.00559
292.5	4.21147	−10.16738
315.0	7.78201	−7.78201
337.5	10.16833	−4.21186
360.0	11.01011	0.0

a. What is the $\pm 3\sigma$ accuracy of the probe without compensation for the pre-travel characteristic?

b. What is the $\pm 3\sigma$ accuracy of the probe after compensation for the pre-travel characteristic using the triangle shown in Fig. 7.33(a)?

c. Plot and compare the expected distribution of errors on the basis of the results found in parts (a) and (b) for the probe with and without pre-travel compensation.

7.3 Suppose that a force sensor constructed as shown in Fig. 7.20(b) is to be used in the force control system shown in the block diagram in Fig. 7.34(a). The sensor deflects 0.1 in. when a static force of 500 lb is applied. An actuator weighing 100 lb is rigidly attached to the sensor (the width of the sensor can be neglected), and the actuator/sensor system can be approximately modeled by the spring–mass–damper system shown in Fig. 7.34(b). If the damping ratio of the system is 0.05, is it likely that a closed-loop bandwidth of 75 Hz can be achieved by the force control system?

7.4 In a computer interface for a rotary resolver, a phase comparator is used as shown in Fig. 7.6. In the phase comparitor, $360°$ of phase shift are measured by a counter as 4000 counts. As indicated in the example in Fig. 7.35, the zero-crossing detectors in the phase comparator detect only transitions from negative to positive voltage, and the counter starts counting at the zero crossing in V_{s_1} and stops at the zero crossing in V_r. The frequency of the sinusoidal resolver excitation voltage V_{s_1} is 2000 Hz, and the sample rate of the closed-loop control system in which the resolver is used as a feedback device is 500 Hz.

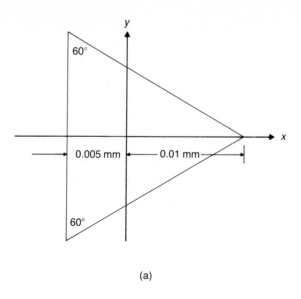

(a)

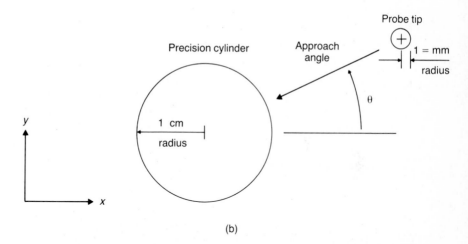

(b)

Figure 7.33 (a) Approximate pre-travel characteristic for probe in Problem 7.2, and (b) measurement of a precision cylinder with touch-trigger probe in Problem 7.2.

a. Assume that the resolver is rotating at a constant speed such that a 360° phase lag in rotor voltage V_r with respect to V_{s_1} occurs in exactly 7.5 feedback samples. If an initial feedback sample is taken just as zero crossings are detected in both the resolver excitation voltage and the rotor voltage inputs to the phase comparitor, calculate the next 8 feedback samples that would be obtained from the counter. (Note that the resolver position changes during the time between the two zero crossings!)

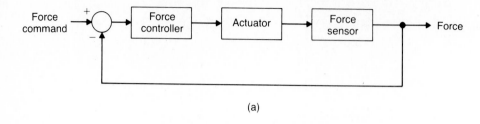

(a)

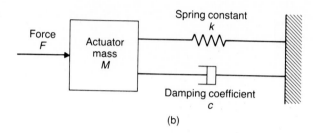

(b)

Figure 7.34 (a) Force control system for Problem 7.3, and (b) approximate dynamic model of actuator/force sensor for Problem 7.3.

    b. What is the error in each of the resolver position feedback samples calculated in part (a) due to the motion of the resolver between zero crossings?

    c. How could this error be compensated for in a computer control system?

7.5 A resolver is directly attached to the screw of a machine drive system as shown in Fig. 7.36. The pitch of the screw is 2 mm per revolution. The phase shift between the resolver's excitation and feedback signals is determined by a counter in the

Figure 7.35 Example of resolver excitation voltage, feedback voltage, and counter output for Problem 7.4.

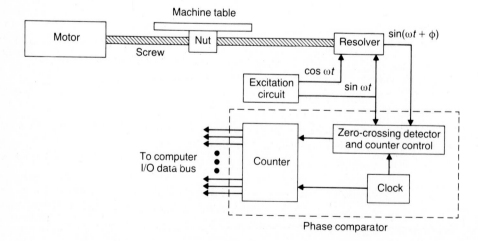

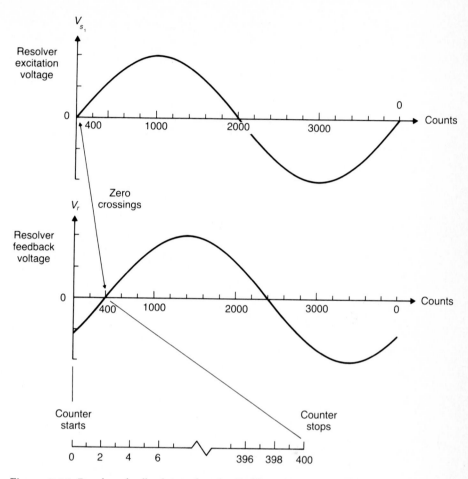

Figure 7.36 Resolver feedback interface for Problem 7.5.

resolver interface that is incremented by 1 every time a clock pulse is generated. The resolver excitation frequency is 1500 Hz, and a 1-μm position resolution is required for accurate control of machine motion. What clock frequency should be used?

7.6   How many circular tracks are required in an absolute optical shaft angle encoder of the form shown in Fig. 7.11 if the shaft angle resolution is to be better than 1°?

7.7   The shaft position resolution provided by an optical encoder and its associated interface is 1/4000 of a shaft revolution. What is the velocity resolution in rpm if shaft position is sampled at a rate of 250 Hz? Can velocity resolution be improved by increasing the sample rate?

7.8   Feedback in a closed-loop motor control system is provided by an incremental optical shaft angle encoder with 4096 grating lines. The quadrature outputs A and B, shown in Fig. 7.13, are generated by photo-detectors offset by one-quarter of the spacing between grating lines. A counter accumulates the pulses shown in Fig. 7.14, and the contents of the counter are sampled every 5.5 ms by a control computer.

a. What is the velocity resolution in rpm?

b. If the counter is reset after each sample, what is in the apparent speed of the motor in 10 successive sample periods when the motor is rotating at a steady speed of 96 rpm? (Assume that the counter is set to zero at the beginning of the first sample period just after a pulse has been generated and counted.)

c. What is the apparent acceleration of the motor between each of the sample periods in part (b)?

7.9   When a motor is rotating at 30 rpm, an 11-Hz ripple is observed in the output voltage of a tachometer, with a gain of 10 volts per 1000 rpm, mounted on the motor shaft. When a velocity control loop is closed around the motor and tachometer, the tachometer output voltage is observed to be nearly constant. This is the case because the actual motor velocity is varied by the controller at 11 Hz so that the ripple in tachometer voltage is nearly canceled out.

a. The amplitude of the ripple is 5 % of the tachometer voltage. Calculate the amplitude of the resulting 11-Hz oscillation in motor shaft position that is superimposed on the 30-rpm mean rate of change in shaft position.

b. To what percentage must the ripple in the output voltage of the tachometer be reduced if the amplitude of the oscillation in shaft position is to be held within 1 minute of arc?

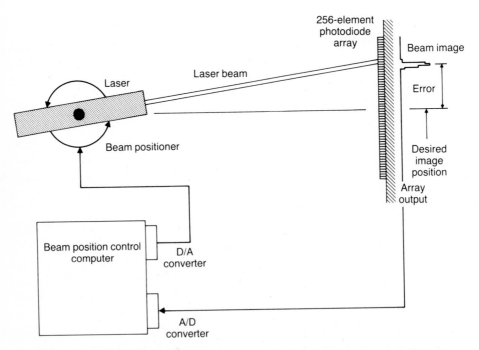

Figure 7.37 Laser beam positioner with photodiode array feedback for Problem 7.10.

7.10 A 256-element linear photodiode array is to be used as a feedback device for aiming a laser beam as illustrated in Fig. 7.37. The output of each element is a voltage proportional to light intensity, and the element outputs appear serially at the output of the array. It takes 20 μs to perform the A/D conversion of each array element output and another 1.2 ms to locate the center of the beam image on the detector. What is the maximum response time that should be expected for closed-loop control of beam position?

Chapter 8

Command Generation in Machine and Process Control

INTRODUCTION

The generation of sophisticated command sequences can be as important as, or more important than, other aspects of control system design. Command generation often represents a major portion of the "intelligence" of a system. For example, industrial robots and computer-controlled machine tools have multiple positioning axes that must move in a coordinated fashion along precisely defined paths to accomplish their programmed tasks. Figures 8.1(a) and 8.1(b) illustrate typical motion profiles that must be stored and generated as position commands in these systems. Practical considerations in these systems (such as large axis inertias and actuator torque limitations) require that axis acceleration and velocity be controlled as well as axis position. The success of machine control systems in general and of many other process control systems depends on the implementation of flexible and powerful command programming, storage, and generation schemes.

Several techniques for command generation in machine and process control are described in this chapter. These can be used with both open-loop and closed-loop control strategies, as well as for the control of multiple process variables. The first technique described is linear interpolation between setpoints at a constant rate of command change. The technique is developed for a single process variable and then extended to synchronized commands for multiple process variables. The second technique involves the use of cubic polynomials for command generation; it allows convenient representation and generation of

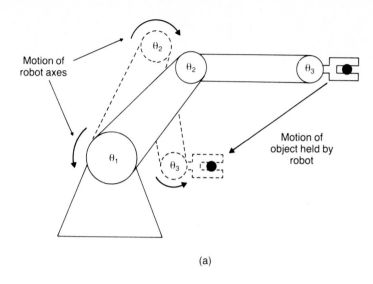

(a)

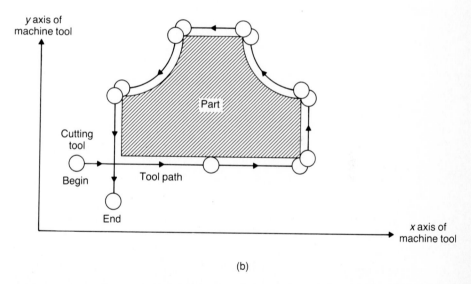

(b)

Figure 8.1 (a) Motion of a three-axis robot while moving an object from point to point, and (b) the path to be followed by a cutting tool in the machining of a part.

complex paths for multi-axis machines with controlled velocity and acceleration. The third technique is specifically developed for open-loop control using incremental actuators such as stepping motors.

8.2
COMMAND GENERATION USING
LINEAR INTERPOLATION

A function that is to be generated can often be stored as a set of points, and linear interpolation can be used to generate intermediate points. In Fig. 8.2(a), for example, a function between time t_0 and time t_k is represented by a set of points $r(t_0), r(t_1), r(t_2), \ldots, r(t_i), \ldots, r(t_k)$. The value of the function at time $t_{i-1} \le t \le t_i$ is obtained by linearly interpolating between $r(t_{i-1})$ and $r(t_i)$ using

$$r(t) = r(t_{i-1}) + [t - t_{i-1}] \frac{[r(t_i) - r(t_{i-1})]}{[t_i - t_{i-1}]} \tag{8.1}$$

Values of the function can be generated at time intervals T by applying this equation at

$$t = t_0 + nT \tag{8.2}$$

where $n = 0, 1, 2, \ldots, N$ and

$$N = (t_k - t_0)/T \tag{8.3}$$

The fractional part of N is truncated so that N is an integer. In general, this results in the final value of the function, $r(t_k)$, not quite being reached because $nT < t_k$. If this is not acceptable, $r(t_k)$ can be generated as r_{N+1} at time $(N + 1)T$, as shown in Fig. 8.2(b).

Example 8.1 Ramp generation It is desired to generate a ramp function $r(t)$ between values r_0 and r_f. The desired rate of change is to be V. If $t_0 = 0$, Eq. (8.1) can be rewritten as

$$r_n = r_0 + nVT$$

This equation can be applied at $n = 0, 1, 2, \ldots, N$, where

$$N = \frac{(r_f - r_0)}{VT}$$

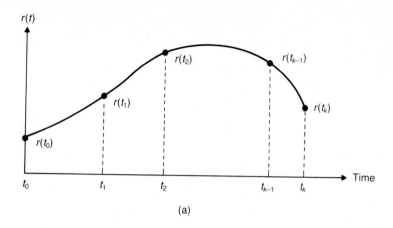

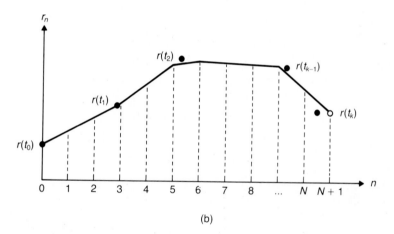

Figure 8.2 (a) A function $r(t)$ represented by a set of points, and (b) a function r_n generated using linear interpolation.

Alternatively, this equation can be written recursively as

$$r_n = r_{n-1} + \Delta r$$

where

$$\Delta r = VT$$

It should be noted that the Δr term can be applied for N sample periods to the input of an incremental closed-loop control such as that shown in Fig. 5.32. The final value r_f will not quite be reached in general, and r_f can be generated as r_{N+1} if necessary. $\square$

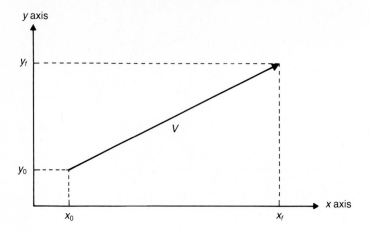

Figure 8.3 Straight-line motion of machine-tool table at velocity V.

Example 8.2 Straight-line motion generation   It is desired to generate commands for the x-axis and y-axis drives of a machine tool for straight-line motion at a velocity V between points (x_0, y_0) and (x_f, y_f), as illustrated in Fig. 8.3. The number of time intervals required to generate the motion is found from

$$N = \frac{[(x_f - x_0)^2 + (y_f - y_0)^2]^{1/2}}{TV}$$

A series of N x–y motion commands can then be generated using

$$x_n = x_{n-1} + \Delta x$$
$$y_n = y_{n-1} + \Delta y$$

where

$$\Delta x = (x_f - x_0)/N$$
$$\Delta y = (y_f - y_0)/N$$

Program 8.1 illustrates this solution using integer arithmetic (except for V and T). Note that truncation errors can occur regardless of whether integer, fractional, or floating-point arithmetic is used. They are often smaller in magnitude with floating-point arithmetic, but they can still accumulate to significant proportions. One way of handling truncation errors (while taking advantage of the rapid computation times of integer arithmetic) is to retain the truncated portions and correct the integers when they accumulate to significance.

```
program Straight_line;
  var
    x0, xf, y0, yf : integer;          {initial and final positions}
    xn, yn : integer;                  {current position}
    dx, dy : integer;                  {position increment}
    v, T : real;                       {velocity and sample period}
    i, n : integer;                    {n=number of motion increments}
  begin
    readln(x0, xf, y0, yf);            {input initial and final positions}
    readln(v, T);                      {input velocity and sample period}
    n := trunc(sqrt((xf - x0) * (xf - x0) + (yf - y0) * (yf - y0)) / v / T);
    dx := (xf - x0) div n;
    dy := (yf - y0) div n;             {source of truncation errors}
    i := 0;
    xn := x0;                          {initial position}
    yn := y0;
    writeln(i, xn, yn);
    for i := 1 to N do
      begin
        xn := xn + dx;                 {incremental position change}
        yn := yn + dy;
        writeln(i, xn, yn)
      end
  end.
```

Program 8.1 Straight-line motion generation.

This approach is illustrated in Program 8.2. Output from both programs is shown for comparison in Table 8.1, where it is apparent that when Program 8.1 is used, the final value of y is 390 rather than the desired value of 400 that is obtained with Program 8.2. □

8.3

COMMAND GENERATION USING CUBIC POLYNOMIALS

Third-order polynomials can be used to store command inputs compactly within a computer memory and to generate complex command sequences in real-time for machine and process control. A function beginning at time t_0 can be broken up into k segments that end at times $t = t_1, t_2, \ldots, t_i, \ldots, t_k$. Each of these segments can then be approximated by using third-order polynomials, often referred to as cubic spline functions, that are continuous with adjacent segments at segment boundaries.

Consider the m-dimensional vector $\mathbf{r}(t)$ representing the commands for a process with m inputs as a function of time t. For example, $\mathbf{r}(t)$ can represent axis position commands for an m-axis machine. The ith segment of $\mathbf{r}(t)$ can be

```
program Corrected_straight_line;
  var
    x0, xf, y0, yf : integer;              {initial and final positions}
    xn, yn : integer;                      {current position}
    dx, dy : integer;                      {position increment}
    v, T : real;                           {velocity and sample period}
    i, n : integer;                        {n=number of motion increments}
    xrem, yrem : integer;                  {remainders}
    xerror, yerror : integer;              {accumulated errors}
  begin
    readln(x0, xf, y0, yf);                {input initial and final positions}
    readln(v, T);                          {input velocity and sample period}
    n := trunc(sqrt((xf - x0) * (xf - x0) + (yf - y0) * (yf - y0)) / v / T);
    dx := (xf - x0) div n;
    dy := (yf - y0) div n;                 {source of truncation errors}
    xrem := (xf - x0) mod n;
    yrem := (yf - y0) mod n;               {remainders for error correction}
    xerror := 0;
    yerror := 0;                           {errors initially zero}
    i := 0;
    xn := x0;                              {initial position}
    yn := y0;
    writeln(i, xn, yn);
    for i := 1 to n do
      begin
        xn := xn + dx;                     {incremental position change}
        yn := yn + dy;
        xerror := xerror + xrem;           {accumulated errors are scaled by n}
        yerror := yerror + yrem;
        if xerror >= n then                {correct x if actual x error > 1}
          begin
            xn := xn + 1;
            xerror := xerror - n           {adjust scaled x error}
          end;
        if yerror >= n then                {correct y if actual y error > 1}
          begin
            yn := yn + 1;
            yerror := yerror - n           {adjust scaled y error}
          end;
        writeln(i, xn, yn)
      end
end.
```

Program 8.2 Straight-line motion generation with truncation error correction.

Table 8.1

Output of Programs 8.1 and 8.2, illustrating accumulated truncation error in Y in Program 8.1. (The final value is 390 rather than 400.) (a) Output of Program 8.1. (b) Output of Program 8.2.

i	X	Y	i	X	Y
1	20	26	1	20	26
2	40	52	2	40	53
3	60	78	3	60	80
4	80	104	4	80	106
5	100	130	5	100	133
6	120	156	6	120	160
7	140	182	7	140	186
8	160	208	8	160	213
9	180	234	9	180	240
10	200	260	10	200	266
11	220	286	11	220	293
12	240	312	12	240	320
13	260	338	13	260	346
14	280	364	14	280	373
15	300	390	15	300	400
	(a)			(b)	

represented by

$$\mathbf{r}(t) = \mathbf{A}_i \mathbf{b}(t) \tag{8.4}$$

where

$$\mathbf{b}(t) = [t^3 \quad t^2 \quad t \quad 1]^T \tag{8.5}$$

and $\mathbf{A}_i$ is the ith spline coefficient matrix that is valid for $t_{i-1} \leq t \leq t_i$. The derivatives of $\mathbf{r}(t)$ along the ith segment are

$$\dot{\mathbf{r}}(t) = \mathbf{A}_i \dot{\mathbf{b}}(t) \tag{8.6}$$

$$\ddot{\mathbf{r}}(t) = \mathbf{A}_i \ddot{\mathbf{b}}(t) \tag{8.7}$$

where

$$\dot{\mathbf{b}}(t) = [3t^2 \quad 2t \quad 1 \quad 0]^T \tag{8.8}$$

$$\ddot{\mathbf{b}}(t) = [6t \quad 2 \quad 0 \quad 0]^T \tag{8.9}$$

Example 8.3 Four-segment motion profile   The motion profiles shown in Fig. 8.4 are generated in four segments using Eq. (8.4) and the following coefficient matrices:

$$\text{Segment 1} \atop (0 \le t \le 5 \text{ s})} \quad \mathbf{A}_1 = \begin{bmatrix} -0.052 & 0.46 & 0 & 0 \\ -0.064 & 0.56 & 0 & 0 \end{bmatrix}$$

$$\text{Segment 2} \atop (5 \le t \le 15 \text{ s})} \quad \mathbf{A}_2 = \begin{bmatrix} 0.001 & -0.065 & 1.275 & 0.125 \\ 0.004 & -0.16 & 2.1 & -1 \end{bmatrix}$$

$$\text{Segment 3} \atop (15 \le t \le 30 \text{ s})} \quad \mathbf{A}_3 = \begin{bmatrix} 5.185 \times 10^{-4} & -2.667 \times 10^{-2} & 0.45 & 5.5 \\ 1.852 \times 10^{-3} & -1.333 \times 10^{-1} & 2.75 & -9.5 \end{bmatrix}$$

$$\text{Segment 4} \atop (30 \le t \le 40 \text{ s})} \quad \mathbf{A}_4 = \begin{bmatrix} 0.0005 & -0.065 & 2.8 & -30 \\ -0.0015 & 0.17 & -6.4 & 82.5 \end{bmatrix}$$

The specific form of Eq. (8.4) in this case is

$$\begin{bmatrix} x(t) \\ y(t) \end{bmatrix} = \begin{bmatrix} c_{x_3} & c_{x_2} & c_{x_1} & c_{x_0} \\ c_{y_3} & c_{y_2} & c_{y_3} & c_{y_0} \end{bmatrix}_i \begin{bmatrix} t^3 \\ t^2 \\ t \\ 1 \end{bmatrix}$$

which represents the polynomials

$$x(t) = c_{x_{3_i}} t^3 + c_{x_{2_i}} t^2 + c_{x_{1_i}} t + c_{x_{0_i}}$$
$$y(t) = c_{y_{3_i}} t^3 + c_{y_{2_i}} t^2 + c_{y_{1_i}} t + c_{y_{0_i}}$$

that are valid between t_{i-1} and t_i. The profiles in Fig. 8.4 are generated by varying time t from 0 to 40 s, identifying on which of the four segments t lies, and then calculating $x(t)$ and $y(t)$ using the appropriate set of coefficients. For example, $t = 10$ s lies on the second segment, and the second set of coefficients must be used for this value of time. The polynomials for x and y in this case are

$$x(t) = 0.001t^3 - 0.065t^2 + 1.275t + 0.125$$

$$y(t) = 0.004t^3 - 0.16t^2 + 2.1t - 1$$

For $t = 10$ s, substituting into the above equation yields

$$x(10) = 7.375 \text{ cm}$$

$$y(10) = 8.0 \text{ cm}$$

Table 8.2 lists the boundary conditions at the end points of the segments shown in Fig. 8.4. Given the position, velocity, acceleration, and so on at each of the end points, one often wants to find the coefficients of the polynomials that satisfy these boundary conditions and interpolate in a smooth, continuous manner over the entire motion sequence. A technique for doing this is presented in the next section. □

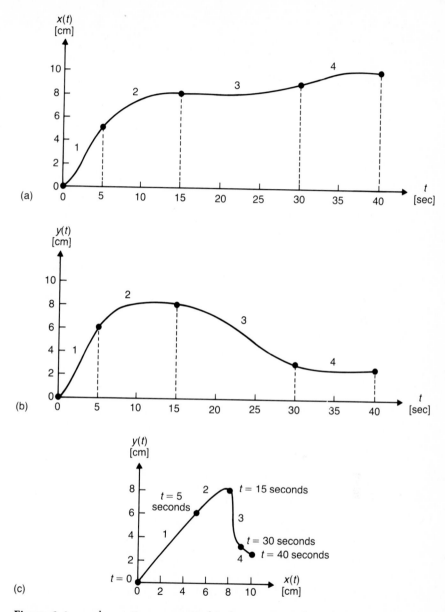

Figure 8.4 x and y motion represented in four segments by cubic functions of time.

Calculation of Cubic Coefficient Matrices

The coefficient matrix $\mathbf{A}_i$ is calculated from four sets of boundary conditions at the end points of the ith segment. In machine systems, matching the position, velocity, and acceleration at the end points is likely to be of interest. This gives a total of six boundary conditions to choose from. Here, position and velocity (the

Table 8.2

Position and velocity at the end points of segments shown in Fig. 8.4

i	t_i[seconds]	x_i[cm]	$\dfrac{dx_i}{dt}$[cm/s]	y_i[cm]	$\dfrac{dy_i}{dt}$[cm/s]
0	0	0.0	0.00	0.0	0.00
1	5	5.0	0.70	6.0	0.80
2	15	8.0	0.00	8.0	0.00
3	30	9.0	0.25	3.0	−0.25
4	40	10.0	0.00	2.5	0.00

function and its first derivative) at both ends of the segment are used as boundary conditions for calculating the coefficient matrices.

Given $\mathbf{r}(t)$ and $\dot{\mathbf{r}}(t)$ at time t_{i-1} and t_i, we can calculate the coefficient matrix for the ith segment using the Hermite method. Combining Eqs. (8.4) and (8.6) at times t_{i-1} and t_i yields

$$\mathbf{R}_i = \mathbf{A}_i \mathbf{B}_i \tag{8.10}$$

where

$$\mathbf{R}_i = [\mathbf{r}(t_{i-1}) \quad \dot{\mathbf{r}}(t_{i-1}) \quad \mathbf{r}(t_i) \quad \dot{\mathbf{r}}(t_i)] \tag{8.11}$$

and

$$\mathbf{B}_i = [\mathbf{b}(t_{i-1}) \quad \dot{\mathbf{b}}(t_{i-1}) \quad \mathbf{b}(t_i) \quad \dot{\mathbf{b}}(t_i)] \tag{8.12}$$

Solving for the coefficient matrix $\mathbf{A}_i$ yields

$$\mathbf{A}_i = \mathbf{R}_i \mathbf{B}_i^{-1} \tag{8.13}$$

When we use this method, only t_i, $\mathbf{r}(t_i)$, and $\dot{\mathbf{r}}(t_i)$ for $i = 0, 1, 2, \ldots, k$ need be stored in the computer memory. We can calculate $\mathbf{r}(t)$ for any time $t_0 \leq t \leq t_k$ by first determining the i for which $t_{i-1} \leq t \leq t_i$ and then calculating $\mathbf{A}_i$ from Eq. (8.13) and $\mathbf{r}(t)$ from Eq. (8.4). If commands are to be generated many times, the $\mathbf{A}_i$ matrices can be precalculated and stored.

Example 8.4 Spline coefficient matrix calculation   From Table 8.2, matrix $\mathbf{R}_2$ in Eq. (8.11) is

$$\mathbf{R}_2 = \begin{bmatrix} 5.0 & 0.7 & 8.0 & 0.0 \\ 6.0 & 0.8 & 8.0 & 0.0 \end{bmatrix}$$

and matrix $\mathbf{B}_2$ is

$$\mathbf{B}_2 = \begin{bmatrix} 5^3 & 3(5^2) & 15^3 & 3(15^2) \\ 5^2 & 2(5) & 15^2 & 2(15) \\ 5 & 1 & 15 & 1 \\ 1 & 0 & 1 & 0 \end{bmatrix}$$

From Eq. (8.13), the matrix $\mathbf{A}_2$ containing the coefficients of the polynomials for $x(t)$ and $y(t)$ that match the boundary conditions in $\mathbf{R}_2$ and interpolate between time t_{i-1} and t_i is

$$\mathbf{A}_2 = \mathbf{R}_2 \mathbf{B}_2^{-1} = \begin{bmatrix} 0.001 & -0.065 & 1.2785 & 0.125 \\ 0.004 & -0.16 & 2.1 & -1 \end{bmatrix}$$

This coefficient matrix generates the second segment of the motion profiles in Fig. 8.4. □

Simplification of Cubic Spline Calculations

To facilitate computer implementation of the cubic spline method, a new normalized time variable u_i can be introduced to simplify calculation of the coefficient matrices and generation of commands. This variable u_i varies from 0 to 1 as t varies from t_{i-1} to t_i on the ith segment of the contour. For $t_{i-1} \le t \le t_i$, t is related to u_i by

$$t = t_{i-1} + (t_i - t_{i-1})u_i \tag{8.14}$$

and

$$\frac{dt}{du_i} = t_i - t_{i-1} \tag{8.15}$$

Equations (8.4) and (8.6) now can be rewritten as

$$\mathbf{r}^*(u_i) = \mathbf{A}_i^* \mathbf{b}^*(u_i) \tag{8.16}$$

$$\dot{\mathbf{r}}^*(u_i) = \mathbf{A}_i^* \dot{\mathbf{b}}^*(u_i) \tag{8.17}$$

where

$$\mathbf{b}^*(u_i) = [u_i^3 \quad u_i^2 \quad u_i \quad 1]^T \tag{8.18}$$

$$\dot{\mathbf{b}}^*(u_i) = [3u_i^2 \quad 2u_i \quad 1 \quad 0]^T \tag{8.19}$$

and $\mathbf{A}_i^*$ is the new coefficient matrix.

Boundary conditions at the end points of the ith segment of the contour are then as follows:

$$\mathbf{r}^*(0) = \mathbf{r}(t_{i-1}) \tag{8.20}$$

$$\dot{\mathbf{r}}^*(0) = \dot{\mathbf{r}}(t_{i-1}) \frac{dt}{du_i} \tag{8.21}$$

$$\mathbf{r}^*(1) = \mathbf{r}(t_i) \tag{8.22}$$

$$\dot{\mathbf{r}}^*(1) = \dot{\mathbf{r}}(t_i) \frac{dt}{du_i} \tag{8.23}$$

The coefficient matrix $\mathbf{A}_i^*$ can now be calculated as follows:

$$\mathbf{A}_i^* = \mathbf{R}_i^* \mathbf{B}^{*-1} \tag{8.24}$$

where

$$\mathbf{R}_i^* = \left[\mathbf{r}(t_{i-1}) \quad \dot{\mathbf{r}}(t_{i-1}) \frac{dt}{du_i} \quad \mathbf{r}(t_i) \quad \dot{\mathbf{r}}(t_i) \frac{dt}{du_i} \right] \tag{8.25}$$

and

$$\mathbf{B}^* = [\mathbf{b}^*(0) \quad \dot{\mathbf{b}}^*(0) \quad \mathbf{b}^*(1) \quad \dot{\mathbf{b}}^*(1)] \tag{8.26}$$

It is important to note that $\mathbf{B}^{*-1}$ is constant and that

$$\mathbf{B}^{*-1} = \begin{bmatrix} 0 & 0 & 1 & 3 \\ 0 & 0 & 1 & 2 \\ 0 & 1 & 1 & 1 \\ 1 & 0 & 1 & 0 \end{bmatrix}^{-1} = \begin{bmatrix} 2 & -3 & 0 & 1 \\ 1 & -2 & 1 & 0 \\ -2 & 3 & 0 & 0 \\ 1 & -1 & 0 & 0 \end{bmatrix} \tag{8.27}$$

Eq. (8.14) can be solved for u_i yielding

$$u_i = \frac{t - t_{i-1}}{t_i - t_{i-1}} \tag{8.28}$$

and the point on the ith segment at time t can now be calculated by using Eqs. (8.16), (8.18), and (8.28).

Example 8.5 Use of normalized spline variable   For the second segment of the motion defined by the end points in Table 8.2,

$$\frac{dt}{du_2} = t_2 - t_1 = 15 - 5 = 10 \text{ s}$$

From Eq. (8.25),

$$\mathbf{R}_2^* = \begin{bmatrix} 5.0 & 7.0 & 8.0 & 0.0 \\ 6.0 & 8.0 & 8.0 & 0.0 \end{bmatrix}$$

From Eqs. (8.24) and (8.27),

$$\mathbf{A}_2^* = \mathbf{R}_2^* \mathbf{B}^{*-1} = \begin{bmatrix} 1.0 & -5.0 & 7.0 & 5.0 \\ 4.0 & -10.0 & 8.0 & 6.0 \end{bmatrix}$$

From Eqs. (8.16) and (8.18), the polynomials in the normalized time variable u_2, which varies from 0 to 1 as time t varies from 5 to 15 s, are

$$x(u_2) = u_2{}^3 - 5u_2{}^2 + 7u_2 + 5$$

$$y(u_2) = 4u_2{}^3 - 10u_2 + 8u_2 + 6$$

At time $t = 10$ s, the corresponding value of u_2 is found from Eq. (8.28) as

$$u_2 = \frac{10 - 5}{15 - 5} = 0.5$$

Substituting this value of u_2 into our equations for x and y yields

$$x(u_2) = 7.375$$

$$y(u_2) = 8.0$$

This result is identical to that obtained in Example 8.3 for $t = 10$ s. □

Example 8.6 Point-to-point three-dimensional motion For three-dimensional motion commands, Eq. (8.16) has the form

$$\begin{bmatrix} x(u_i) \\ y(u_i) \\ z(u_i) \end{bmatrix} = \begin{bmatrix} a_{x_3}^* & a_{x_2}^* & a_{x_1}^* & a_{x_0}^* \\ a_{y_3}^* & a_{y_2}^* & a_{y_1}^* & a_{y_0}^* \\ a_{z_3}^* & a_{z_2}^* & a_{z_1}^* & a_{z_0}^* \end{bmatrix}_i \begin{bmatrix} u_i{}^3 \\ u_i{}^2 \\ u_i \\ 1 \end{bmatrix}$$

If it is desired to generate the point-to-point motion shown in Fig. 8.5 over a period of 10 s using one segment, then with zero velocity at the end points of the motion,

$$\mathbf{r}(t_0) = \begin{bmatrix} 0 \\ 0 \\ 0 \end{bmatrix} \quad \mathbf{r}(t_1) = \begin{bmatrix} 1 \\ 1 \\ -1 \end{bmatrix} \quad \frac{d\mathbf{r}(t_0)}{dt} = \begin{bmatrix} 0 \\ 0 \\ 0 \end{bmatrix} \quad \frac{d\mathbf{r}(t_1)}{dt} = \begin{bmatrix} 0 \\ 0 \\ 0 \end{bmatrix}$$

Then, from Eq. (8.15),

$$\frac{dt}{du_1} = 10 - 0 = 10 \text{ s}$$

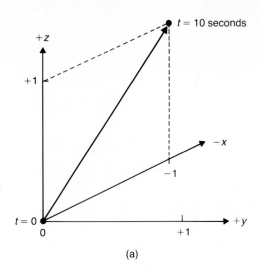

(a)

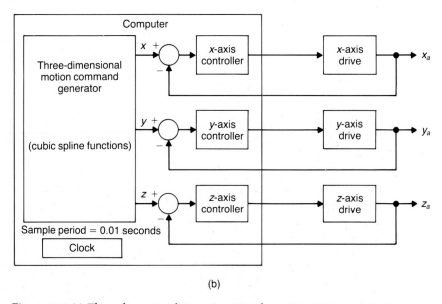

(b)

Figure 8.5 (a) Three-dimensional (x, y, z) motion from $(0, 0, 0)$ to $(-1, 0, 1)$; (b) three-axis computer control with cubic spline motion command generator.

and from Eq. (8.25),

$$
\mathbf{R}_1^* = \begin{bmatrix} 0 & 0(10) & 1 & 0(10) \\ 0 & 0(10) & 0 & 0(10) \\ 0 & 0(10) & -1 & 0(10) \end{bmatrix} = \begin{bmatrix} 0 & 0 & 1 & 0 \\ 0 & 0 & 0 & 0 \\ 0 & 0 & -1 & 0 \end{bmatrix}
$$

Finally, the spline coefficient matrix is obtained via Eqs. (8.24) and (8.27).

$$
\mathbf{A}_1^* = \begin{bmatrix} 0 & 0 & 1 & 0 \\ 0 & 0 & 0 & 0 \\ 0 & 0 & -1 & 0 \end{bmatrix} \begin{bmatrix} 2 & -3 & 0 & 1 \\ 1 & -2 & 1 & 0 \\ -2 & 3 & 0 & 0 \\ 1 & -1 & 0 & 0 \end{bmatrix}
$$

$$
= \begin{bmatrix} -2 & 3 & 0 & 0 \\ 0 & 0 & 0 & 0 \\ 2 & 3 & 0 & 0 \end{bmatrix}
$$

The equations for the individual elements of $\mathbf{r}(u_1)$ are then

$$
x(u_1) = -2u_1^3 + 3u_1^2
$$
$$
y(u_1) = 0
$$
$$
z(u_1) = 2u_1^3 - 3u_1^2
$$

If the computer control for the machine is structured as shown in Fig. 8.5(b) and has a sample period T of 0.01 s, it is desirable to generate a new command every 0.01 s. The increment Δu_1 that corresponds to T is found, using Eq. (8.15), to be

$$
\Delta u_1 = T\left(\frac{du_1}{dt}\right) = \frac{0.01 \text{ s}}{10 \text{ s}} = 0.001
$$

Examples of the commands generated are given in Table 8.3. Only 11 out of 1000 are listed.)

Straight lines, parabolas, and third-order paths of any duration can be exactly represented by a single cubic polynomial, but in general, sinusoids and other more complex paths cannot be exactly represented. Table 8.4 shows the accuracy achieved when a cubic spline function, calculated from the position and velocity at the end points, is used to represent a two-dimensional circular path generated by $x(t) = R \cos(\omega t)$ and $y(t) = R \sin(\omega t)$. The vector error at the midpoint of the

Table 8.3

Commands Generated in Example 8.6.
(Only 11 out of 1000 are listed.)

n	t	u_1	x	y	z
0	0.0	0.0	0.000	0.000	−0.000
100	1.0	0.1	0.028	0.000	−0.028
200	2.0	0.2	0.104	0.000	−0.104
300	3.0	0.3	0.216	0.000	−0.216
400	4.0	0.4	0.352	0.000	−0.352
500	5.0	0.5	0.500	0.000	−0.500
600	6.0	0.6	0.648	0.000	−0.648
700	7.0	0.7	0.784	0.000	−0.784
800	8.0	0.8	0.896	0.000	−0.896
900	9.0	0.9	0.972	0.000	−0.972
1000	10.0	0.1	1.000	0.000	−1.000 □

spline is given as a function of radius for segments of various arc lengths. As can be observed from the table, the accuracy of the representation increases greatly as arc length and hence spline length decrease. A cubic spline function representing a 90° arc on a circle with a 2-in. (5.08-cm) radius has a vector error of 0.030 in. (0.0762 cm) at its midpoint, whereas a spline function representing a 45° arc has an error of 0.002 in. (0.00508 cm). In this case, doubling the number of splines increases the accuracy by a factor of 15.

Table 8.4

Error at Midpoint of Spline for Various Arc Lengths

ARC LENGTH FIT	SPLINES PER CIRCLE	ERROR (AT MIDPOINT/ UNIT RADIUS)
15°	24	0.00002 in./in. (m/m)
30	12	0.00019
45	8	0.00098
60	6	0.00307
90	4	0.01521
120	3	0.04655

8.4

POINT-TO-POINT MOTION GENERATION USING SPLINE FUNCTIONS

Controlled acceleration is crucial for precision control and reliable operation in many positioning systems. All machine drives have velocity and acceleration limitations, and the cubic spline method is a convenient way to generate point-to-point motion while observing these limits. In this section, several methods of point-to-point motion generation for a single axis are presented and are then extended to synchronized motion of multiple axes.

Single Axis, Single Spline

As was illustrated in Example 8.6, point-to-point motion for a machine axis can easily be generated by using a single cubic spline function. Consider the general case shown in Fig. 8.6(a), where the position of the x axis of a machine is to be changed from x_0 at time t_0 to x_f at time t_f. The axis velocity is to be zero before and after the move. The machine accelerates from the starting point x_0 up to some maximum velocity that is reached at the midpoint of the move. It then immediately begins to decelerate, stopping at x_f.

If a single cubic spline function is to represent the entire move, it must satisfy the following boundary conditions:

$$\text{At } u = 0: \qquad x^*(0) = x_0 \quad \text{and} \quad \frac{dx^*(0)}{du} = 0$$

$$\text{At } u = 1: \qquad x^*(1) = x_f \quad \text{and} \quad \frac{dx^*(1)}{du} = 0$$

Substituting these boundary conditions into Eq. (8.24) yields

$$A^* = [-2\Delta x \quad 3\Delta x \quad 0 \quad x_0] \tag{8.29}$$

where

$$\Delta x = x_f - x_0 \tag{8.30}$$

It can be observed that the velocity has maximum magnitude at $u = 0.5$, a point halfway between x_0 and x_f, and that

$$\left(\frac{dx(t)}{dt}\right)_{\text{max}} = \left(\frac{dx^*(u)}{du}\right)_{\text{max}} \frac{du}{dt} = \frac{3}{2}\frac{|\Delta x|}{\Delta t} \tag{8.31}$$

where

$$\Delta t = t_f - t_0 \tag{8.32}$$

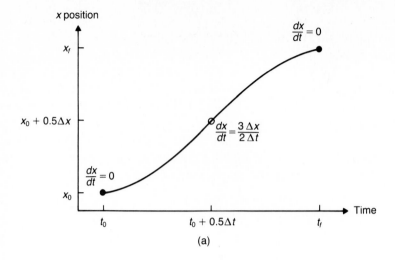

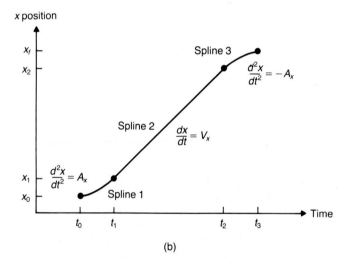

Figure 8.6 (a) Single axis point-to-point motion generated using a single cubic spline function; (b) single axis point-to-point motion generated using three cubic spline functions.

If V_x is the maximum velocity to be allowed, the minimum motion duration Δt_v associated with this limit can be found from Eq. (8.31) as

$$\Delta t_v = \frac{3}{2} \frac{|\Delta x|}{V_x} \tag{8.33}$$

For small values of Δx, Eq. (8.33) gives a correspondingly small value of Δt_v. The machine must accelerate to V_x and then decelerate to a stop during this short time interval, and the acceleration rates required may be high. Some minimum time Δt_a must therefore be allowed when moving a distance of Δx to prohibit accelerations greater than some limit A_x. Acceleration is zero at $s = 0.5$ and is maximum in magnitude at the end points. The maximum acceleration is therefore

$$\left(\frac{d^2x(t)}{dt^2}\right)_{max} = \left(\frac{d^2x^*(x)}{du^2}\right)_{max}\left(\frac{du}{dt}\right)^2 = \frac{|6\,\Delta x|}{\Delta t^2} \tag{8.34}$$

Substituting the maximum allowable acceleration into Eq. (8.34) and solving for the minimum motion duration, we find that

$$\Delta t_a = \left(\frac{6|\Delta x|}{A_x}\right)^{1/2} \tag{8.35}$$

To satisfy both the velocity limit and the acceleration limit, the value of Δt should be chosen as

$$\Delta t = \max(\Delta t_v, \Delta t_a) \tag{8.36}$$

Thus, when we are using one spline function to move a distance of Δx, choosing the greater of Δt_v or Δt_a will result in the fastest possible move without exceeding the velocity and acceleration limits V_x and A_x.

Single Axis, Three Splines, Constant Acceleration

It can be observed from Eq. (8.31) that the average velocity along the single spline is two-thirds of the maximum velocity reached only at the center point of the spline. The time required for the move is thus half again that which would be required if it were possible to make the entire move at the maximum velocity. A more efficient method for point-to-point motion uses three spline functions to represent the move. The first spline function accelerates the axis at the maximum rate until the maximum velocity is reached, the second spline function maintains the maximum velocity, and the third spline function decelerates the machine to a stop at the maximum rate. Figure 8.6(b) shows the path produced when this method is used. The three spline functions can be found from Δx, V_x, and A_x as follows:

$$\Delta t_1 = t_1 - t_0 = \frac{V_x}{A_x} \tag{8.37}$$

$$\Delta t_2 = t_2 - t_1 = \frac{|\Delta x|}{V_x} - \Delta t_1 \tag{8.38}$$

$$\Delta t_3 = t_3 - t_2 = \Delta t_1 \tag{8.39}$$

The boundary conditions for use in Eq. (8.24) are

$$x(t_0) = x_0 \qquad\qquad \dot{x}(t_0) = 0$$

$$x(t_1) = x_0 + \frac{1}{2} A_x \, \Delta t_1^{\,2} \, \text{sign}(\Delta x) \qquad\qquad \dot{x}(t_1) = V_x \, \text{sign}(\Delta x)$$

$$x(t_2) = x_f - \frac{1}{2} A_x \, \Delta t_3^{\,2} \, \text{sign}(\Delta x) \qquad\qquad \dot{x}(t_2) = V_x \, \text{sign}(\Delta x)$$

$$x(t_3) = x_f \qquad\qquad \dot{x}(t_3) = 0$$

It is possible for the calculation of Δt_2 in Eq. (8.38) to result in a negative number. This occurs when

$$|\Delta x| < \frac{V_x^{\,2}}{A_x} \tag{8.40}$$

and it indicates that the distance to be moved is not long enough to allow acceleration to the maximum velocity. If this is the case, the constant-velocity segment should be eliminated, and Eqs. (8.37), (8.38), and (8.39) should be modified to

$$\Delta t_1 = \left(\frac{|\Delta x|}{A_x} \right)^{1/2} \tag{8.41}$$

$$\Delta t_2 = 0 \tag{8.42}$$

$$\Delta t_3 = \Delta t_1 \tag{8.43}$$

Single Axis, Three Splines, Decreasing Acceleration

Many machine drives are capable of high acceleration at low velocities but of only little acceleration at high velocities. It is convenient in this case to modify the previous method by making the acceleration a maximum of A_x at time t_0 but decreasing it linearly to zero at time t_1 when the maximum velocity is reached. The modified spline functions can be found from

$$\Delta t_1 = 2 \frac{V_x}{A_x} \tag{8.44}$$

$$\Delta t_2 = \frac{|\Delta x|}{V_x} - \frac{4}{3} \Delta t_1 \tag{8.45}$$

$$\Delta t_3 = \Delta t_1 \tag{8.46}$$

with the boundary conditions

$$x(t_0) = x_0 \qquad\qquad \dot{x}(t_0) = 0$$

$$x(t_1) = x_0 + \frac{1}{3} A_x \, \Delta t_1{}^2 \, \mathrm{sign}(\Delta x) \qquad \dot{x}(t_1) = V_x \, \mathrm{sign}(\Delta x)$$

$$x(t_2) = x_f - \frac{1}{3} A_x \, \Delta t_3{}^2 \, \mathrm{sign}(\Delta x) \qquad \dot{x}(t_2) = V_x \, \mathrm{sign}(\Delta x)$$

$$x(t_3) = x_f \qquad\qquad \dot{x}(t_3) = 0$$

Here Δt_2 is negative when

$$|\Delta x| < \frac{8}{3} \frac{V_x^2}{A_x} \tag{8.47}$$

If this is the case, the constant-velocity segment should be eliminated and

$$\Delta t_1 = \left(\frac{3}{2} \frac{|\Delta x|}{A_x}\right)^{1/2} \tag{8.48}$$

$$\Delta t_2 = 0 \tag{8.49}$$

$$\Delta t_3 = \Delta t_1 \tag{8.50}$$

Multiple Axes, Single Spline

The methods we have described can be used to individually control the axes of mutiple-axis machines according to their individual acceleration and velocity limits, which may be quite different depending on the axis design and the actuator technology used. This implies that when two or more axes begin to move together, there is no guarantee that they will reach their respective end positions at the same time. However, there is often a need for coordinated point-to-point motion. For example, a machine with x, y, and z motions in a cartesian coordinate system might be required to move in a straight line from one point to another.

The single-spline method of motion generation can be modified easily for application in multiple-axis systems. If there are m axes, then let

$$\Delta x_i = x_{f_i} = x_{0_i} \tag{8.51}$$

$$\Delta t_{v_i} = \frac{3}{2} \frac{|\Delta x_i|}{V_i} \tag{8.52}$$

$$\Delta t_{a_i} = \left(\frac{6|\Delta x_i|}{A_i}\right)^{1/2} \tag{8.53}$$

where $i = 1, 2, \ldots, m$, and V_i and A_i are the ith-axis velocity and acceleration limits. In this case, Eq. (8.33) and (8.35) can be applied for each axis, and the shortest allowable Δt for all axes is found from

$$\Delta t = \max(\Delta t_{v_1}, \Delta t_{a_1}, \Delta t_{v_2}, \Delta t_{a_2}, \ldots, \Delta t_{v_m}, \Delta t_{a_m}) \qquad (8.54)$$

Multiple Axes, Three Splines, Constant Acceleration

Figure 8.7(a) shows an example of two-axis machine motion where significantly different velocity and acceleration profiles result from Eqs. (8.37), (8.38), and (8.39) and the motion of the two axes is not synchronized. For synchronized motion as shown in Figure 8.7(b), one axis is accelerated at its maximum limit to a velocity that is less than its maximum limit. Simultaneously, the second axis accelerates at less than its maximum limit but reaches its maximum velocity. Both axes then move at constant velocity until they are simultaneously decelerated to a stop. In an analysis similar to that given previously for a single axis, it can be shown that to achieve maximum acceleration on the ith axis and maximum velocity on the jth axis, the following relationships are required:

$$\Delta t_{1_{ij}} = \frac{|\Delta x_i|\, V_j}{|\Delta x_j|\, A_i} \qquad (8.55)$$

$$\Delta t_{2_{ij}} = \frac{|\Delta x_j|}{V_j} - \Delta t_{1_{ij}} \qquad (8.56)$$

$$\Delta t_{3_{ij}} = \Delta t_{1_{ij}} \qquad (8.57)$$

for $i = 1, 2, \ldots, m$ and $j = 1, 2, \ldots, m$.

Again, if $\Delta t_{2_{ij}}$ is negative, these time variables must be modified so that the constant-velocity segment is of zero duration and

$$\Delta t_{1_{ij}} = \left(\frac{|\Delta x_i|}{A_i}\right)^{1/2} \qquad (8.58)$$

$$\Delta t_{2_{ij}} = 0 \qquad (8.59)$$

$$\Delta t_{3_{ij}} = \Delta t_{1_{ij}} \qquad (8.60)$$

The minimum time to move axes i and j, given the i-axis acceleration limit and the j-axis velocity limit, is therefore

$$\Delta t_{ij} = \Delta t_{1_{ij}} + \Delta t_{2_{ij}} + \Delta t_{3_{ij}} \qquad (8.61)$$

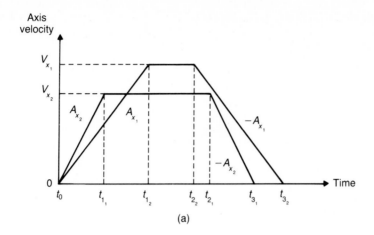

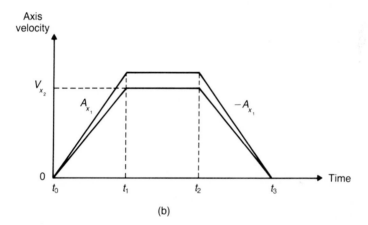

Figure 8.7 (a) Non-synchronized and (b) synchronized point-to-point motion of two axes.

The shortest time for the move that satisfies all velocity and acceleration limits, while maintaining synchronized motion of n axes, is then

$$\Delta t = \max_{i=1 \to m} \left[\max_{j=1 \to m} \Delta t_{ij} \right] \tag{8.62}$$

The critical combination of axes is that combination of axis i and axis j for which $\Delta t_{ij} = \Delta t$. The individual components of Δt are $\Delta t_{1_{ij}}$, $\Delta t_{2_{ij}}$, and $\Delta t_{3_{ij}}$, and the boundary conditions for any axis k are

$$x_k(t_0) = x_{k_0} \qquad\qquad \dot{x}_k(t_0) = 0$$

$$x_k(t_1) = x_{k_0} + \frac{1}{2} A_i \, \Delta t_1^{\,2} \, \frac{\Delta x_k}{|\Delta x_i|} \qquad \dot{x}_k(t_1) = V_j \frac{\Delta x_k}{|\Delta x_j|}$$

$$x_k(t_2) = x_{k_f} - \frac{1}{2} A_i \, \Delta t_1^{\,2} \, \frac{\Delta x_k}{|\Delta x_i|} \qquad \dot{x}_k(t_2) = V_j \frac{\Delta x_k}{|\Delta x_j|}$$

$$x_k(t_3) = x_{k_f} \qquad\qquad \dot{x}_k(t_3) = 0$$

Multiple Axes, Three Splines, Decreasing Acceleration

As we have noted, it can be beneficial to reduce the acceleration demanded as velocity approaches the maximum value. For synchronized motion of m axes under these conditions,

$$\Delta t_{1_{ij}} = 2 \frac{|\Delta x_i|}{|\Delta x_j|} \frac{V_j}{A_i} \tag{8.63}$$

$$\Delta t_{2_{ij}} = \frac{|\Delta x_j|}{V_j} - \frac{4}{3} \Delta t_{1_{ij}} \tag{8.64}$$

$$\Delta t_{3_{ij}} = \Delta t_{1_{ij}} \tag{8.65}$$

If $\Delta t_{2_{ij}}$ is negative, the following equations should be used

$$\Delta t_{1_{ij}} = \left(\frac{3}{2} \frac{|\Delta x_i|}{A_i} \right)^{1/2} \tag{8.66}$$

$$\Delta t_{2_{ij}} = 0 \tag{8.67}$$

$$\Delta t_{3_{ij}} = \Delta t_{1_{ij}} \tag{8.68}$$

The boundary conditions for any axis k are then

$$x_k(t_0) = x_{k_0} \qquad\qquad \dot{x}_k(t_0) = 0$$

$$x_k(t_1) = x_{k_0} + \frac{1}{3} A_i \, \Delta t_1^{\,2} \, \frac{\Delta x_k}{|\Delta x_i|} \qquad \dot{x}_k(t_1) = V_j \frac{\Delta x_k}{|\Delta x_j|}$$

$$x_k(t_2) = x_{k_f} - \frac{1}{3} A_i \, \Delta t_3^{\,2} \, \frac{\Delta x_k}{|\Delta x_i|} \qquad \dot{x}_k(t_2) = V_j \frac{\Delta x_k}{|\Delta x_j|}$$

$$x_k(t_3) = x_{k_f} \qquad\qquad \dot{x}_k(t_3) = 0$$

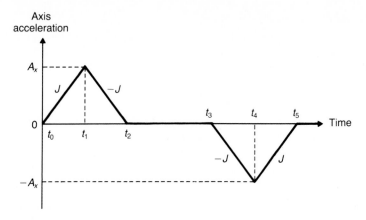

Figure 8.8 Five-spline motion profile with constant rate of change of acceleration (constant jerk).

8.5
CONTROLLED RATE OF CHANGE OF ACCELERATION

In precision machine control systems, large step changes in commanded acceleration can excite vibrations in the machine's mechanical structures. These vibrations can lead to position inaccuracies during high-speed manipulation and, in some cases, to the failure of high-precision position transducers to reliably measure position. Control of the rate of change of acceleration can often be a more cost-effective solution to these problems than redesigning mechanical structures to minimize vibration and improve accuracy.

Acceleration can be conveniently commanded using spline functions so that high-speed manipulation can occur without exciting machine structures. For example, Fig. 8.8 shows an acceleration profile for point-to-point motion consisting of five cubic spline functions, four of which have a constant rate of change in acceleration (constant jerk) J. The objective is to accelerate up to a constant velocity and later decelerate to a stop without step changes in acceleration. A variation is to use fourth-order polynomials to achieve parabolic acceleration and deceleration profiles with three spline functions.

8.6
FEEDFORWARD CONTROL USING CUBIC
SPLINE FUNCTIONS

The accuracy with which a process follows computer-generated commands depends largely on the process dynamics and controllers. Each process variable typically has its own controller and actuator system. Although a variety of types of controllers can be implemented, proportional and proportional-plus-integral control are perhaps the most common. Both have limitations that result in dynamic deviations of the process outputs from their desired values. The use of strictly proportional control often results in following errors for ramp inputs. These errors can be reduced by increasing proportional gains, but most processes become unstable or undesirably oscillatory with high gain. The addition of integral control can result in zero steady-state following errors, but system stability is sacrificed and other undesirable behaviors may result.

Feedforward control has already been discussed in Chapter 4, and offers an alternative that can be adapted easily to spline command generation methods. When used in conjunction with proportional feedback control, feedforward control can significantly reduce following errors. Feedforward control is accomplished by modifying the spline coefficients to offset commands by an amount that compensates for the following errors inherent in process control systems. Assume that the closed-loop dynamics of a controlled process can be approximated by

$$\mathbf{G}^{-1}\dot{\mathbf{c}}(t) + \mathbf{c}(t) = \mathbf{r}'(t) \tag{8.69}$$

where $\mathbf{c}(t)$ represents the process outputs, $\mathbf{G}$ is a diagonal gain matrix, and $\mathbf{r}'(t)$ represents the modified command inputs to the controller. To make the error $\mathbf{r}(t) - \mathbf{c}(t)$ zero, the modified position command $\mathbf{r}'(t)$ is generated from

$$\mathbf{r}'(t) = \mathbf{r}(t) + \mathbf{G}^{-1}\dot{\mathbf{r}}(t) \tag{8.70}$$

Using Eqs. (8.4) and (8.6),

$$\mathbf{r}'(t) = \mathbf{A}_i\mathbf{b}(t) + \mathbf{G}^{-1}\mathbf{A}_i\dot{\mathbf{b}}(t) \tag{8.71}$$

However, as can be observed from Eqs. (8.5) and (8.8),

$$\dot{\mathbf{b}}(t) = \mathbf{Q}\mathbf{b}(t) \tag{8.72}$$

where

$$\mathbf{Q} = \begin{bmatrix} 0 & 3 & 0 & 0 \\ 0 & 0 & 2 & 0 \\ 0 & 0 & 0 & 1 \\ 0 & 0 & 0 & 0 \end{bmatrix} \tag{8.73}$$

Then

$$\mathbf{r}'(t) = [\mathbf{A}_i + \mathbf{G}^{-1}\mathbf{A}_i\mathbf{Q}]\mathbf{b}(t) \tag{8.74}$$

or

$$\mathbf{r}'(t) = \mathbf{A}_i' \mathbf{b}(t) \tag{8.75}$$

where

$$\mathbf{A}_i' = \mathbf{A}_i + \mathbf{G}^{-1} \mathbf{A}_i \mathbf{Q} \tag{8.76}$$

The matrix A_i' is easily calculated because of the many zero terms in $\mathbf{G}$ and $\mathbf{Q}$. Equation (8.75) now can be used in place of Eq. (8.4) in generating commands. Therefore, via knowledge of the process gains, feedforward control may be incorporated directly into the spline coefficients, and low proportional controller gains may be used in the system, resulting in a relatively stable control system with most of the following error eliminated.

Example 8.7 Inclusion of command feedforward   Assume that the process G_p in the block diagram shown in Fig. 8.9(a) can be approximately modeled by

$$\frac{dc(t)}{dt} = k_p m(t)$$

A proportional controller with gain k_c is used to control the process, and the approximate closed-loop model

$$\tau \frac{dc(t)}{dt} + c(t) = r(t)$$

results, where the closed-loop time contant is

$$\tau = \frac{1}{k_c k_p}$$

When the single cubic spline method described in Section 8.4 is to be used to change the setpoints of the process from r_0 to $r_0 + \Delta r$, the command input has the form

$$r(u) = -2\Delta r u^3 + 3\Delta r u^2 + r_0$$

where

$$u = t/\Delta t$$

The error

$$e(t) = r(t) - c(t)$$

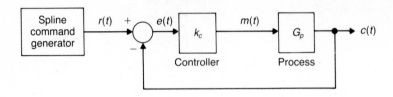

(a)

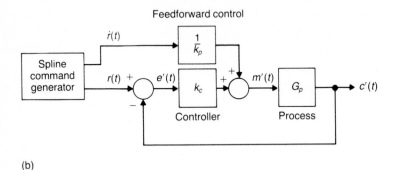

(b)

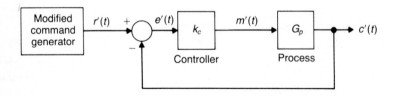

(c)

Figure 8.9 Process with spline function command generator.
(a) Without feedforward control; (b) with command feedforward control;
(c) with command feedforward control included in command generator.

due to the process dynamics modeled above is plotted in Fig. 8.10 for various
closed-loop time constants.

If the model is exact, these errors can be completely eliminated by using
feedforward control as shown in Fig. 8.9(b) and (c). The modified command input
that compensates for these errors is obtained from Eq. (8.70) or Eq. (8.75) as

$$r'(u) = -2\Delta r u^3 + \left(3\Delta r - \frac{6\tau\Delta r}{\Delta t}\right)u^2 + \frac{6\tau\Delta r}{\Delta t}u + r_0 \quad \square$$

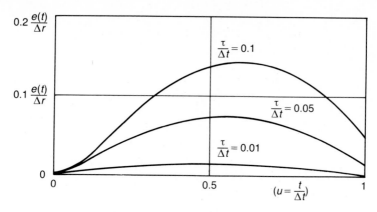

Figure 8.10 Following error for a setpoint change using the single cubic spline method without feedforward control.

8.7

OPEN-LOOP POSITION CONTROL USING STEPPING MOTORS

Open-loop control is an inexpensive and easily implemented control strategy that can be effective when process behavior is well known and not seriously affected by disturbances. A device that is ideal for open-loop control of position is the **stepping motor**. The variable-reluctance stepping motor, one commonly used type, consists of a multiple-phase stator and a rotor with multiple teeth as illustrated in Fig. 8.11. For this type of motor, the number of steps per revolution is found from

$$N = \text{(number of teeth)} \times \text{(number of stator phases)} \qquad (8.77)$$

The excitation of the stator phases in the proper sequence to cause the motor to step is controlled by electronic circuitry packaged as the "translator" indicated in Fig. 8.12. The **translator** accepts as its input a series of pulses, one pulse corresponding to one step of the motor. The motor velocity therefore depends on the frequency of the input pulse trains. A second signal determines the direction in which the step is to occur: clockwise or counterclockwise.

The torque generated per phase is determined by the electromagnetic characteristics of the motor. Excitation of a single phase results in a torque/ rotation curve such as that shown in Fig. 8.13. It is important to note that there is a maximum holding torque beyond which the motor cannot hold its position.

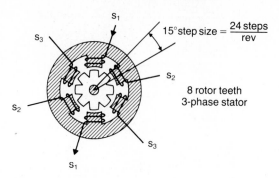

Figure 8.11 Variable-reluctance stepping motor.

Because the stator cannot distinguish between individual rotor teeth, load torque and acceleration must be carefully considered when designing a system that uses stepping motors. In general, the low torque characteristics of electromagnetic stepping motors restricts their use in high-load applications. Torque also tends to drop off at high stepping frequencies.

Computer software and hardware can be used to generate direction signals and a pulse train to the translator, causing the motor to rotate the desired number of steps in the desired direction. An example is shown in Fig. 8.14. Whenever a pulse is received by the translator and the direction signal is high, the motor advances one step in the positive direction. The motor moves one step in the negative direction when a pulse is received and the direction signal is low.

The problem of motion generation for stepping motors is the formation of appropriate pulse trains to serve as the inputs to the translator circuit. These pulse trains have to conform to the physical limitations of the stepping motor with regard to acceleration and deceleration limits and maximum stepping frequency. Stepping motors typically do not respond well to irregular pulse trains. Because

Figure 8.12 Computer-controlled stepping motor.

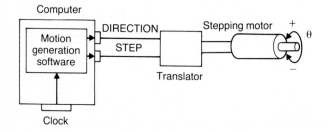

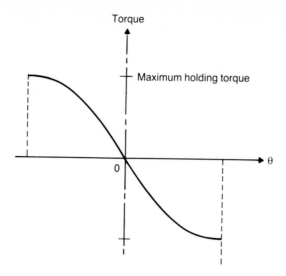

Figure 8.13 Stepping motor torque characteristics.

of the high instantaneous accelerations commanded, sporadic pulse spacing can cause the motor to misstep and even stall.

Several methods are available for generating pulse trains, including pulse rate multiplication and the use of digital differential analysis (DDA). The DDA method has been found to give excellent results as far as the uniformity of the pulse train is concerned. First, we will show how velocity profiles can be generated for a single axis. Then we will extend the concept to include techniques for generating acceleration profiles and controlling two or more axes.

Figure 8.14 Incremental changes in stepping motor position as a result of computer commands.

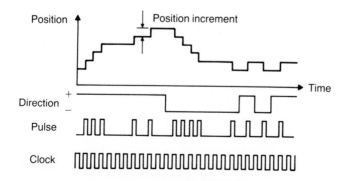

Constant-Velocity Profile

A simple case of motion generation using the DDA method is illustrated by the velocity profile shown in Fig. 8.15(a). It is necessary to generate a constant-frequency pulse train for each of the velocities. The method for generation of constant velocities is illustrated in Fig. 8.15(b). The number stored in the velocity register V is added to the displacement register P every time there is a 0 to 1 transition in the clock[1]. If there is an overflow from P, a pulse is sent to the motor. The rate at which there are overflows from P depends on the value of V and on the clock frequency f. For a fixed clock frequency, the output pulse rate can be changed by altering the value of V.

One way to implement Fig. 8.15(b) is to interpret the contents of register P as a signed fraction representing a fraction of a step. When this value becomes greater than 1 or less than -1, the stepping motor should be indexed one step in the plus or minus direction, respectively. This condition can be detected by detecting overflow of the P register, the sign of which determines the direction in which to step. Program 8.3 illustrates this implementation using the hardware shown in Fig. 8.16 and the logic-level I/O procedures defined in Section 5.6.

Example 8.8 Velocity profile   Say we want to implement Fig. 8.15(b), given the following information:

> Total rotation desired = 1000 steps
> Desired velocity = 51 steps/second
> Clock frequency = 10,000 Hz
> P and V registers = 16-bit signed fractions

The desired value in the V register is

$$v = \frac{51 \text{ steps}}{s} \times \frac{1 \text{ s}}{10,000 \text{ adds}} = 0.0051 \text{ steps/add}$$

Converting this number to a signed 16-bit binary fraction (15 bits plus sign) yields

$$V = 167 \times 2^{-15} = 0.000000010100111_2 = 0.0050964355 \text{ steps/add}$$

Note that

$$V \neq v$$

This error is due to the need to represent V with a fixed number of bits.

---

[1] The unit "add" will be used in calculations throughout this section. It represents a unit of time due to the fact that an addition occurs each time there is a 0 to 1 transition in the clock.

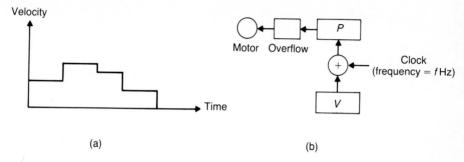

Figure 8.15 Generation of constant-velocity profiles. (a) Constant-velocity profiles; (b) method for generation of constant-velocity profiles.

With this value of V, the desired number of clock periods or additions required to output 1000 steps is

$$n = \frac{1000 \text{ steps}}{\dfrac{167 \times 2^{-15} \text{ steps}}{1 \text{ add}}} = 196215.6 \text{ adds}$$

Here it is necessary to round any fractional number of clock periods up to the next highest integer. Hence, the number of 0 to 1 clock transitions (additions) that must actually be used is

$$N = 196216 \text{ adds } \square$$

Example 8.9 Effect of word size   To investigate the effect of decreasing the number of bits used in P and V, consider Example 8.8 implemented with 8-bit signed fractions (8-bit words) rather than 16-bit signed fractions (16-bit words). The desired value of V is again

$$v = \frac{51 \text{ steps}}{1 \text{ s}} \times \frac{1 \text{ s}}{10000 \text{ adds}} = 0.0051 \text{ steps/add}$$

but in this case, when this value is converted to an 8-bit signed fraction, the result is

$$V = 1 \times 2^{-7} = 0.0000001_2 \text{ steps/add}$$

This is the smallest nonzero velocity that can be commanded. The actual velocity generated is

$$V = 1 \times 2^{-7} \text{ steps/add} = 78.125 \text{ steps/s}$$

```
program Constant_velocity;
  var
    P, V : real;                          {position and velocity registers}
    iP : integer;                         {integer portion of P:  +1,0 or -1}
    i, n,  : integer;                     {n is total number of adds}
    clock : boolean;                      {clock is logic input channel 1}
begin
  readln(V, n);                           {velocity is a fraction [steps/add]}
  DLC(false, 1);                          {initialize step output to 0}
  P := 0.0;                               {initialize position fraction to zero}
  for i := 1 to n do                      {add n times}
    begin
      repeat                              {wait for clock=0}
        LDC(clock, 1)
      until not clock;
      repeat                              {wait for 0 to 1 clock transition}
        LDC(clock, 1)
      until clock;
      P := P + V;                         {add velocity to position}
      iP := trunc(P);                     {check for -1 < P < 1}
      case iP of
        1 :                               {step in the positive direction}
          begin
            DLC(true, 2);                 {direction is logic output 2}
            DLC(true, 1);                 {step is logic output 1}
            DLC(false, 1);
            P := P - 1.0                  {reduce position to fraction}
          end;
        0 :                               {step not required}
          begin
          end;
        -1 :                              {step in the negative direction}
          begin
            DLC(false, 2);                {direction is logic output 2}
            DLC(true, 1);                 {step is logic output 1}
            DLC(false, 1);
            P := P + 1.0                  {reduce position to fraction}
          end
      end
    end
end.
```

Program 8.3 Constant-velocity profile (uses logic I/O procedures LDC and DLC).

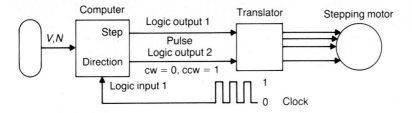

Figure 8.16 Stepping motor controller.

The velocity error in this case is very large because of the lack of velocity resolution that results from a very short word length.

It should be noted that the actual velocity generated depends on the desired velocity chosen, the clock frequency, and the velocity resolution. In this case, the velocity resolution is

$$V_{min} = 0.0000001_2 \text{ steps/add} = 78.125 \text{ steps/s}$$

Any integer multiple of V_{min} can be generated exactly as long as

$$-1 \le V < 1 \text{ step/add}$$

Decreasing the clock frequency adversely affects the maximum velocity that can be generated. For this example, the maximum positive stepping rate is

$$V_{max} = 0.1111111_2 \text{ steps/add} = 9922 \text{ steps/s}$$

whereas for a clock frequency of 1000 Hz,

$$V_{max} = 992.2 \text{ steps/s} \quad \square$$

Constant-Acceleration Profile

Acceleration often must be limited when stepping motors are used so that torque limits are not exceeded. For example, it may be necessary to accelerate the motor up to speed and later decelerate to a halt, as shown in the velocity profile in Fig. 8.17(a). In the figure, the desired velocity increases uniformly from zero, is held constant, and then decreases uniformly back to zero. The means for achieving such a profile is shown in Fig. 8.17(b). For every 0 to 1 transition in the clock, A is added to V and V is added to P. V starts at an initial value of zero and increases to a final value at t_1 that depends on the value in A. The rate at which P overflows increases at a uniform rate. At the end of the acceleration segment, the acceleration register A is set to zero. At the end of the constant-velocity segment,

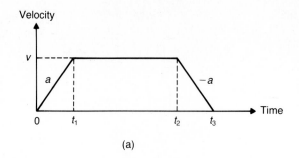

(a)

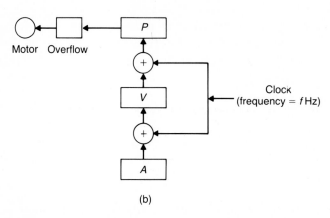

(b)

Figure 8.17 Generation of constant-acceleration profiles.
(a) Constant-acceleration profiles; (b) method for
generation of constant-acceleration profiles.

t_2, A is made negative and additions are continued until the velocity is again zero at t_3. The acceleration a, velocity v, and total number of steps to be traveled can be specified for the profile.

Example 8.10 Acceleration profile   A stepping motor to be controlled has a position resolution of 200 steps per revolution. The inertia of the motor and load is 0.05 in.-lb$_f$-s^2/rad. The maximum torque rating of the motor is 20 in.-lb$_f$. The motor and load are to be accelerated for 0.1 s to a speed of 100 rpm, operated at 100 rpm for 2 s, and then decelerated back to a stop in 0.1 s. The rate of deceleration is therefore the same as the initial acceleration rate.

The clock frequency is to be 1000 Hz. The registers are all 16-bit signed fractions, and it is desired to change only register A to generate the profile. It is therefore necessary to determine the values to be placed in register A and the number of 0 to 1 clock transitions that are to occur during each segment of the profile.

At the beginning of the first segment, the displacement register P and the velocity register V are set to zero. The desired value to be placed in the acceleration register A to accelerate the motor to 100 rpm in 0.1 s is then

$$a_1 = \frac{100 \text{ rpm}}{0.1 \text{ s}} \times \frac{200 \text{ steps}}{1 \text{ rev}} \times \frac{1 \text{ min}}{60 \text{ s}} \times \left(\frac{1 \text{ s}}{1000 \text{ adds}}\right)^2$$

$$= 3.3333 \times 10^{-3} \text{ steps/add}^2$$

From the given motor and load inertia and the maximum motor torque, the maximum acceleration is

$$a_{max} = \frac{T_{max}}{I} = \frac{20 \text{ in.-lb}_f}{0.05 \dfrac{\text{in.-lb}_f\text{-s}^2}{1 \text{ rad}}} = \frac{400 \text{ rad}}{1 \text{ s}^2}$$

or

$$a_{max} = 400 \frac{1 \text{ rad}}{\text{s}^2} \times \frac{1 \text{ rev}}{2\pi \text{ rad}} \times \frac{200 \text{ steps}}{1 \text{ rev}} \times \left(\frac{1 \text{ s}}{1000 \text{ adds}}\right)^2$$

$$= 12.732 \times 10^{-3} \text{ steps/add}^2$$

The acceleration desired is therefore within the rated capability of the motor. In the 16-bit signed fraction representation (15 bits plus sign), the value in register A should be

$$A_1 = 109 \times 2^{-15} \text{ steps/add}^2$$

The velocity desired after 0.1 s is

$$V = 100 \text{ rpm} \times \frac{200 \text{ steps}}{1 \text{ rev}} \times \frac{1 \text{ min}}{60 \text{ s}} \times \frac{1 \text{ s}}{1000 \text{ adds}} = 0.333 \text{ steps/add}$$

The number of clock periods or adds at which A is to be maintained to achieve this value is

$$n_1 = \frac{v}{A_1} = 100.0 \text{ adds}$$

and

$$N_1 = 100 \text{ adds}$$

The actual velocity reached after 100 adds is then

$$V_2 = N_1 A_1 = 10900 \times 2^{-15} \text{ steps/add} = 99.79 \text{ rpm}$$

At this point, register A is set to $A_2 = 0$, and the number of adds for the second segment is

$$N_2 = 2 \text{ s} \times \frac{1000 \text{ adds}}{1 \text{ s}} = 2000 \text{ adds}$$

For the third segment it is necessary to decelerate to a stop at the same rate for the first segment. Hence,

$$A_3 = -A_1 = -109 \times 2^{-15} \text{ steps/add}^2$$

and

$$N_3 = N_1 = 100 \text{ adds}$$

To summarize, the value of A for each segment and the number of adds for each segment are as follows:

$$A_1 = 109 \times 2^{-15} \qquad A_2 = 0 \qquad A_3 = -109 \times 2^{-15}$$

$$N_1 = 100 \qquad\qquad N_2 = 2000 \qquad N_3 = 100 \qquad\qquad \square$$

Example 8.11 Profile specification   In Example 8.10, a velocity profile was specified in terms of time. Suppose instead that it is desired to accelerate and decelerate at one-half the maximum rate of 400 rad/s^2 and that the total change in position is to be 5 revolutions.

Referring to the results of Example 8.10, we find here that

$$a_1 = \frac{1}{2} a_{max} = 6.3662 \times 10^{-3} \text{ steps/add}^2$$

and that, after rounding to a 15-bit fraction,

$$A_1 = 209 \times 2^{-15} \text{ steps/add}^2$$

The number of clock periods or adds required to reach 100 rpm is then

$$n_1 = \frac{100 \text{ rpm}}{\dfrac{209 \times 2^{-15} \text{ steps}}{1 \text{ add}^2}} \times \frac{200 \text{ steps}}{1 \text{ rev}} \times \frac{1 \text{ min}}{60 \text{ sec}} \times \frac{1 \text{ s}}{1000 \text{ adds}} = 52.3 \text{ adds}$$

Rounding off to the nearest integer yields

$$N_1 = 52 \text{ adds}$$

The velocity after 52 adds is

$$V_2 = N_1 A_1 = 10868 \times 2^{-15} \text{ steps/add}$$

The number of steps generated during the 52 adds is found from

$$p_2 = A_1 \sum_{i=1}^{N_1} i = (209 \times 2^{-15}) \sum_{i=1}^{52} i = 8.8 \text{ steps}$$

Thus 16 steps are generated during the first and third segments of the profile, leaving 984 steps to be generated during the second segment to make up the total of 5 revolutions. The number of clock periods or adds required in this segment is then

$$n_2 = \frac{984 \text{ steps}}{V_2} = 2966.8 \text{ adds}$$

Because it is important to generate all 984 steps, this value should be rounded up to the next higher integer.

$$N_2 = 2967 \text{ adds}$$

To summarize, the initial values of P and V are zero. The value of A at the beginning of each segment and the number of adds for each segment are as follows:

$$A_1 = 209 \times 2^{-15} \qquad A_2 = 0 \qquad A_3 = -209 \times 2^{-15}$$

$$N_1 = 52 \qquad\qquad N_2 = 2967 \qquad N_3 = 52 \qquad\qquad \square$$

Generation of Straight-Line Motion

The DDA method can be extended to two or more axes to achieve continuous path control. For generating straight-line motion in the x-y plane, the slope of the line gives the desired velocity ratio between the two axes. The two axes can be operated at constant velocity or constant acceleration, but as long as the velocity ratio is maintained, the desired straight line will be generated. For the case shown in Fig. 8.18 when both axes are moving at constant velocity, the method can be implemented as shown in Fig. 8.19. The equations of motion for a vector velocity V are

$$\Delta x = L \cos \theta \tag{8.78}$$

$$\Delta y = L \sin \theta \tag{8.79}$$

$$V_x = V \cos \theta \tag{8.80}$$

$$V_y = V \sin \theta \tag{8.81}$$

Figure 8.18 Straight-line motion in the x-y plane.

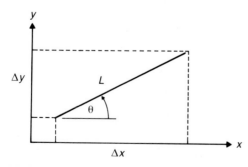

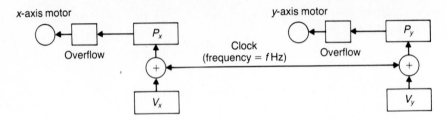

x-axis motor   y-axis motor

Overflow   Clock
(frequency = *f* Hz)   Overflow

Figure 8.19 Method for generating constant-velocity straight-line motion in the *x-y* plane.

Example 8.12 *x-y* Motion   The *x* and *y* stepping drives on an *x-y* positioning table both have a position resolution of 100 steps/cm. It is desired to move a distance of 50 cm at 100 cm/min at a 30° angle. 16-bit signed fractions are used with a clock frequency of 1000 Hz.

The number of clock periods or additions required is

$$N = 50 \text{ cm} \times \frac{1 \text{ min}}{100 \text{ cm}} \times \frac{60 \text{ s}}{1 \text{ min}} \times \frac{1000 \text{ adds}}{1 \text{ s}} = 30{,}000 \text{ adds}$$

The *x* and *y* velocities must then be

$$V_x = \frac{50 \cos(30°) \text{ cm}}{30{,}000 \text{ adds}} \times \frac{100 \text{ steps}}{1 \text{ cm}} = 4730 \times 2^{-15} \text{ steps/add}$$

$$V_y = \frac{50 \sin(30°) \text{ cm}}{30{,}000 \text{ adds}} \times \frac{100 \text{ steps}}{1 \text{ cm}} = 2731 \times 2^{-15} \text{ steps/add} \quad \square$$

Figure 8.20 Generation of a circular contour.

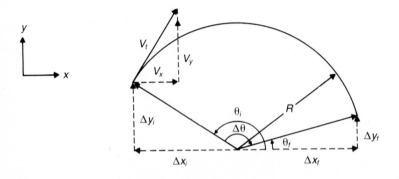

Generation of Circular Motion

For the more complicated case of generating circular motion in the x-y plane, the velocity of each axis is proportional to the position along the other axis, as shown in Fig. 8.20. The equations of motion for this case are

$$\Delta x = R \cos \theta \tag{8.82}$$

$$\Delta y = R \sin \theta \tag{8.83}$$

$$V_x = -\frac{d\theta}{dt} \Delta y \tag{8.84}$$

$$V_y = \frac{d\theta}{dt} \Delta x \tag{8.85}$$

The method for generating circular motion is shown in Fig. 8.21. The value of the velocity registers must be changed at every overflow to conform to the changing velocity defined in Eqs. (8.84) and (8.85). This can be accomplished by multiplying each step generated by the rotational velocity $d\theta/dt$ and adding it to the y velocity register in the case of a step in x or subtracting it from the x velocity register in the case of a step in y. Because each step has a value of ± 1, the multiplication is not necessary and $d\theta/dt$ can be added or subtracted directly to or from the appropriate velocity register. Program 8.4 illustrates how the technique can be implemented using hardware such as that shown in Fig. 8.22.

Example 8.13 Circular motion Suppose a $45°$ arc is to be generated at a speed of 100 cm/min with a radius of 5 cm. As in Example 8.12, the position resolution is 100 steps/cm and 16-bit signed fractions are to be used with a clock rate of 1000 Hz. The initial angle is $0°$.

Figure 8.21 Method for generating circular motion in the x-y plane.

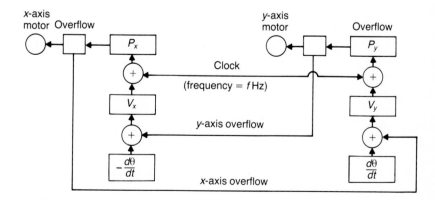

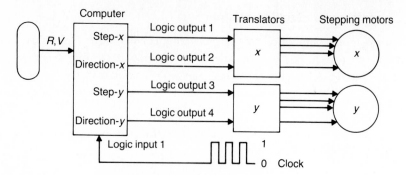

Figure 8.22 Controller for two stepping motors.

Program 8.4 Circular-motion generator (uses logic I/O procedures LDC and DLC).

```
program Circular_motion;
  var
    Px, Py, Vx, Vy : real;              {position and velocity registers}
    iPx, iPy : integer;                 {integer position register portion}
    r, v : real;                        {radius [steps], velocity [steps/add]}
    i, n : integer;
    clock : boolean;                    {clock is logic input channel 1}
begin
  readln(r, v);                         {input radius and velocity}
  DLC(false, 1);                        {initialize step outputs to 0}
  DLC(false, 3);
  Px := 0.0;                            {initialize position and velocity}
  Py := 0.0;
  Vx := 0.0;                            {start at THETA=0}
  Vy := v;
  n := round(6.28318 / (v / r));        {adds for complete circle}
  for i := 1 to n do
    begin
      repeat                            {wait for clock=0}
        LDC(clock, 1)
      until not clock;
      repeat                            {wait for 0 to 1 clock transition}
        LDC(clock, 1)
      until clock;
      Px := Px + Vx;                    {add velocity to position}
      Py := Py + Vy;
      iPx := trunc(Px);                 {check for -1 < Px < 1}
      case iPx of
        1 :                             {step in the positive x direction}
          begin
            DLC(true, 2);               {x direction is logic output 2}
            DLC(true, 1);               {x step is logic output 1}
            DLC(false, 1);
            Px := Px - 1.0;             {reduce x position to fraction}
            Vy := Vy + v / r            {increment y velocity}
          end;
```

```
      0 :
       ;                                {step in x not required}
     -1 :                               {step in the negative x direction}
       begin
         DLC(false, 2);                 {x direction is logic output 2}
         DLC(true, 1);                  {x step is logic output 1}
         DLC(false, 1);
         Px := Px + 1.0;                {reduce x position to fraction}
         Vy := Vy - v / r               {decrement y velocity}
       end
     end;
     iPy := trunc(Py);                  {check for -1 < Py < 1}
     case iPy of
      1 :                               {step in the positive y direction}
       begin
         DLC(true, 4);                  {y direction is logic output 4}
         DLC(true, 3);                  {y step is logic output 2}
         DLC(false, 3);
         Py := Py - 1.0;                {reduce y position to fraction}
         Vx := Vx - v / r               {decrement x velocity}
       end;
      0 :
       ;                                {step in y not required}
     -1 :                               {step in the negative y direction}
       begin
         DLC(false, 4);                 {y direction is logic output 4}
         DLC(true, 3);                  {y step is logic output 2}
         DLC(false, 3);
         Py := Py + 1.0;                {reduce y position to fraction}
         Vx := Vx + v / r               {increment x velocity}
       end
     end
   end
 end.
```

It is first necessary to calculate the initial values of all the registers. Here the angular velocity is

$$\frac{d\theta}{dt} = \frac{V}{R} = \frac{\dfrac{100 \text{ cm}}{1 \text{ min}}}{5 \text{ cm}} + \frac{1 \text{ min}}{60 \text{ sec}} \times \frac{1 \text{ s}}{1000 \text{ adds}} = 11 \times 2^{-15} \text{ rad/add}$$

Also,

$$\Delta x_i = 5 \text{ cm} \times \frac{100 \text{ steps}}{1 \text{ cm}} = 500.0 \text{ steps}$$

$$\Delta y_i = 0.0 \text{ steps}$$

The initial values of the registers are then

$$P_{x_i} = 0 \text{ steps}$$

$$V_{x_i} = -\frac{d\theta}{dt}\,\Delta y_i = 0 \text{ steps/add}$$

$$P_{y_i} = 0 \text{ steps}$$

$$V_{y_i} = \frac{d\theta}{dt}\,\Delta x_i = \frac{11 \times 2^{-15} \text{ rad}}{1 \text{ add}} \times 500 \text{ steps} = 5500 \times 2^{-15} \text{ steps/add}$$

Note that P_{x_i} is zero because this register holds only residual fractions of a step.

Finally, it is necessary to calculate the number of 0 to 1 clock transitions (additions) required to generate the 45° arc. This is found from the angular velocity as follows:

$$N = \frac{\theta_f - \theta_i}{d\theta/dt} = \frac{45°}{\dfrac{11 \times 2^{-15} \text{ rad}}{1 \text{ add}}} \times \frac{2\pi \text{ rad}}{360°} = 2340 \text{ adds} \quad \square$$

When the DDA method is used for generating circular motion, inaccuracies occur that are primarily due to truncation errors and to the implied rectangular integration approximation used. These errors can be held within limits by using higher clock frequencies, using larger registers, and re-initializing periodically to reduce error buildup. An alternative technique is to use a chordal approximation with the end points of straight-line segments directly computed from sine and cosine functions. It is also possible to use the sine- and cosine-generating functions given in Table 3.1 to generate successive axis positions between which straight-line motion can be generated. These techniques may give better results, but in all cases, careful consideration must be given to the precision with which computations are carried out and to the rate at which errors build up.

8.8

SUMMARY

In this chapter a number of techniques have been presented for command generation for both closed-loop and open-loop computer control. Although control of machine motion has been emphasized, the techniques presented are applicable to generating commands for control of a wide variety of processes and actuators. Emphasis has also been placed on control of the commanded rates of change of process variables with process limitations in mind. When this precaution is taken, the process is not commanded to operate in limiting modes and the resulting response is smooth, predictable, and governed by the generated commands rather than process limits and process dynamics. Ramps and paraboloids are used rather than step inputs, for example, because most processes cannot follow a step without dynamic errors.

Linear and spline interpolation techniques have been introduced, along with a number of methods for point-to-point motion command generation with controlled velocity and acceleration. The DDA method for the control of stepping motors has been described and illustrated for generating straight-line and circular motion. The method can be implemented in software or hardware. Microelectronic circuits that implement this technique for stepping motor control are commercially available and are easily interfaced to microprocessors.

BIBLIOGRAPHY

Barnhill, R. E., and R. F. Riesenfeld. *Computer Aided Geometric Design*. New York, NY: Academic Press, 1974.

Bollinger, J., and N. Duffie, Computer Algorithms for High Speed Continuous-Path Robot Manipulation," *Annals of the CIRP*, Vol. 28/1, 1979, pp. 391–395.

Craig, J. J. *Introduction to Robotics*. Reading, MA: Addison-Wesley, 1986.

Faux, I. D., and M. J. Pratt. *Computational Geometry for Design and Manufacturing*. New York, NY: Wiley, 1979.

Mortenson, M. *Geometric Modeling*. New York, NY: Wiley, 1985.

Foley, J. D., and A. Van Dam. *Fundamentals of Interactive Computer Graphics*. Reading, MA: Addison-Wesley, 1982.

Fu, K. S., R. C. Gonzalez, and C. S. G. Lee. *Robotics: Control, Sensing, Vision, and Intelligence*. New York, NY: McGraw-Hill, 1987.

Koren, Y. *Robotics for Engineers*. New York, NY: McGraw-Hill, 1985.

Kuo, B. C. (ed). *Theory and Applications of Stepping Motors*. St. Paul, MN: West, 1974.

Paul, R. *Robot Manipulators: Mathematics, Programming and Control*. Boston, MA: MIT Press, 1981.

Rogers, D. F., and J. A. Adams. *Mathematical Elements for Computer Graphics*. New York, NY: McGraw-Hill, 1976.

Taft, C. K., R. G. Gauthier, and T. J. Harned. *Stepping Motor System Design and Analysis, 9/e*. Durham, NH: University of New Hampshire, 1985.

PROBLEMS

8.1 Motion commands are to be generated for an x-y positioning machine that cause it to move from (2,4) to point (16,10) at constant velocity as shown in Fig. 8.23. Develop equations for $x(t)$ and $y(t)$ that linearly interpolate between these points.

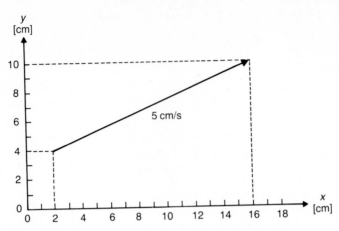

Figure 8.23 Straight-line, point-to-point motion for Problems 8.1 and 8.2.

8.2   Consider the motion shown in Fig. 8.23.
   a. Develop recursion equations for x_n and y_n that incorporate constant increments Δx and Δy.
   b. If the sample frequency of the computer control for the machine is 250 Hz, for how many sample periods should the equations in part (a) be used?
   c. If there is very high precision in the representation of x_n, y_n, Δn, and Δy in the control computer, what is the error in final position that results from using the equations in part (a) for the number of samples in part (b)?
   d. If the precision of representation of x_n, y_n, Δx, and Δy is μm, what is the error in final position? How does this compare with your result in part (c)?

8.3   Plot against time t, using the results of Example 8.3. Also plot $z(t)$ against $x(t)$ and $y(t)$.

8.4   What is the maximum commanded velocity (velocity vector magnitude) in Example 8.3?

8.5   It is desired to accelerate a motor up to a speed of 3000 in 50 ms, to operate it at that speed for 5 s, and then to decelerate it to a stop in 50 ms. The shaft position profile, $\theta(t)$, is to be represented by the following three cubic polynomials. Determine the spline coefficients.

$$0 \le t \le t_1: \qquad \theta(t) = a_3 t^3 + a_2 t^2 + a_1 t + a_0$$
$$t_1 \le t \le t_2: \qquad \theta(t) = b_3 t^3 + b_2 t^2 + b_1 t + b_0$$
$$t_2 \le t \le t_3: \qquad \theta(t) = c_3 t^3 + c_2 t^2 + c_1 t + c_0$$

8.6   It is desired to move the θ axis of a robot from an angle of 20° to an angle of 60°. The maximum velocity of rotation is to be 5°/s, and the robot is to be at rest at the beginning and end of the motion.
   a. Calculate the coefficients of a single cubic polynomial for $\theta(t)$ that can be used to represent this motion.
   b. What is the maximum acceleration, and when does it occur?

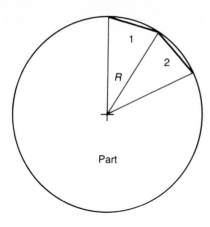

Figure 8.24 Chordal approximation
for cutting the circular part in
Problem 8.7.

8.7 A laser machining system is to cut a circular part using the chordal approximation shown in Fig. 8.24. The accuracy is to be 0.001 in., and the radius of the part, R, is to be no more than 5 in. If clockwise motion at 300 in./min is to start at the top of the circle, what are the straight-line interpolation equations for the first three chordal segments of the motion?

8.8 Repeat Problem 8.7, but consider the case shown in Fig. 8.25 where a hole is to be cut in a part and its radius, R, is to be no less than 5 in.

8.9 Write a computer program to use the coefficient matrices given in Example 8.3 to verify the profiles shown in Fig. 8.4.

Figure 8.25 Chordal approximation for cutting the hole
in Problem 8.8.

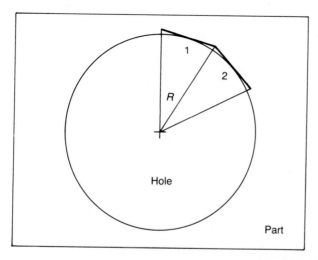

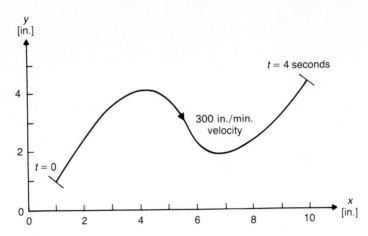

Figure 8.26 Constant-velocity motion segment for Problem 8.10.

8.10 Calculate the coefficient matrix A in Eq. (8.4) assuming constant velocity along the motion segment shown in Fig. 8.26. Compare the result to the plot in the figure. Determine the source of any differences.

8.11 Calculate the coefficients for the cubic polynomials in normalized time variables u_1, u_2, u_3, and u_4 for the four segments with end-point data listed in Table 8.2. Calculate Δu_1, Δu_2, Δu_3, and Δu_4 for a sample period of 0.005 s.

8.12 The three-dimensional motion profile shown in Fig. 8.27 illustrates the path to be followed by a robot welding a seam on an automobile part. Represent the seam using cubic spline functions.

8.13 An approximately sinusoidal motion profile is shown in Fig. 8.28, together with entry and exit segments that provide smooth transitions into and out of the profile. The motion is to be represented by four spline functions, two of which are repeated as often as desired while approximately sinusoidal motion is required. Calculate coefficients for the four spline functions that are required to represent these segments.

8.14 An r-θ-z robot moves a part from point 1 to point 2 as illustrated in Fig. 8.29. Synchronized, nearly straight-line motion is desired from point to point in 2.8 s.
    a. Use a multiple-axis, single-spline motion representation to generate the required motion.
    b. Plot the motion generated in the x-y, y-z, and z-x planes and identify deviations from straight-line motion in cartesian (x, y, z) space.

8.15 Derive the equations for single-axis, point-to-point motion using the three spline constant-acceleration and decreasing-acceleration methods. Compare the total times required for the motion $(\Delta t = \Delta t_1 + \Delta t_2 + \Delta t_3)$.

8.16 Use the multiple-axis, three-spline, constant-acceleration method to compute the spline functions that generate synchronized motion of the r-θ-z robot in Fig. 8.29

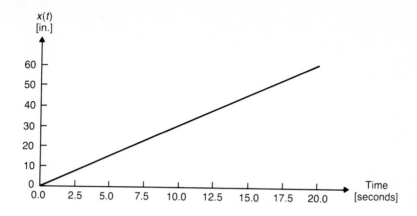

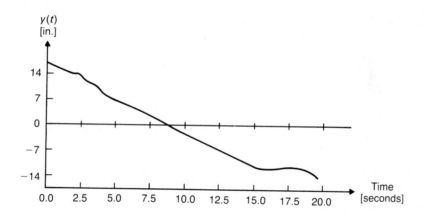

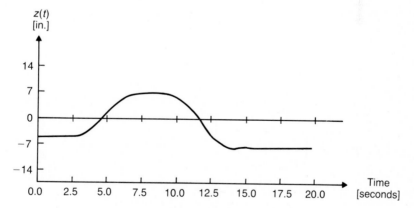

Figure 8.27 Motion profile for seam welding in Problem 8.12.

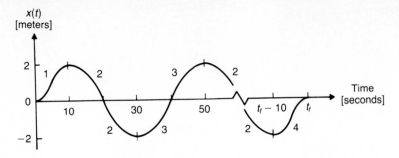

Figure 8.28 Approximately sinusoidal profile with entry and exit segments for Problem 8.13.

Figure 8.29 Robot and point-to-point motion for Problem 8.14.

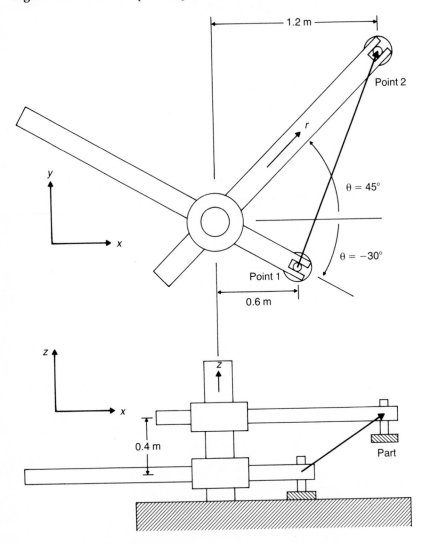

from point 1 to point 2. The maximum acceleration and velocity for each of the axes are as follows:

Axis	Maximum velocity	Maximum acceleration
r	10 m/min	15 cm/s^2
θ	2 rev/s	4 rev/s^2
z	6 m/min	20 cm/s^2

Plot the acceleration, velocity, and position generated for each axis.

8.17 Repeat Problem 8.16 using the multiple-axis, three-spline, decreasing-acceleration method for synchronized point-to-point motion. Plot the accelerations generated, and compare them and the time required to complete the motion to your results in problem 8.16.

8.18 Repeat Problem 8.16 using the multiple-axis, single-spline method for synchronized point-to-point motion. Plot the accelerations generated and compare them and the time required to complete the motion to your results in Problem 8.16.

8.19 Repeat Problem 8.16 using non-synchronized motion and the single-axis, three-spline, constant-acceleration method. Compare the cartesian (x, y, z) coordinates of the motion generated with your results in Problem 8.16.

8.20 To minimize vibration in a very compliant machine structure, fifth-order polynomials are to be used to generate motion with constant jerk (no step changes in acceleration). The acceleration profile is to be that shown in Fig. 8.30. Use normalized time variables to develop the polynomials.

8.21 Calculate and plot a spline function for moving a motor shaft one revolution in 0.1 s with zero velocity at the beginning and end of the motion.

8.22 Repeat Problem 8.21, but apply feedforward compensation to the spline coefficients for a closed-loop system consisting of an integrating process with a gain of 2.874 and a proportional controller with a gain of 5.8. Compare the plot with your result in Problem 8.21.

Figure 8.30 Acceleration profile for Problem 8.20.

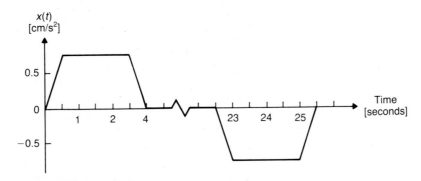

8.23 Add feedforward compensation to the coefficient matrix given in Example 8.3 for the second segment of the motion profile shown in Fig. 8.4. Assume that the x- and y-axis drives have been tuned so that they are approximately characterized as first-order systems with unity gain and a 3-Hz bandwidth. (*Hint*: The bandwidth in rad/s is $1/\tau$ for a first-order system.)

8.24 Write a simulation program and plot the x- and y-axis following error for the motion profile defined by the spline coefficient matrices in Example 8.3 for an overdamped, second-order, closed-loop system with unity gain and time constants of 0.1 s and 0.008 s. Ignore the shortest time constant and add command feedforward control to the spline coefficient matrices. Plot the new following error, and evaluate the effect of the addition of feedforward control.

8.25 Suppose that the motion profiles defined in Example 8.3 and Fig. 8.4 represent commands generated for an x-y positioning system. Each axis can be modeled as a second-order system with unity gain, a natural frequency of 200 rad/s, and a damping ratio of $\zeta = 0.7$. Develop and evaluate a method for combined velocity and acceleration command feedforward control.

8.26 A stepping motor with 720 steps/revolution is commanded using the DDA method illustrated in Fig. 8.15(b). Eight-bit registers are used for position and velocity. At time zero, the word 0.001111_2 is loaded into the velocity register. The position register is initially zero. At each rising edge of a 1000-Hz clock, the velocity register is added to the position register.
   a. What is the velocity of the motor in revolutions per minute?
   b. What is the velocity resolution of the system?
   c. What is the maximum positive velocity of the system? What is its maximum negative velocity?

8.27 A 10,000-Hz clock is used with the method shown in Fig. 18.15(b) to drive a 360-step/rev stepping motor at 100 rpm. The velocity resolution is to be approximately 1 rpm.
   a. Show the initial values of the registers in binary form.
   b. What is the velocity resolution?
   c. What is the maximum speed in revolutions per minute that can be achieved?

8.28 Consider the system shown in Fig. 8.31(a) and the desired velocity profile shown in Fig. 8.31(b).
   a. What should be in A between times 0 and t_1 to accelerate the motor at a rate of 500 rad/s^2? How many clock periods are required to reach the desired speed of 200 rpm?
   b. What should be in A between times t_1 and t_2? How many clock periods are required if the motor is to be run at 200 rpm for 20 s?
   c. What is the actual speed in rpm that results from part (a)?
   d. What should be in V at time t_1 to ensure that a velocity of 200 rpm is generated?
   e. What should be in A between times t_2 and t_3 to decelerate the motor to a stop at a rate of 500 rad/s^2? How many clock periods are required?
   f. How many revolutions does the motor turn?
   g. How many clock periods are required between times t_1 and t_2 to ensure that exactly 100 revolutions are made during the whole profile?

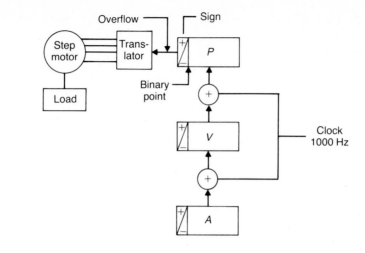

(a)

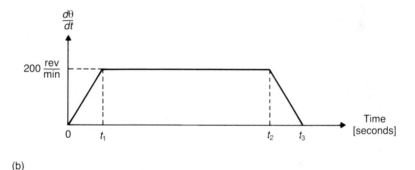

(b)

Figure 8.31 (a) Step-generating algorithm for Problem 8.28. *Note:* (1) P, V, and A are 7-bit fractions plus a sign bit (2's-complement); (2) the stepping motor has a resolution of 180 steps per revolution; (3) P and V are initially zero; (4) only register A is to be changed. (b) Profile to be generated in Problem 8.28.

8.29 A stepping motor with a position resolution of 200 steps/revolution is to follow the velocity profile shown in Fig. 8.32. Determine the series of values to be placed in the acceleration register of Fig. 8.17(b) and the number of clock periods for which each is to be used. A 10,000-Hz clock is to be used. Give your answers in base 10.

8.30 To prevent sudden pressure pulsations in a pipe through which flow is controlled by a valve operated by a stepping motor, the valve is to be opened and closed slowly. The motor has a resolution of 30 steps per revolution, it takes 5 revolutions to open and close the valve, and the valve is to be opened or closed in 2 s. The computer that sends stepping commands to the motor has access to a 240-Hz clock. Develop an algorithm that will open and close the valve at the rate desired.

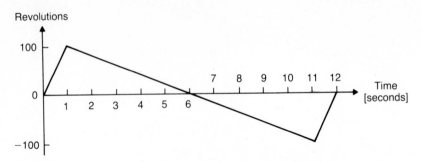

Figure 8.32 Velocity profile for stepping motor in Problem 8.29.

8.31 Develop a stepping motor control algorithm of the form shown in Fig. 8.19 to generate the motion shown in Fig. 8.23. Assume that the clock frequency to be used is 50 kHz.

8.32 Stepping motors are used on the x and y axes of a positioning table. The motors have resolutions of 400 steps/rev, torque ratings of 0.2 oz-in. and drive screws with a pitch of 10 rev/in. The inertia of the x axis is 0.01 oz-in.-s^2, and the inertia of the y axis is 0.006 oz-in.-s^2. Develop a control algorithm to drive the table with the synchronized x-y motion shown in Fig. 8.23.

8.33 Develop a stepping motor control algorithm that generates the motion shown in Fig. 8.29, given the following velocity and acceleration limits.

Axis	Maximum velocity	Maximum acceleration
R	10 m/min	0.5 cm/s^2
θ	2 rev/s	0.8 rev/s^2
Z	6 m/min	4 cm/s^2

8.34 Develop a stepping motor control algorithm of the form shown in Fig. 8.21 that generates a circle with a 5-in. radius in the x-y plane in 1 min.

Chapter 9

Sequential Control Using Programmable Logic Controllers

9.1

INTRODUCTION

Many control applications exist that require the switching (on or off) of various outputs as a function of the condition (on or off) of a number of input devices. This type of control is referred to as **switching control** or **logic control**. The relative simplicity of switching control makes it attractive for use in automatic machines and processes where the requirement is for the machine or process to follow a set sequence of operations. One example is a transfer line wherein each station performs a certain operation on a part, after which the processed part is transferred to the next station and replaced by an unprocessed part from the previous station. Another example is a process in which various dry bulk materials are weighed, combined, mixed, and ultimately discharged. The use of switching control to cause a machine or process to go through a set sequence of operations gives rise to the term **sequential control**.

Sequential control can be implemented in a number of ways, including electro-mechanical relays and various pneumatic, fluidic, and solid-state devices. This chapter, however, focuses on digital computer systems that are primarily dedicated to the implementation of switching control. These special-purpose computers are called **programmable logic controllers**.

The characteristic structure of programmable logic controllers is illustrated in Fig. 9.1. Provision is made to accept information from various input switching devices such as push-buttons, limit switches, and relay contacts. Also provided are terminals to which output devices such as solenoids, relay coils, and indicator lights can be connected. There are no direct-wired connections between the

369

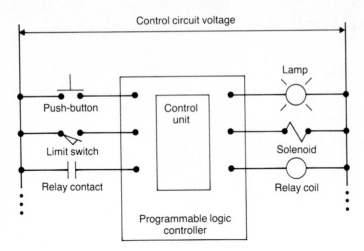

Figure 9.1 Characteristic structure of a programmable logic controller.

inputs and outputs. Instead, the switched conditions of the inputs are converted to logic-level signals that are input to a digital computer in the control unit. A program stored in the computer specifies which outputs should be energized on the basis of the present and past states of the input signals. Logic-level outputs from the control unit are converted to the voltage levels required to energize or de-energize the various output devices.

In light of the preceding discussion, it is clear that the role of the control unit is to continuously scan its program, setting outputs on or off depending on the switch conditions of the inputs. The controller must therefore be provided with a program that defines the desired switching sequences. This chapter considers approaches that can be utilized in programming programmable logic controllers. Emphasis is placed on the organizing control requirements in such a way as to obtain a set of Boolean logic equations defining the tasks to be performed. Although various commercial programmable logic controllers require alternative programming techniques such as ladder diagrams or English-like logic statements, these must eventually lead to a set of logical operations that can be performed by the computer in the control unit. Boolean logic is therefore fundamental both in programming programmable logic controllers and in understanding their operation.

In this chapter, fundamental concepts of Boolean algebra are discussed first, together with the implications of the serial nature of the digital computer for controller performance. This approach is then related to the traditional ladder diagram approach. Techniques for designing sequential logic are discussed next, along with examples of sequential system design. Finally, the problems of programming programmable logic controllers and monitoring their operation are examined.

9.2

BOOLEAN ALGEBRA

The mathematical basis for switching control is **Boolean algebra**. If switching requirements can be specified in the form of **Boolean equations**, then the programmable logic controller becomes a computer that "solves" Boolean equations. It is not always necessary to program the programmable logic controller directly in Boolean equations, but it is instructive to consider how a digital computer can be used to solve Boolean equations.

Variables in Boolean algebra can have only the value 0 or the value 1. Logical functions containing operations such as AND, OR, EXCLUSIVE OR, and NOT (already described in Section 5.3) are easily expressed using Boolean algebra, and equations can be constructed and evaluated that consist of any number of Boolean variables and operations.

Theorems of Boolean Algebra

A number of Boolean algebra theorems are given in Table 9.1. Of particular interest are the distributive laws, listed as entries 14 and 15 in the table, and DeMorgan's Laws, listed as entries 19 and 20, which can be useful in simplifying Boolean equations representing sequential logic. This is of lesser importance in programmable logic controllers than would be the case with "hard-wired" solid-state or relay logic, because the cost of additional instructions in a program is usually significantly less than the cost of additional solid-state or relay hardware.

Two noteworthy conventions are followed in Table 9.1. First, the AND function $A \cdot B$ is often written as simply AB. Hence

$$A \cdot B = AB \tag{9.1}$$

Table 9.1

Theorems of Boolean Algebra

1. $A + B = B + A$	11. $\overline{\overline{A}} = A$
2. $AB = BA$	12. $(A + B) + C = A + (B + C)$
3. $A + 0 = A$	13. $(AB)C = A(BC)$
4. $A \cdot 0 = 0$	14. $AB + AC = A(B + C)$
5. $A + 1 = 1$	15. $(A + B)(A + C) = A + BC$
6. $A \cdot 1 = A$	16. $A + AB = A$
7. $A + A = A$	17. $A + \overline{A}B = A + B$
8. $AA = A$	18. $A(A + B) = A$
9. $A + \overline{A} = 1$	19. $\overline{AB} = \overline{A} + \overline{B}$
10. $A\overline{A} = 0$	20. $\overline{A + B} = \overline{A} \cdot \overline{B}$

Second, the AND function takes precedence over the OR function in evaluating equations. Therefore

$$A + BC = A + (BC) \tag{9.2}$$

The theorems in Table 9.1 can be proved in several ways, as illustrated in the following examples.

Example 9.1 Proof and application of theorem 15   One approach to proving the theorems in Table 9.1 uses truth tables. We will illustrate it by proving Theorem 15,

$$(A + B)(A + C) = A + BC$$

Table 9.2(a) shows the value of the left side of this equation for all possible values of A, B, and C. Table 9.2(b) shows the value of the right side of the equation for the same values of A, B, and C. Because the results are the same for all possible values of A, B, and C, we can conclude that the theorem is correct.

      To illustrate the usefulness of this theorem, consider the solid-state logic circuits shown in Fig. 9.2, which determine the status of a space vehicle that can be controlled by either an astronaut (A) or one of two on-board computers ($C1$ and $C2$). Assume that computer $C1$ is also used for communications and that computer $C2$ is also used for logging data. The vehicle status is determined as follows:

$$
\begin{array}{lll}
\text{(Excellent)} & E = A + C1 \cdot C2 \\
\text{(Good)} & G = A + C1 \\
\text{(Fair)} & F = A + C2
\end{array}
$$

Table 9.2

Truth Tables Proving $(A + B)(A + C) = A + BC$

A	B	C	A + B	A + C	(A + B)(A + C)	A	B	C	BC	A + BC
0	0	0	0	0	0	0	0	0	0	0
0	0	1	0	1	0	0	0	1	0	0
0	1	0	1	0	0	0	1	0	0	0
0	1	1	1	1	1	0	1	1	1	1
1	0	0	1	1	1	1	0	0	0	1
1	0	1	1	1	1	1	0	1	0	1
1	1	0	1	1	1	1	1	0	0	1
1	1	1	1	1	1	1	1	1	1	1

                                (a)   (b)

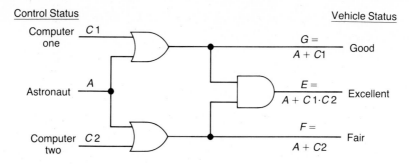

Figure 9.2 Logic for determining the status of a space vehicle.

where the values (0 or 1) of A, $C1$, and $C2$ represent whether or not the astronaut and computers are functional. The circuit in Fig. 9.2 takes advantage of Theorem 15 by using the alternative logic:

$$E = (A + C1)(A + C2) = G \cdot F$$

which minimizes the number of solid-state logic gates in the circuit. ◻

Example 9.2 Algebraic manipulation and simplification A second approach to proving the theorems listed in Table 9.1 involves algebraic manipulation and simplification. With Theorem 16, for example, Theorems 6, 14, 5, and 6 can be applied, in that order, to construct the following proof:

$$
\begin{aligned}
A + AB &= A \cdot 1 + AB && \text{(Theorem 6)} \\
&= A(1 + B) && \text{(Theorem 14)} \\
&= A \cdot 1 && \text{(Theorem 5)} \\
&= A && \text{(Theorem 6)}
\end{aligned}
$$

Therefore Theorem 16 proved, and

$$A + AB = A \quad ◻$$

Boolean Equations for Switching Logic

If the switching-logic functions to be performed by a programmable logic controller can be expressed in the form of Boolean equations, the digital computer in the control unit of the programmable logic controller can be used to solve these equations in much the same way as a general-purpose digital computer is used to solve algebraic equations. That is, the values of the unknown variables in the equations can be determined from the values of the known variables.

Consider an example wherein PB1 and PB2 are variables whose values are determined by the state (0 = OFF, 1 = ON) of two push buttons that are interfaced to the programmable logic controller. LAMP is a variable whose value will be used to determine the state of an indicator lamp that is interfaced to the programmable logic controller. If the relationship between these variables is determined by the equation

$$LAMP = PB1 + PB2 \tag{9.3}$$

then the value of the unknown variable LAMP at any instant is exactly determined by the values of the known variables PB1 and PB2 according to the OR function. Eq. (9.3) establishes a *continuous* relationship between variables PB1, PB2, and LAMP.

Next, consider a slightly more complex equation:

$$LAMP = (LAMP + PB1) \cdot \overline{PB2} \tag{9.4}$$

Again, PB1 and PB2 are known variables associated with push-button inputs to the programmable logic controller. The state of the output LAMP as determined by Eq. (9.4) is shown in Table 9.3. The first entry is the most interesting because it indicates that when both inputs are 0 (OFF), LAMP remains at its previous state, which may be 0 or 1. It is only when an input is activated that the state of LAMP can be changed. The effect of Eq. (9.4) is to allow a momentary activation of push-button input PB1 to energize the output LAMP and a momentary activation of PB2 to de-energize LAMP. The variable LAMP therefore serves as a memory of which push-button was activated last. When both are activated, the output is de-energized.

As is the case with algebraic equations, solving for multiple Boolean variables requires multiple Boolean equations. For example, consider the following set of equations, which describe three indicator lamp outputs.

$$LAMP1 = \overline{PB1} \cdot \overline{PB2} \tag{9.5}$$

$$LAMP2 = PB1 \cdot PB2 \tag{9.6}$$

$$LAMP3 = \overline{LAMP1 + LAMP2} \tag{9.7}$$

Table 9.3

Output LAMP Defined by Eq. (9.4)

PB1	PB2	LAMP
0	0	LAMP
0	1	0
1	0	1
1	1	0

Table 9.4

Outputs Defined by Eqs. (9.5), (9.6), and (9.7)

PB1	PB2	LAMP1	LAMP2	LAMP3
0	0	1	0	0
0	1	0	0	1
1	0	0	0	1
1	1	0	1	0

Table 9.4 shows the state of the three variables LAMP1, LAMP2, and LAMP3 as a function of the possible values of PB1 and PB2. It can be seen from Table 9.4 that LAMP1 is 1 when neither button is pushed, LAMP3 is 1 when one button is pushed, and LAMP2 is 1 when both buttons are pushed. Even though the value variables LAMP1 and LAMP2 are used to energize indicator lamps interfaced as outputs to the programmable logic controller, there is no reason why these variables cannot be used in further equations such as Eq. (9.7).

9.3

DIGITAL SOLUTION OF BOOLEAN EQUATIONS

The digital computer in the control unit of a programmable logic controller is capable of performing the logic operations necessary to solve a set of Boolean equations. A single bit is all that is required in the computer memory to store the value of each variable in the equations. The state of a particular variable can be determined at any time by testing the value of its memory bit. If a variable is associated with an input to the programmable controller, the state of its memory bit is forced to the state of that input. If a variable is associated with an output from the programmable logic controller, its memory bit is used to determine the state of that output. Variables not associated with inputs or output can be created at will by the programmer as a matter of programming necessity or convenience.

Boolean Assignment Statements

If the set of Boolean equations defining the logic functions to be performed by a programmable logic controller can be transformed into a set of **Boolean assignment statements**, then a solution to the Boolean equations can be obtained. However, the *discrete* nature of this method has serious implications that will be discussed in the next section.

If the Boolean equations to be solved are restricted to having a single Boolean variable on the left side of each equation, we obtain the equivalent assignment statement by replacing " = " by " ← " in each equation. For example, the equivalent assignment statements for the system with two input pushbuttons and three lamp outputs defined by Eqs. (9.5), (9.6), and (9.7) are

$$\text{LAMP1} \leftarrow \overline{\text{PB1}} \cdot \overline{\text{PB2}} \tag{9.8}$$

$$\text{LAMP2} \leftarrow \text{PB1} \cdot \text{PB2} \tag{9.9}$$

$$\text{LAMP3} \leftarrow \overline{\text{LAMP1} + \text{LAMP2}} \tag{9.10}$$

These statements are interpreted as follows:

In Eq. (9.8), the function $\overline{\text{PB1}} \cdot \overline{\text{PB2}}$ is evaluated and the result (0 or 1) is assigned to LAMP1. In other words, the value of the variable on the left side of the statement is set to the value of the expression on the right side of the statement. Equation (9.9) is evaluated in a similar manner, and LAMP2 is set accordingly. The results of Eqs. (9.8) and (9.9) are then used in Eq. (9.10) to determine the value of LAMP3.

The values of PB1 and PB2 are predetermined because they are input variables. It is obvious that an input variable should not appear on the left side of an assignment statement; its value is already determined externally and is not the result of logic evaluation. The values of output variables that are determined by the evaluation of assignment statements can be directly transferred to programmable logic controller's outputs.

The equivalent assignment statement to Eq. (9.4) is

$$\text{LAMP} \leftarrow (\text{LAMP} + \text{PB1}) \cdot \overline{\text{PB2}} \tag{9.11}$$

Because the variable LAMP appears on the right side of the statement, its *previous value* is used in the evaluation of the right side to determine the value of the left side. The result of this evaluation is then the *new value* of the LAMP. The statement in Eq. (9.11) can be programmed directly in a high-level language, or it can be programmed in assembly language, perhaps as follows:

```
START     LOAD         #0
LOOP      STORE        LAMP
          OUTPUT       LAMPOUT
          INPUT        PB1IN
          OR           LAMP
          STORE        LAMP
          INPUT        PB2IN
          COMPLEMENT
          AND          LAMP
          JUMP         LOOP
```

Clearly, a single evaluation of an assignment statement such as Eq. (9.11) is not sufficient to monitor the state of PB1 and PB2 and continually determine the value of LAMP. Rather, repeated evaluation of Eq. (9.11) is required to produce a continuing solution of Eq. (9.4). This feature is present in the foregoing program. Some of the effects of this repetitive solution will be considered.

Scan Time

The foregoing program illustrates the step-by-step nature of the digital computer in making logical computations and performing operations such as testing the state of the switches and turning the lamp on and off. It is necessary to pass repeatedly through the program in order to implement the desired logic function. One pass through the program can be thought of as one operating cycle of the programmable logic controller. During each cycle, all inputs to the programmable controller are sampled, logic statements are evaluated, and outputs are set accordingly. Such a cycle is often referred to as a **scan**, and the time required for one cycle is called the **scan time**. The scan time is an indication of how often each logic function is evaluated.

The step-by-step nature of the digital computer—and hence of the programmable logic controller—is fundamentally different from the operation of other sequential logic hardware. Instead of being continuously solved (as would be the case with solid-state or relay logic), a given logic function is evaluated only once per scan. To appreciate this difference more clearly, consider the following:

$$A = \bar{A} \tag{9.12}$$

$$A \leftarrow \bar{A} \tag{9.13}$$

Equation 9.12 has no solution because there is no value of A that will satisfy it. The assignment statement in Eq. (9.13), however, is perfectly valid.

The following instructions can be inserted into a larger program that implements the logic for control of a given system:

```
LOOP
          ⋮
     LOAD          A
     COMPLEMENT
     STORE         A
     OUTPUT        AOUT
          ⋮
     JUMP          LOOP
```

Given an initial value of A before the first scan, A will have the opposite value after the first scan. After each successive scan, A will have a value opposite the value it had on the preceding scan. The output generated, shown in Fig. 9.3, has a

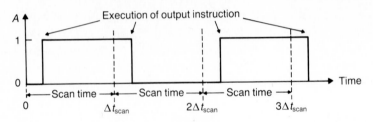

Figure 9.3 Plot of output A generated by $A \leftarrow \bar{A}$.

period of twice the scan time, and can be used for measurement of the scan time. It also serves to highlight the non-continuous nature of the solution of logic functions by a programmable logic controller.

Of course, the great operating speed of the digital computer means that for many applications, the scan time is so short that the solution of logic functions by the programmable logic controller can be assumed to be nearly instantaneous. It is only when a large number of logic functions are to be performed, events of very short duration must be detected, or very fast response times are required that it becomes important to consider the scan time.

If the duration of the shortest event that must be detected by a programmable logic controller is Δt and the scan time for the programmed logic is Δt_{scan}, then there is a requirement that

$$\Delta t_{\text{scan}} < \Delta t \tag{9.14}$$

If the scan time is longer than an event that must be detected, it is possible that the event will occur between the evaluation, in successive scans, of the logic functions designed to detect that event. In this case the event will not be detected. This situation, which is illustrated in Fig. 9.4, is potentially serious if detection of the event is crucial to proper system operation.

The scan time for a given logic program can be estimated by counting the number of operations that must be performed by the programmable logic controller. If Δt_{op} is the approximate time required by the control unit to perform an operation (AND, OR, NOT, assignment, and so on) and there are N operations in the program, the scan time for the program can be approximated as

$$\Delta t_{\text{scan}} \approx N \Delta t_{\text{op}} \tag{9.15}$$

Equations (9.14) and (9.15) define a limit of

$$N_{\text{max}} < \frac{\Delta t}{\Delta t_{\text{op}}} \tag{9.16}$$

where N_{max} is the maximum number of operations that can be programmed.

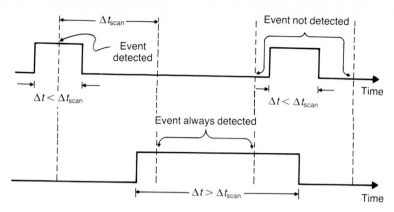

Figure 9.4 Detection of events shorter than and longer than the scan time.

This simple analysis demonstrates that the complexity of sequential logic that can be implemented on a programmable logic controller is limited by both the speed of the digital computer in the control unit, which determines Δt_{op}, and the requirements of the control application, which determines Δt. For many applications, Δt is large and hence is not a critical factor.

Example 9.3 Event duration and scan time As an example of an application where scan time can be a factor, consider a switch actuated by a rotating cam as shown in Fig. 9.5. If θ is the dwell angle of the cam in degrees and ω is the rotational speed of the cam in revolutions per minute, then the time during which the switch is activated by the cam is

$$\Delta t = \frac{60\theta}{360\omega} \text{ s}$$

Figure 9.5 Switch activated by cam rotating at speed ω with dwell angle θ.

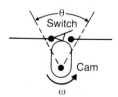

If the dwell angle is $30°$ and the cam is rotating at 2000 rpm, the length of this event is

$$\Delta t = \frac{30}{6(2000)} = 0.0025 \text{ s}$$

The scan time of a programmable logic controller required to detect this event at some point in its programmed switching logic must therefore be shorter than 2.5 ms. If the time required to perform a typical logic operation is 10 μs for this controller, then, from Eq. (9.16), the limit on the number of operations that can be performed in one scan is

$$N_{max} < \frac{0.0025}{10 \times 10^{-6}}$$

or

$$N_{max} < 250$$

This is a significant limitation. If, for example, an average of 5 operations per Boolean logic statement were to be assumed, then only about 50 such statements could be included in the program. Should a scan time of greater than 2.5 ms be obtained, there is a danger that an activation of the switch will be missed. If one of the requirements of sequential control is to detect *every* actuation of the switch, the control system fails to meet this requirement. There are several possibilities for redesign:

1. Optimizing the logic to reduce the number of operations and thereby reduce the scan time.

2. Increasing the length of time during which the switch is open.

3. Including logic to detect switch activation in more than one place in the program.

4. Using a different programmable logic controller with a faster rate of logic processing to reduce scan time.

5. Employing solid-state or other sequential logic hardware for this application rather than using a programmable logic controller.

It should be noted that many programmable logic controllers operate with a preset scan time. In such a system, the scan time is not directly dependent on the number of logic operations performed and is controlled by a clock instead. After each scan, the controller waits until the clock indicates that it is time to do the next scan. It is important to ensure that the scan time is long enough so that all the programmed logic operations can be performed in each scan. Hence the timing considerations for this type of system are similar to those for systems with a scan time that is fixed by program length, as described above. □

9.4

LADDER DIAGRAMS

Sequential control systems traditionally have been designed using **ladder diagrams**. This is a result of the extensive use of electro-mechanical relays in these systems in the past and the need to develop a standardized technique for specifying relay logic circuits. In fact, many commercial programmable logic controllers have facilities for programming with ladder diagrams. Ladder diagrams entered by the programmer are translated into equivalent Boolean operations by the programmable logic controller so that the required logic functions can be performed.

Components of Ladder Diagrams

The basic components described schematically in ladder diagrams are control relay coils, relay contacts, inputs such as push-buttons and limit switches, and actuators such as solenoids and indicator lights. Figure 9.6(a) shows the symbols used in this section for various components of ladder diagrams.

Control Relays

Figure 9.6(b) is a schematic drawing of a **control relay**. The relay coil consists of a moving core and an electromagnet. When the relay coil is de-energized, the spring pulls the core downward; the core moves upward when the coil is energized. The second part of the relay contains switching contacts that open and close as a function of core displacement. The **normally open contacts** shown in Fig. 9.6(b) complete a circuit only when the relay coil is energized. The term *normally* refers to the state of the contact when the coil is de-energized. The **normally closed contacts** complete a circuit only when the relay coil is de-energized. In general, control relays can be purchased with as many contacts as the application demands. The symbols for a control relay coil and for the normally open and normally closed relay contacts are shown in Fig. 9.6(a).

Solenoids

Solenoids are electromagnetic actuators. They are commonly used for actuating pneumatic and hydraulic valves that control fluid flow into cylinders. The symbol for a solenoid is shown in Fig. 9.6(a). Figure 9.7(a) shows a two-position, solenoid-actuated valve used to control a spring-returned pneumatic cylinder. The direction of air flow through the valve is determined by the state of the solenoid on the valve. When the solenoid is de-energized, the valve position adjacent to the spring is in effect. Air is forced out of the cylinder and exhausted to the atmosphere as the piston is forced to the left by the piston return spring.

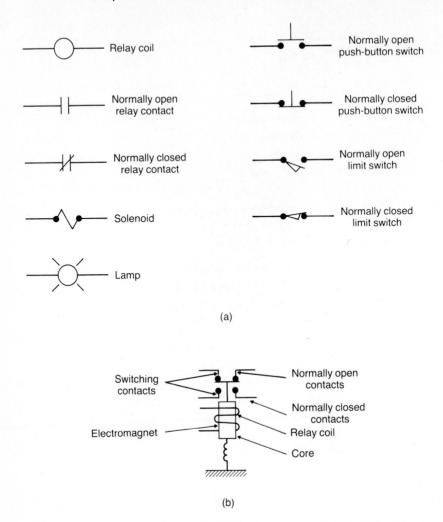

Figure 9.6 (a) Symbols used in ladder diagrams; (b) schematic drawing of control relay.

Conversely, when the solenoid is actuated, the valve position adjacent to solenoid is in effect and the pressure of the air supply forces the piston to the right.

Figure 9.7(b) shows a similar valve for a hydraulic cylinder. Pressurized hydraulic fluid (typically oil) is supplied to the right side of the cylinder when the solenoid is de-energized, and the piston moves to the left. When the solenoid is energized, the direction of fluid flow is reversed and the piston moves to the right. In both cases, hydraulic fluid flowing out of the cylinder is returned through the valve to a reservoir.

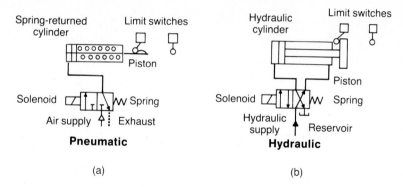

Figure 9.7 Two-position, solenoid-actuated valves and cylinders.
(a) Pneumatic; (b) Hydraulic.

Lamp

Lamps are often used in sequential control systems to indicate the state of the machine or process to the operator. The symbol for a lamp is given in Fig. 9.6(a).

Push-Button Switches

The momentary-contact **push-button switch** allows a circuit to be completed or opened as the switch is operated manually. The normally open push-button switch completes the circuit between the two terminals when the button is pushed and opens the circuit when the button is released. The normally closed switch opens the circuit between the two terminals when it is pushed and completes the circuit when it is released. The symbols for normally open and normally closed push-button switches are shown in Fig. 9.6(a).

Limit Switches

A second type of switch is the **limit switch**, which is actuated through mechanical contact with an actuating arm. The symbols for normally open and normally closed limit switches are shown in Fig. 9.6(a).

The normally open limit switch completes the circuit between the two terminals when the switch is actuated and opens the circuit when the switch is deactuated. The normally closed limit switch opens the circuit when it is actuated and completes the circuit when it is released.

Electrical Circuit Diagrams

Electrical circuit diagrams for relay logic can be constructed by drawing two vertical lines. These lines represent electrical leads with the control circuit voltage

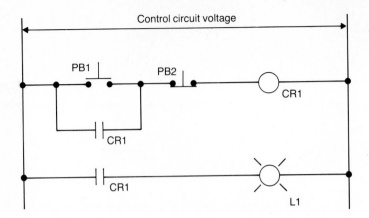

Figure 9.8 Indicator lamp controlled by two push-buttons.

applied between them. Horizontal lines appear on the diagram as various components are added. The result is called a ladder diagram because of its appearance.

An illustrative electrical circuit diagram is shown in Fig. 9.8. When the control circuit voltage is initially applied, there is no current path either to the control relay coil or to the lamp, and neither is energized. When push-button switch PB1 is pushed, coil CR1 is energized and the two normally open contacts close. One of these energizes the lamp L1 while the other establishes a parallel path around the PB1 contact. This contact is called a holding contact because it holds the circuit in a completed state after PB1 is released. The lamp remains energized until push-button PB2 is pushed, opening the relay coil circuit and opening the relay contacts.

Logic Functions

The logic functions performed by relay logic are determined by the switching contact configurations in the ladder diagram. Figure 9.9 shows the contact configurations used to obtain the AND, OR, EXCLUSIVE OR, and NOT logic functions described in Section 5.3.

AND

Two normally open relay contacts in series are shown in Fig. 9.9(a). This combination produces an AND function, because both relay coil A and relay coil B must be energized in order to complete the circuit.

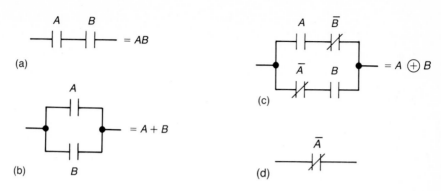

Figure 9.9 Contact configurations for various logic functions. (a) AND; (b) OR; (c) EXCLUSIVE OR; (d) NOT.

OR

In Figure 9.9(b), a parallel arrangement of two normally open contacts is given. This is an OR function because energizing coil A, coil B, or both completes the circuit.

EXCLUSIVE OR

The EXCLUSIVE OR function is shown in Fig. 9.9(c). The circuit is completed when either coil A or coil B is actuated, but not both.

NOT

A single normally closed contact is shown in Fig. 9.9(d). This is a NOT function because energizing coil A results in an open circuit, whereas de-energizing coil A provides a closed circuit.

Obtaining Boolean Equations from Ladder Diagrams

When the sequential logic for a switching control system has been specified using a ladder diagram, it may be necessary to be able to convert the ladder diagram into a set of Boolean equations if a programmable logic controller is to be used rather than relay logic hardware. This conversion can be carried out automatically by programmable logic controllers that allow direct entry of ladder diagrams, but it is also readily accomplished by inspecting switching contact configurations. By breaking these combinations into AND, OR, and NOT subgroups, it is possible to develop a Boolean equation for each control relay, solenoid, and lamp in the ladder diagram.

For example, the ladder diagram in Fig. 9.8 describes the activation of control relay CR1 and lamp L1. The circuit for CR1 has a normally open push-button PB1 and a normally open contact CR1 in an OR configuration. This group is in series with, and hence forms an AND function with, a normally closed push-button $\overline{PB2}$ (a NOT function). The Boolean equation describing this circuit is

$$CR1 = (CR1 + PB1) \cdot \overline{PB2} \tag{9.17}$$

The lamp circuit is much simpler, and

$$L1 = CR1 \tag{9.18}$$

Converting Boolean Equations into Ladder Diagrams

Just as a set of Boolean equations can be obtained to describe a ladder diagram, an equivalent ladder diagram can be developed from a set of Boolean equations. In this case, it is usually possible to construct one "rung" in the ladder diagram from each Boolean equation. Equations are restricted to a single entity on the left that represents the device to be actuated (such as control relay, solenoid, or light). The elements on the right are either control relay contacts or input switches (such as push-buttons or limit switches). Groups inside parentheses are converted first; inverted (NOT-ed) variables specify normally closed contacts, and all other variables specify normally open contacts.

Example 9.4 Conversion to ladder diagram   Suppose that the following equations are to be represented in ladder diagram form:

$$CR1 = (CR1 + (PB1 \cdot PB2)) \cdot \overline{LS3}$$

$$SOL1 = \overline{CR1} \cdot \overline{LS1} \cdot LS2$$

CR1 is a control relay, SOL1 is a solenoid, PB1 and PB2 are push-button switches, and LS1, LS2, and LS3 are limit switches. The equivalent ladder diagram is shown in Fig. 9.10. □

Example 9.5 Conversion to ladder diagram   Another example will illustrate the fact that ladder diagrams may be a somewhat less convenient representation than Boolean equations. Consider the equation

$$CR1 = (START + CR1) \cdot (\overline{LS1 + LS2 + LS3})$$

This conversion is most easily accomplished if we realize that by applying Theorem 20 in Table 9.1, we get

$$\overline{LS1 + LS2 + LS3} = \overline{LS1} \cdot \overline{LS2} \cdot \overline{LS3}$$

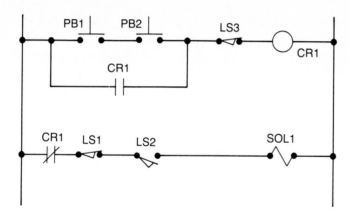

Figure 9.10 Ladder diagram for Example 9.4.

Figure 9.11 Two ladder diagrams for Example 9.5.

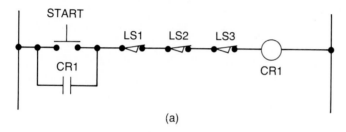

(a)

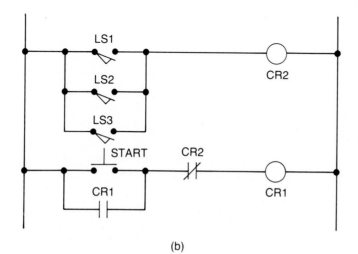

(b)

The ladder diagram is then that shown in Fig. 9.11(a). Otherwise, because only the inversion of switching contacts (and not that of groups of switching contacts) can be constructed using relay logic hardware, a dummy control relay CR2 must be introduced as shown in the alternative implementation in Fig. 9.11(b). The sole function of CR2 is to perform the inversion, and there are now two defining equations:

$$CR2 = LS1 + LS2 + LS3$$

$$CR1 = (START + CR1) \cdot \overline{CR2} \quad \square$$

9.5

SEQUENTIAL LOGIC DESIGN

This section describes several approaches to organizing control requirements in such a way as to provide information useful for the programming of programmable logic controllers. The basic objective is to express the appropriate control relationships in terms of Boolean equations. Although it is possible to obtain control relationships in ways other than through the formulation of Boolean equations (such as through the use of ladder diagrams), Boolean equations offer a fundamental approach that is likely to enjoy increasing acceptance with the passage of time. We have already seen in the previous section that a ladder diagram can be constructed easily once a set of Boolean equations has been developed.

Three general approaches to obtaining Boolean equations will be discussed. The first is to obtain the appropriate equations from a written statement of the switching functions to be accomplished. The second approach utilizes a switching table, which indicates the switched conditions of outputs and inputs during a sequence of operations. The third approach is to work from a flowchart or state-transition diagram that defines the machine or process operations that must take place. We will illustrate each approach by applying it to determine the Boolean equations required to meet specified control requirements for a particular machine or process.

Off-Dominant Design of Holding Functions

In designing sequential logic, we often need functions that provide a memory of past inputs and sequences so that the next step in the sequence can be taken at the appropriate time. This type of function has already been discussed with regard to Eq. (9.6), where the variable LAMP remained on or off depending on which push-button was last pushed. Control relay CR1 in Fig. 9.8 serves the same function.

To investigate further how these holding functions are constructed using Boolean algebra, consider the following equations:

$$A = \overline{B + A[0]} \tag{9.19}$$

$$B = \overline{A + A[1]} \tag{9.20}$$

where A, B, $A[0]$, and $A[1]$ are Boolean variables. Readers familiar with solid-state logic will recognize Eqs. (9.19) and (9.20) as constituting a flip-flop constructed from NOR (NOT OR) functions.

Substituting Eq. (9.20) for B in Eq. (9.19) yields

$$A = (A + A[1]) \cdot \overline{A[0]} \tag{9.21}$$

Table 9.5 shows the state of variable A as a function of variables $A[1]$ and $A[0]$. It is obvious from the table that $A[1]$ is the function that results in $A = 1$ and that $A[0]$ is the function that results in $A = 0$. Equation (9.21) is called **off-dominant** because when both $A[1]$ and $A[0]$ are 1, the result is still $A = 0$. The variable A can be 1 only when $A[0] = 0$. This feature is often useful in the control of machines and processes where it is usually safer to have an output off when a conflict in input conditions exists. A simple case is when START and STOP buttons are pushed simultaneously.

Timing Functions

Time delays and other **timing functions** are often required in sequential logic systems. There may, for example, need to be a pause between one event and another for a machining operation to take place or for some other stage in a process to be performed. It may be more convenient to allow a fixed amount of time for such an event than to detect its completion with a transducer. Hence programmable logic controllers usually provide facilities for timing functions.

Table 9.5

State of Variable A resulting from
$$A = (A + A[1]) \cdot \overline{A[0]}$$

$A[1]$	$A[0]$	A
0	0	A
0	1	0
1	0	1
1	1	0

Table 9.6

Time Delay Input TIMER[Δ*t*] as a
Function of Timer Control Output
TIMER

SEQUENCE	TIMER	TIMER[Δ*t*]
0. Timer off	0	0
1. Timer on	1	0
2. Time out	1	1
0. Timer off	0	0

Timing functions are most easily handled if a timing device is visualized as being connected to the programmable logic controller. A timer control signal, generated by the programmable logic controller, activates the timer, and a timer input to the programmable logic controller indicates whether the desired time delay has occurred. Table 9.6 summarizes this approach. TIMER is the controller output, and timing starts when TIMER is changed from 0 to 1. After the specified time delay Δ*t*, the controller input, TIMER[Δ*t*], changes from 0 to 1. TIMER can be reset to 0 at any time *without delay* by setting TIMER = 0.

Example 9.6 Time delay   If a lamp is to be turned on 10 s after a switch is activated, then, if the time delay for a timer is set to 10 s, the following equations can be used to define the control logic:

$$TIMER = SWITCH$$

$$LAMP = TIMER[10 \text{ s}] \quad \square$$

Specification of Control Requirements

The basic control objectives of a sequential control system can generally be specified by using written statements of required actions. This is a good starting point for any design, and, for uncomplicated systems, the written statements can be directly translated into Boolean equations.

Example 9.7 Design of system logic   Consider an example wherein a sump receives waste water on an intermittent basis and must be pumped out from time to time when the water level becomes high. It is desired to design a control system to begin pumping automatically when the level reaches some upper limit and to continue pumping until some lower limit is reached. Level detection is accomplished by the use of a float rod and two limit switches, LS1 and LS2, as illustrated in Fig. 9.12. These will be input to a programmable logic controller,

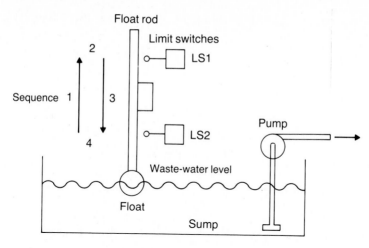

Figure 9.12 Waste-water sump with pump and float rod.

and an output—PUMP—will be used to control the pump motor. The problem is to obtain a Boolean equation that defines output PUMP in terms of inputs LS1 and LS2.

The operation of the pump can be more specifically described as follows:

1. The pump is to be turned on (PUMP = 1) and should stay on when limit switch LS1 is actuated (LS1 = 1).

2. The pump should be turned off (PUMP = 0) and stay off when limit switch LS2 is actuated (LS2 = 1).

It can be seen in Fig. 9.12 that a memory function is required, because limit switch LS1 will be de-actuated immediately as the water level falls but the pump must remain on until LS2 is reached. Equation (9.21) should therefore be used to develop an equation for the output PUMP from the foregoing statements of system requirements. From statement 1, LS1 = 1 is required for PUMP = 1. Therefore

$$\text{PUMP}[1] = \text{LS1}$$

From statement 2, LS2 = 1 is required for PUMP = 0, so

$$\text{PUMP}[0] = \text{LS2}$$

For this application, Eq. (9.21) becomes

$$\text{PUMP} = (\text{PUMP} + \text{PUMP}[1]) \cdot \overline{\text{PUMP}[0]}$$

or, by substitution into the above equation,

$$\text{PUMP} = (\text{PUMP} + \text{LS1}) \cdot \overline{\text{LS2}}$$

This is the desired Boolean equation describing the state of the output PUMP as a function of the inputs LS1 and LS2. ☐

Switching Tables

Basic system functions can be directly expressed in written statements, but as systems grow more complex it becomes desirable to present the sequence that occurs in a more manageable form. A **switching table** can help identify those input combinations that produce changes in outputs. Hence it can help us develop Boolean equations.

The first stage in the construction of a switching table is to identify the steps in the sequence that constitute a complete automatic cycle of the system. In the pump control example in Fig. 9.12, step 1 is the filling of the sump with no pumping taking place. Step 2 is the actuation of limit switch LS1. Step 3 is the emptying of the sump by pumping, and step 4 is the actuation of LS2. These four steps describe one complete cycle for the system.

The next stage is to develop a table that shows the state of the two inputs utilized in the automatic cycle and the single output for each of the different steps in the cycle. We begin constructing this table˙ (See Table 9.7) by providing columns for each input and output and rows for each of the steps. The purpose of the two additional columns (S and D) will be explained later.

The switching table for the waste-water pumping example (Fig. 9.12) is completed as follows. During step 1, neither limit switch is actuated and the pump is OFF. Thus the row for step 1 in Table 9.7 is 0,0,0. Limit switch LS1 is actuated at step 2 and the pump is turned ON. The row for step 2 is therefore, 1,0,1. During step 3, neither limit switch is actuated but the pump must continue to be ON. The resulting row is 0,0,1. Finally, in step 4, LS2 is encountered and the pump must be turned OFF. The row for step 4 is therefore 0,1,0. The next row is step 1, the beginning of a new cycle. The switching table now shows the

Table 9.7

Switching Table for Waste-Water Pumping Cycle

SEQUENCE	LS1	LS2	PUMP	S	D
1. Sump filling	0	0	0		
2. LS1 actuated	1	0	1	*	
3. Sump emptying	0	0	1		
4. LS2 actuated	0	1	0	*	
1. Sump filling	0	0	0		

state of the two inputs and the one output for all the events that make up one complete automatic cycle.

The procedure for obtaining Boolean equations from switching tables is as follows:

1. First *identify* those steps that cause switching in output values. The column labeled S (switching) in the switching table is used for this purpose. This column is marked to indicate those steps in the sequence where switching in output values occurs.

2. Then *identify* those steps that cause switching and possess a combination of current input states and previous output states *duplicated* in other steps. For all steps identified, place the number of the step that causes switching in the column labeled D [duplicated). Note that the previous output states must be used because new output states may be generated in the current step.

3. *Dummy variables* must be created to resolve all switching problems that have been identified. A column is created in the switching table for each dummy variable, and steps in the sequence are identified that can conveniently be used to change the state of the dummy variable. The state of these dummy variables can then be used along with input and output states to differentiate each step that causes switching from all other steps in the sequence.

4. *Boolean equations* are written for each output and dummy variable by identifying the input, output, and dummy variable states required to turn each output or dummy variable ON or OFF. Equation (9.21) can be applied if a holding function is required.

Example 9.8 Design using a switching table   Applying the foregoing procedure to the pump example, we mark steps 2 and 4 in Table 9.7 with asterisks in the S column to indicate that a state change in the output PUMP occurs in these steps. No steps in the table have input states identical to either step 2 or step 4, and no dummy variables are required.

An equation for the output PUMP can now be developed. From the switching table it can be seen that $LS1 = 1$ can be used to set $PUMP = 1$, and $LS2 = 1$ can be used to set $PUMP = 0$. The state of PUMP must be held constant between these events, and applying Eq. (9.21) again yields

$$PUMP = (PUMP + LS1) \cdot \overline{LS2} \quad \square$$

Flowcharts

As the complexity of system requirements increases, a **flowchart** can be considered. This is particularly true when there are a large number of alternative sequences or "branching" paths in the logic. The switching tables can be clumsy

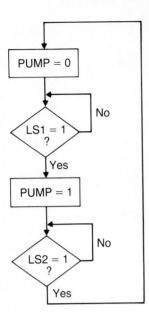

Figure 9.13 Flowchart for
waste-water pumping
cycle.

for this purpose because they efficiently represent only a single repetitive
sequence. Flowcharts allow a considerable amount of decision making and
branching to be represented in a way that tends to make the development of
Boolean equations easier.

A flowchart for the pump control example is shown in Fig. 9.13. In the
convention used, changes in output states are shown in rectangular boxes. These
are reached through diamond-shaped decision boxes that test input and output
states. The flowchart clearly indicates the sequence of events in the control
function and shows which conditions must be present to move to the next step in
the sequence.

State Diagrams

The **state diagram** allows convenient illustration of switching sequences, even
when many alternative sequences are possible. A state diagram for our automatic
pumping cycle is shown in Fig. 9.14. System states are represented by circles.
Output values are determined by the system state. Figure 9.14 reveals that
PUMP = 0 in state 1 and PUMP = 1 in state 2. Output values are not changed
without changing to a new system state.

The arrows in the state diagram show which system state changes are
allowed and what conditions are required for each state change. No state change
occurs along a given arrow unless the conditions shown on the arrow are

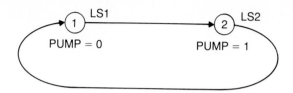

Figure 9.14 State diagram for waste-water pumping cycle.

satisfied. When these conditions are satisfied, the system state is assumed to change instantaneously to the new state. In Fig. 9.14, a state change from state 1 to state 2 can occur only when LS1 = 1. Similarly, the system state can return to state 1 only when LS2 = 1.

One strategy for writing Boolean equations using state diagrams is to create a dummy variable for each state. For an arbitrary state S, S = 1 when the system is in state S, and S = 0 when the system is not in state S. We can construct an equation S[1] by using the arrows pointing toward state S and an equation S[0] by using the arrows pointing out of state S. These can then be combined via Eq. (9.21) to produce an equation for S.

Example 9.9 Design using a state-transition diagram   In the example given in Fig. 9.14 there are two states, S1 and S2. The conditions required to turn these states ON are

$$S1[1] = S2 \cdot LS2$$

$$S2[1] = S1 \cdot LS1$$

In other words, state S1 is turned ON when the system is in state S2 and LS2 is activated. State S2 is turned ON when the system is in state S1 and LS1 is activated.

The arrows pointing out of state S1 and S2 are used to determine the conditions required to turn these states OFF. In the example, these are

$$S1[0] = S2 \cdot LS1$$

$$S2[0] = S1 \cdot LS2$$

The system state changes from S1 to S2 when LS1 is activated and the system is in state S1. Similarly, the state changes from S2 to S1 when LS2 is activated and the system is in state S2.

Equations describing S1 and S2 are then obtained by applying Eq. (9.21), which results in

$$S1 = (S1 + S2 \cdot LS2)(\overline{S2 \cdot LS1})$$

$$S2 = (S2 + S1 \cdot LS1)(\overline{S1 \cdot LS2})$$

Once the state transitions have been described, the outputs can be defined as functions of the states. From the state diagram for the automatic pumping cycle shown in Fig. 9.14, we note that the output PUMP is ON in state 2. The appropriate equation is

$$PUMP = S2$$

It is obvious that equations obtained by this method are not necessarily optimal in terms of the number of equations, variables, and logical operations that must be performed. It is often possible to reduce these equations by inspection.

For this example, we can see that the inputs alone can be used to define the state transitions. Hence, the equations we derived can be rewritten as

$$S1 = (S1 + LS2) \cdot \overline{LS1}$$

$$S2 = (S2 + LS1) \cdot \overline{LS2}$$

Substituting PUMP into the equation for S2 yields

$$PUMP = (PUMP + LS1) \cdot \overline{LS2}$$

The equation for S1 can then be discarded, because the variable S1 does not appear in the equation for PUMP. □

In this approach, equations are obtained directly from the state diagram with little effort, and the equations and the diagram are therefore very closely related. This can help both in ensuring the reliability of the system (an important safety consideration) and in modifying and debugging the logic sequence. The cost of additional equations may be high when implemented with relay or solid-state logic hardware, but it is likely to be inconsequential when a programmable logic controller is used. This is because the only effects of increased program size on a programmable logic controller are increased memory requirements and increased scan time—effects that are insignificant in many applications.

It should be noted that when logic equations are solved by a digital computer, unanticipated results can be obtained if serious consideration is not given to the sequential nature of the solution. Among the possible side effects is the likelihood that more than one state will be active for short periods of time. This is because the logic operations that deactivate the old state are not performed simultaneously with those that activate the new state. It is therefore possible, depending on the order of operations, that a state transition may not be completed until part way through the next scan. This can cause erroneous state transitions in either scan, the activation of multiple outputs, blown fuses, and other problems. Care must always be taken that multiple states are not erroneously activated. These problems can be avoided by including additional logic and by carefully considering the order in which equations are solved.

Example 9.10 State-transition diagram with timer   Suppose that in the system represented by the state-transition diagram in Fig. 9.15, it is possible for both PB1 and TIMER[Δt] to be true at the beginning of a scan. In this case, the statements

$$S1 \leftarrow (S1 + S2 \cdot PB2 + S3 \cdot PB2) \cdot \overline{(S2 \cdot PB1 + S3 \cdot TIMER[\Delta t])}$$

$$S2 \leftarrow (S2 + S1 \cdot PB1) \cdot \overline{(S1 \cdot PB2)}$$

$$S3 \leftarrow (S3 + S1 \cdot TIMER[\Delta t]) \cdot \overline{(S1 \cdot PB2)}$$

$$TIMER \leftarrow S1$$

can result in transitions to both S2 and S3, states that should be mutually exclusive. To prevent such erroneous transitions, these statements can be modified as follows:

$$S2 \leftarrow (S2 + S1 \cdot \overline{S3} \cdot PB1) \cdot \overline{S1}$$

$$S3 \leftarrow (S3 + S1 \cdot \overline{S2} \cdot TIMER[\Delta t]) \cdot \overline{S1}$$

$$S1 \leftarrow (S1 + S2 \cdot PB2 + S3 \cdot PB2) \cdot \overline{(S2 + S3)}$$

$$TIMER \leftarrow S1 \quad \square$$

Example 9.11 System with duplicated input conditions   This example illustrates the problem of duplicated input conditions that require different output states and the use of a time delay in the logic sequence. Two logic designs will be obtained for the system. The first will be obtained in this example by using a switching table and the second will be obtained in Example 9.12 by using a state-transition diagram.

Figure 9.15 State-transition diagram for Example 9.10.

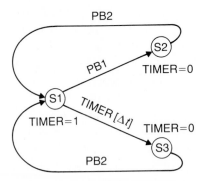

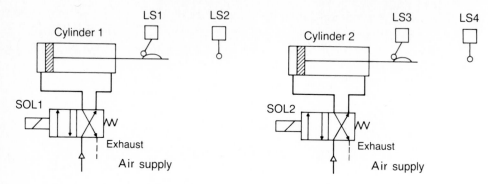

Figure 9.16 Two-cylinder system.

A two-cylinder pneumatic system powered by two-position, solenoid-actuated valves is shown in Fig. 9.16. The required automatic cycle is to be as follows:

1. A momentary-contact push-button, START, is depressed.
2. SOL1 is activated and the piston rod of cylinder 1 advances to limit switch LS2.
3. SOL1 is deactivated and the rod retracts to LS1.
4. A time delay of 2 s occurs.
5. SOL2 is activated and the piston rod of cylinder 2 advances to LS4.
6. SOL2 is deactivated and the rod retracts to LS3.
7. The system remains at rest awaiting a new START signal.

A push-button, STOP, is also to be provided that will cause the system to return to the rest condition at any time during the automatic cycle.

A switching table can now be developed for this example. Table 9.8 shows the inputs START, LS1, LS2, LS3, LS4, and TIMER[2 s] as well as the outputs SOL1, SOL2, and timer control TIMER. A brief description is given for each step in the sequence. Three additional columns are shown: S, D, and DUMMY. Their use is explained in the following discussion. The input STOP is not shown because it must remain OFF during the sequence. The STOP function will be included in the logic at a later stage in its development.

The filling in of Table 9.8 proceeds as described previously. For step 0, the at-rest condition, both solenoids are de-energized and only switches LS1 and LS3 are actuated. In step 1, the cycle is initiated by the actuation of START, and SOL1 is energized. Step 2 is the transition from LS1 to LS2. START and LS1 are no longer actuated, but SOL1 must continue to be energized. The remaining entries

Table 9.8

Switching Table for Two-Cylinder System

SEQUENCE	START	LS1	LS2	LS3	LS4	TIMER[2 s]	SOL1	SOL2	TIMER	S	D	DUMMY
0. At rest	0	1	0	1	0	0	0	0	0		5	0
1. START actuated	1	1	0	1	0	0	1	0	0	*		0
2. Transition to LS2	0	0	0	1	0	0	1	0	0			0
3. LS2 actuated	0	0	1	1	0	0	0	0	0	*		1
4. Transition to LS1	0	0	0	1	0	0	0	0	0			1
5. LS1 actuated	0	1	0	1	0	0	0	0	1	*	5	1
6. Timer on [2 s]	0	1	0	1	0	0	0	0	1			1
7. Time out	0	1	0	1	0	1	0	0	1	*		1
8. Transition to LS4	0	1	0	0	0	0	0	1	0	*		1
9. LS4 actuated	0	1	0	0	1	0	0	1	0	*		0
10. Transition to LS3	0	1	0	0	0	0	0	0	0			0
0. At rest	0	1	0	1	0	0	0	0	0		5	0

in the table are arrived at in a similar manner. Steps 6 and 7 describe the time delay. The timer is activated with the output TIMER in step 5. The input TIMER[2 s] appears after the desired time delay of 2 s in step 7. The timer can be deactivated conveniently in step 8.

The first step in the development of Boolean equations from the switching table is to identify the steps in the sequence that cause switching of output states. The column headed S is used to mark these steps. An asterisk is shown at steps 1, 3, 5, 7, 8, and 9 because SOL1 switches from 0 to 1 at step 1 and from 1 to 0 at step 3, TIMER changes from 0 to 1 at step 5, and so forth.

Next, entries are made in the column headed D. Each step with a current input and previous output combination that duplicates a step identified in column S as causing switching is marked with the number of that step. There is no duplication of the input combination at step 1 in the table. Likewise, there are no duplications with steps 3, 7, 8, and 9. However, steps 0 and 6 have input states that are identical to step 5. The control system cannot, from input conditions alone, distinguish between the at-rest state, the time delay, and the activation of LS1.

Because the input conditions alone cannot resolve this switching problem, output conditions must be considered. Step 6 does not pose a problem because TIMER is ON just before step 6 and OFF just before steps 0 and 5. TIMER can therefore be used to differentiate between these events. However, all outputs are in the OFF state immediately preceding step 0. Therefore, output conditions cannot provide information that will distinguish between step 0 and step 5, which causes switching. The column labeled D is therefore marked with 5 for these steps.

The remaining way to resolve the switching problem is to add a dummy variable to the system. In the column headed DUMMY in Table 9.8, a 0 is entered for step 10, just before step 0, and a 1 is entered at step 4, just before step 5. When the switched conditions of the variable DUMMY are now combined with input and output information, the problem is resolved. The variable DUMMY must be turned ON and OFF, just the same as an output. Because DUMMY can be turned ON at any point between steps 10 and 4, limit switch LS2 can be conveniently used to set DUMMY = 1. Similarly, it is convenient to set DUMMY = 0 when limit switch LS4 is actuated. The column headed DUMMY can now be filled in as shown, and the switching table is complete.

Boolean equations now can be written for SOL1, SOL2, TIMER, and DUMMY. In the case of SOL1, the key event required for energization is the actuation of the START push-button. This would constitute the minimum information necessary to turn on the solenoid. However, some interlocking may be desirable. Requiring that both cylinders be fully retracted in order to initiate a cycle would mean that the actuation of LS1 and LS3 was also necessary. Under the assumption that it is desirable to check for the full retraction of both cylinders, the equations for SOL1[1] and SOL1[0] are

$$SOL1[1] = START \cdot LS1 \cdot LS3$$

$$SOL1[0] = LS2 + STOP$$

Note that it is convenient to include the STOP push-button in the above equation. A holding feature is required, and the substitution of the foregoing equations into Eq. (9.21) yields the required Boolean equation for SOL1:

$$SOL1 = (SOL1 + START \cdot LS1 \cdot LS3) \cdot \overline{LS2} \cdot \overline{STOP}$$

The time delay input, TIMER[2 s], can be used to energize SOL2, and limit switch LS4 can be used to de-energize SOL2. Including the STOP push-button, the equations for SOL2 are

$$SOL2[1] = TIMER[2 \ s]$$

$$SOL2[0] = LS4 + STOP$$

and

$$SOL2 = (SOL2 + TIMER[2 \ s]) \cdot \overline{LS4} \cdot \overline{STOP}$$

The equation for the timer control, TIMER, is

$$TIMER = LS1 \cdot DUMMY$$

Finally, the equation for the variable DUMMY is developed from

$$DUMMY[1] = LS2$$

$$DUMMY[0] = LS4 + STOP$$

$$DUMMY = (DUMMY + LS2) \cdot \overline{LS4} \cdot \overline{STOP}$$

Note that STOP must be included in the equation for DUMMY to return it to the OFF state in the event that STOP is activated.

This example brings out some of the advantages of using a switching table. First, it presents a clear picture of all the switching events taking place in one cycle. This feature, in turn, makes it easy to detect switching problems and offers information in a convenient form to help solve any such problems. The basic idea underlying the use of the switching table to develop Boolean equations is to start as simply as possible, using only the inputs to the system. It is necessary build to whatever level of complexity is required to solve the overall switching problem.

□

Example 9.12 Using a state diagram for Example 9.11   A State diagram can also be used to obtain Boolean equations defining the control for the two-cylinder system in Example 9.11 and Fig. 9.16. Six system states can be identified from the specification of system operation. These are

State 1, the at-rest state

State 2, translation from LS1 to LS2

State 3, translation from LS2 to LS1

State 4, time delay of 2 s

State 5, translation from LS3 to LS4

State 6, translation from LS4 to LS3

The system returns to state 1 after state 6. The state diagram shown in Fig. 9.17 can now be constructed for the two-cylinder example.

All outputs are de-energized in state 1, the at-rest state. The only way to leave the at-rest state is for START to be activated. LS1 and LS3 are also required to ensure that both cylinders are retracted before a cycle begins. Transition to state 2 occurs when START, LS1, and LS3 are actuated. SOL1 is energized in state 2. When LS2 is actuated, the system enters state 3, wherein all outputs are again de-energized.

The timer is turned on by the actuation of LS1, and the time delay, state 4, begins. Transition from state 4 to state 5 occurs when the timer runs out and SOL2 is energized. Actuation of LS4 and, later, actuation of LS3 cause transitions to state 6 and back to the at-rest state, respectively.

An alternative route back to the at-rest state, state 1, is shown for each state. These routes indicate the action required by actuation of the STOP button. The system can therefore return directly to state 1 at any point in its operation.

Figure 9.17 State diagram for two-cylinder system.

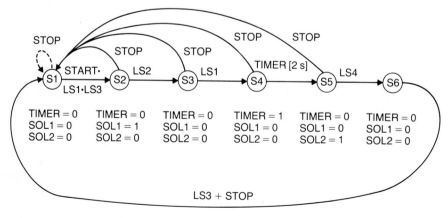

Inclusion of LS1 and that of LS2 in the transition from state 1 to state 2 allows both cylinders to return fully retracted before a new cycle can start. Ideally, for reasons of safety, STOP should cause all motion to cease. However, that is not possible in this system.

Boolean equations can now be written for each state. The equation for state 2 is obtained using Eq. (9.21) as follows:

$$S2[1] = S1 \cdot START \cdot LS1 \cdot LS3$$

$$S2[0] = S3 \cdot LS2 + S1 \cdot STOP$$

$$S2 = (S2 + S1 \cdot START \cdot LS1 \cdot LS3) \cdot \overline{S3 \cdot LS2} \cdot \overline{S1 \cdot STOP}$$

The equation for state 3 is obtained from

$$S3[1] = S2 \cdot LS2$$

$$S3[0] = S4 \cdot LS1 + S1 \cdot STOP$$

$$S3 = (S3 + S2 \cdot LS2) \cdot \overline{S4 \cdot LS1} \cdot \overline{S1 \cdot STOP}$$

Obtaining equations for states 4, 5, and 6 in a similar manner, results in

$$S4 = (S4 + S3 \cdot LS1) \cdot \overline{S5 \cdot TIMER[2 \text{ s}]} \cdot \overline{S1 \cdot STOP}$$

$$S5 = (S5 + S4 \cdot TIMER[2 \text{ s}]) \cdot \overline{S6 \cdot LS4} \cdot \overline{S1 \cdot STOP}$$

$$S6 = (S6 + S5 \cdot LS4) \cdot \overline{S1 \cdot LS3} \cdot \overline{S1 \cdot STOP}$$

The equation for state 1 is more complex, but it is greatly simplified if a transition from state 1 to state 1 can be added for the STOP button. This transition, which is shown in Fig. 9.17, clearly does not change the operation of the system. The equation for state 1 is obtained as follows:

$$S1[1] = S1 \cdot STOP + S2 \cdot STOP + S3 \cdot STOP + S4 \cdot STOP$$
$$+ S5 \cdot STOP + S6 \cdot (LS3 + STOP)$$

This can be rewritten as

$$S1[1] = S6 \cdot LS3 + (S1 + S2 + S3 + S4 + S5 + S6) \cdot STOP$$

However, because the system should always be in one of its states at any given time,

$$S1 + S2 + S3 + S4 + S5 + S6 = 1$$

Table 9.9

Boolean Equations for Two-Cylinder System. (a) From Switching Table. (b) From State Diagram.

$SOL1 = (SOL1 + START \cdot LS1 \cdot LS3) \cdot \overline{LS2} \cdot STOP$	$S1 = (S1 + S6 \cdot LS3 + STOP) \cdot S2 \cdot START \cdot LS1 \cdot LS3$
$SOL2 = (SOL2 + TIMER[2\ s]) \cdot \overline{LS4} \cdot STOP$	$S2 = (S2 + S1 \cdot START \cdot LS1 \cdot LS3) \cdot S3 \cdot LS2 \cdot S1 \cdot STOP$
$DUMMY = (DUMMY + LS2) \cdot \overline{LS4} \cdot STOP$	$S3 = (S3 + S2 \cdot LS2) \cdot \overline{S4 \cdot LS1} \cdot S1 \cdot STOP$
$TIMER = LS1 \cdot DUMMY$	$S4 = (S4 + S3 \cdot TIMER[2\ s]) \cdot \overline{S6 \cdot LS4} \cdot S1 \cdot STOP$
	$S5 = (S5 + S4 \cdot TIMER[2\ s]) \cdot \overline{S6 \cdot LS4} \cdot S1 \cdot STOP$
	$S6 = (S6 + S5 \cdot LS4) \cdot \overline{S1 \cdot LS3} \cdot S1 \cdot STOP$
	$SOL1 = S2$
	$SOL2 = S5$
	$TIMER = S4$
(a)	(b)

Therefore,

$$S1[1] = S6 \cdot LS3 + STOP$$

Also,

$$S1[0] = S2 \cdot START \cdot LS1 \cdot LS3$$

STOP should not be included in the foregoing equation because the system must remain in state S1 if STOP is activated. Then, from Eq. (9.21),

$$S1 = (S1 + S6 \cdot LS3 + STOP) \cdot \overline{S2 \cdot START \cdot LS1 \cdot LS3}$$

Output equations are obtained from this system in a straightforward manner from the state diagram. Here,

$$TIMER = S4$$

$$SOL1 = S1$$

$$SOL2 = S5$$

We can use Table 9.9 to compare the results of the switching-table approach with those of the state-diagram approach for this example. It is apparent that the switching-table approach resulted in four equations, whereas the state-diagram approach resulted in nine equations. Although some simplification could be performed on the equations of the state-diagram approach, the scan time of this approach is likely to be approximately twice that of the switching-table approach. This may not be a critical consideration for this system, because the number of equations is small and the speed requirements of the system are likely to be low. The clarity of the equations obtained by the state-diagram approach and the ease with which they were obtained may make this approach preferable. □

9.6

PROGRAMMING, MONITORING, AND DIAGNOSIS

Programmable logic controllers typically are capable of accepting commands from an operator that change, update, and monitor the control logic program. Controllers installed in production machines may not have all of these capabilities, but development versions enable the operator to enter logic statements or ladder diagrams, observe the operation of the controller, diagnose faulty logic, and then modify the logic to correct faults or implement design changes. Some of the features found in various programmable logic controllers that allow these programming, monitoring, and diagnostic functions to be performed are discussed in the following paragraphs.

Programming Languages

There is little standardization in programming languages and display formats for programmable logic controllers. Among the programming languages found in commercial controllers are

- Ladder diagrams
- Logic statements
- English statements
- Machine-language instructions

Some units may allow only one form of entry, such as ladder diagrams or Boolean logic statements. Others may allow several and may also allow additional functions (such as arithmetic statements, subroutine calls, and closed-loop control) to be performed.

Programming Devices

Some form of programming device is required to allow the programmer to interact with the programmable logic controller. These programming devices include

- CRT/keyboard
- Special programming panels
- Logic displays
- Tape unit
- ROM (read-only memory)

If development functions are to be carried out, a keyboard or programming panel and a device to display programmed logic are required. For systems used in nondevelopment, production applications, it is only necessary to be able to enter the program into the programmable logic controller. This is typically done by using a tape unit or by inserting a preprogrammed read-only memory (ROM).

Monitoring and Diagnosis

As with any computer control system, there exists in the programmable logic controller the possibility of programmer error or hardware or software failure. A number of devices have been developed to detect these faults, including external monitors and internal diagnostic software. The programmable logic controller is usually self-sufficient and can handle its control functions without external hardware. However, some controllers make provision for interfacing with an external "watch-dog" device that can serve in a monitoring or supervisory role. Such an **external monitor** can be programmed to accomplish tasks such as monitoring the operations of the controller, reporting any irregularities, providing information to the operator at appropriate times, offering a diagnostic service in case of operating difficulties, and stopping the machine or process in the event of an emergency.

One simple but effective technique for detecting faults is to monitor the scan time of the programmable logic controller. In many applications, the scan time varies less than 20 % from its mean value. If a "watch-dog timer" is employed to measure the time required for each scan, then when allowable limits are exceeded, it can be assumed that a fault occurred. The programmable logic controller then can be shut down according to a predetermined sequence designed to protect the machine or process.

Status lights can be employed in several ways to allow an operator to monitor a programmable logic controller's operation. Lights can be used on input and output interfaces to indicate their current state (ON or OFF). They can also be used to indicate the presence of power in various parts of the system and hence to indicate when a power failure has occurred. Status lights that are controlled by programmed logic can provide an indication for the operator of the state of important internal variables. By inspecting the state of the machine or process, together with the state of the programmable logic controller and interfaces as indicated by status lights, an operator can often determine the source of a fault in the programmable controller or its external hardware.

Some programmable logic controllers are capable of displaying the state of programmed logic in real time as the controller is operating. The ability to observe the changes in state of inputs, outputs, and timers is as useful as the ability to force them into a desired state. In this way, portions of the control logic can be checked out and faults detected in either programmed logic or input and output devices. Because a **real-time display** is often created by the programmable logic controller itself, a certain amount of its computing capacity may be consumed by this function. The effect is to increase the scan time when such a display is in operation. This fact needs to be taken into consideration when scan time is a critical factor in the control logic.

Simulation

The value of **simulation** should not be overlooked in developing and verifying sequential logic. Simulation of the machine or process to be controlled is a useful step in the effort to prevent damage to the real machine or process during the development and verification of control logic. A panel with indicator lights and switches is often enough to allow basic tests of programmed logic to be performed manually.

External software can also be used to verify control logic. The control program can be tested outside the programmable logic controller (usually on a more general-purpose computer), and its operation under expected conditions can be simulated. Switching sequences that result from given input conditions can be recorded and displayed so that faulty programmed logic can be corrected before implementation on the programmable logic controller.

9.7

SUMMARY

A significant portion of this chapter has been devoted to illustrating approaches for obtaining the switching relationships required for sequential control of machines and processes. These approaches should be thought of as available tools. They should be used in the same manner as all tools; they should be selected judiciously and used with a certain amount of imagination. Emphasis was placed on obtaining Boolean equations, because having the switching requirements specified in this form provides sufficient information for programming even if the programming does not involve the direct use of Boolean equations. It was shown that Boolean equations can be used to develop ladder diagrams as an alternative form of logic specification.

Where switching requirements are straightforward and readily verbalized, the sensible approach is to develop Boolean equations directly from written statements of the switching to be done. On the other hand, the use of switching tables, flowcharts, or state diagrams further structures the switching requirements and is beneficial in cases of greater complexity. Switching tables are useful where a set sequence of operations is involved. Flowcharts and state diagrams become more useful when several alternative sequences are possible.

With regard to the severity of switching problems, we must recognize that the relative ease of designing sequential control logic can depend heavily on the way in which the machine or process itself is designed. Control requirements should be included as one of the many considerations taken into account at the time of the initial design of the basic machine or process. It is counterproductive to inadvertently create control problems that must later be solved by increased complexity in control logic.

The major difference between programmable logic controllers and other types of sequential control hardware is the use of a computer program rather than "hard-wired" connections to implement control logic. This feature allows an operator to utilize a given programmable logic controller in many different sequential control applications simply by supplying the proper input hardware, output hardware and program software for that application. At a later date, modifications can be readily made in the control logic by changing the computer program rather than altering hardware connections. Although the effects of scan time pose an additional problem for the designer, the flexibility of programmable logic controllers have led to their extensive use in machinery and process control applications. Programmable logic controllers are often augmented with closed-loop control functions, and as a result, there is often little fundamental difference between a programmable logic controller and a general-purpose control computer.

BIBLIOGRAPHY

Beckwith, T., N. Buck, and R. Marangoni. *Mechanical Measurements, 3/e.* Reading, MA: Addison-Wesley, 1981.

Doebelin, E. O. *Control System Principles and Design.* New York, NY: Wiley, 1985.

Fletcher, W. I. *An Engineering Approach to Digital Design.* Englewood Cliffs, NJ: Prentice-Hall, 1980.

Harrison, H. L. and J. G. Bollinger. *Introduction to Automatic Controls.* New York, NY: Harper and Row, 1969.

Hill, F. J., and G. R. Peterson. *Digital Systems: Hardware Organization and Design, 2/e.* New York, NY: Wiley, 1978.

Hill, F. J., and G. R. Peterson. *Introduction to Switching Theory and Logical Design, 3/e.* New York, NY: Wiley, 1978.

Mano, M. *Digital Design.* Englewood Cliffs, NJ: Prentice-Hall, 1983.

Short, K. L. *Microprocessors and Programmed Logic.* Englewood Cliffs, NJ: Prentice-Hall, 1980.

Stone, H. S. *Microcomputer Interfacing.* Reading, MA: Addison-Wesley, 1982.

Wakerly, J. *Microcomputer Architecture and Programming.* New York, NY: Wiley, 1981.

PROBLEMS

9.1 To anticipate potential jamming, a warning device is to be actuated (ALARM $= 1$) if the pressure in a hydraulic press exceeds some upper limit *before* the press is fully closed. High pressure is detected by a pressure switch (PS $= 1$). Press closure is detected by a limit switch (LS $= 1$). Develop a Boolean equation for ALARM.

9.2 An operation C is to take place when switch S is actuated along with either push-button A or push-button B. However, the operation is not to take place while both push-buttons are actuated. Develop a Boolean equation for C.

9.3 A building security system will unlock a door and allow an employee to enter when his or her CARD is inserted in a cardreader, his or her CLEARANCE is valid, and it

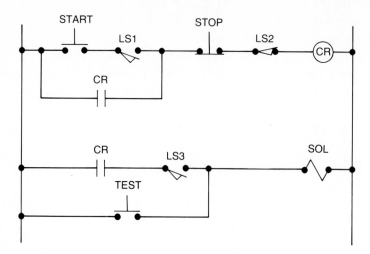

Figure 9.18 Ladder diagram for Problem 9.4.

is not EVENING, NIGHT, a WEEKEND, or a HOLIDAY. The unlocking action is generated by deactivating a signal called LOCK, which is to stay deactivated until 5 s after the CARD was inserted. LOCK is also unconditionally deactivated by a signal called EMERGENCY. Write a Boolean equation describing the signal LOCK as a function of the terms in capital letters in this problem statement. Incorporate the term TIMER[5 s] that becomes true 5 s after TIMER is set true and becomes false immediately upon TIMER being set false. Also write a Boolean equation for TIMER.

9.4   Obtain a set of Boolean equations that represent the ladder diagram in Fig. 9.18.

9.5   For safety reasons, an operator is required to actuate two push-buttons in order to close a stamping press. The intention is to ensure that both of the operator's hands are out of the press when it closes. The press is equipped with the relay logic circuit shown in Fig. 9.19, which incorporates timing relays that prevent the operator from taping one push-button down and making the press faster but unsafe to operate.

The normally-closed timing relay contacts labeled TR1[1 s] and TR2[1 s] in Fig. 9.19 open 1 s after the associated relay coil is energized. These contacts close immediately when the relay coil is de-energized. The normally open contacts labeled TR1 and TR2 close and open immediately. A solenoid labeled SOL1 is used to actuate the hydraulic press. The push-buttons are PB1 and PB2. It is desired to replace the relay logic with a programmable logic controller that is programmed using Boolean logic statements. Develop a set of Boolean equations for TR1, TR2, and SOL1 that describe the logic in the ladder diagram.

9.6   When push-button A is actuated, one of three solenoids is to be energized, depending on the states of two limit switches. The circuit requirements are shown

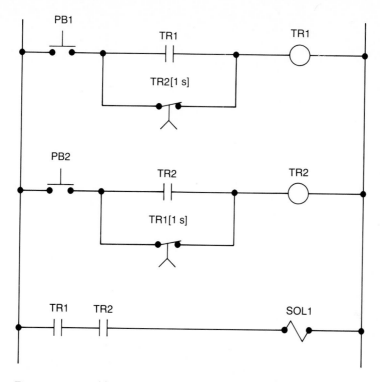

Figure 9.19 Ladder diagram for Problem 9.5, including timing relays for safe press operation.

in the flowchart in Fig. 9.20. Obtain Boolean equations for each of the solenoids, and draw an equivalent ladder diagram.

9.7 Use the truth-table approach to prove DeMorgan's Laws, entries 19 and 20 in Table 9.1.

9.8 Two clamps must be securely closed before a manufacturing operation can begin on a workpiece, and they must remain closed while the operation is being performed. The following equation describes the logic designed for controlling the operation:

MOTOR
$$= (\text{MOTOR} + \text{START} \cdot \text{CLAMP1} \cdot \text{CLAMP2}) \cdot \overline{\text{STOP}} \cdot \text{CLAMP1} \cdot \text{CLAMP2}$$

a. Simplify this equation.
b. Translate the result of part (a) directly into an assembly-language program (using AND, OR, COMPLEMENT, and so on) implementing the logic on a

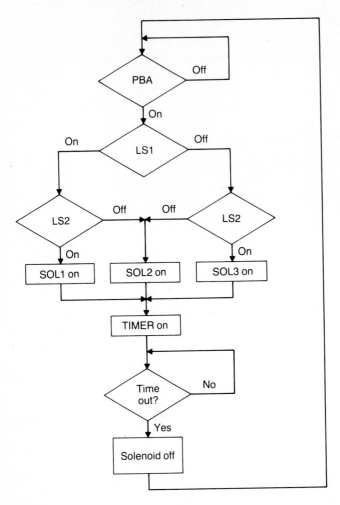

Figure 9.20 Flowchart for Problem 9.6.

computer. Use bit 0 (the least significant bit) of the following I/O interface addresses:

Signal	Address	Type
MOTOR	0	output
START	1	input
STOP	2	input
CLAMP1	3	input
CLAMP2	4	input

c. If each instruction in the program in part (b) takes 1 μs to execute, what is the scan time of the logic?

9.9 Repeat part (b) of Problem 9.8, but add outputs LAMP1 and LAMP2, which are activated when CLAMP1 and CLAMP2 are activated, respectively. Also, all signals are to be input and output in parallel according to the following interface assignments:

Signal	Address	Bit	Type
MOTOR	0	0	output
LAMP1	0	1	output
LAMP2	0	2	output
START	1	0	input
STOP	1	1	input
CLAMP1	1	2	input
CLAMP2	1	3	input

9.10 The nuclear reactor shown in Fig. 9.21 is used for irradiating samples of food, building products, organic tissue, and the like. To protect its operators from overdoses of radiation, a locking mechanism has been installed on the cover of the

Figure 9.21 Nuclear reactor for Problem 9.10.

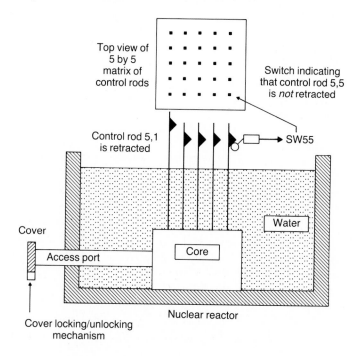

access port to the reactor that prevents it from being opened when the reactor is active. The reactor is active if *any one* of its 5-by-5 matrix of control rods is in a *retracted* position. In order to sense this condition, a limit switch has been placed on each control rod (inputs SW11, SW12, ..., SW45, SW55) that is activated only when the control rod is *not retracted*.

The locking mechanism is to be unlocked (output UNLOCK) *only* when *all* 25 control rod sensors are activated and a switch (input SWITCH) has been activated. In addition, when a control rod is in a retracted position, appropriate row and column lamps are illuminated to help indicate what corrective action is to be taken (outputs ROW1, ROW2, ..., COL1, COL2, ...). Portions of the computer software used to implement this logic are as follows:

```
INPUT    SW11      and    INPUT    SW11
STORE    TEMP             NOT
INPUT    SW12             STORE    TEMP
AND      TEMP             INPUT    SW12
STORE    TEMP             NOT
INPUT    SW13             OR       TEMP
AND      TEMP             STORE    TEMP
STORE    TEMP             INPUT    SW13
INPUT    SW14             :
  :                       INPUT    SW15
INPUT    SW15             NOT
AND      TEMP             OR       TEMP
STORE    TEMP             OUTPUT   ROW1
INPUT    SW21
  :
INPUT    SW55
AND      TEMP
STORE    TEMP
INPUT    SWITCH
AND      TEMP
OUTPUT   UNLOCK
```

If each instruction in the *entire* program takes an average of 4 μs on the programmable logic controller (computer) to be used to control the system, how long must SWITCH be activated to ensure that UNLOCK is actuated? What is the scan time of the logic?

9.11 The relay logic circuit for a computer-controlled x-y table is defined by the ladder diagram in Fig. 9.22(a). The limit switches are used to detect plus and minus x and y overtravel. ENABLE is activated by the computer and is similar to START in the manual mode. The control panel for the operator is shown in Fig. 9.22(b). It is desired to eliminate the relay hardware for the first rung of the ladder diagram and to implement the appropriate logic in the computer where x- and y-axis servo-control functions are already implemented. The second and third rungs represent x and y drive power circuit enabling functions and are not to be modified.

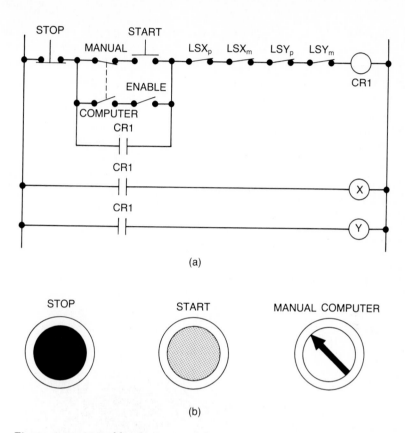

(a)

(b)

Figure 9.22 (a) Ladder diagram and (b) switches on operator's control panel for Problem 9.11.

a. Develop a Boolean equation describing the first rung of the ladder diagram.
b. Show a wiring diagram and computer interfaces for implementing this logic. Use as a starting point the basic hardware shown in Fig. 9.23.
c. Design computer software to implement the logic in part (a), using the hardware in part (b).
d. Determine the scan time if the instruction execution times on the computer are approximately as follows:

Instruction	Time
COMPLEMENT	2×10^{-6} s
LOAD, AND, OR, etc.	4×10^{-6} s
STORE	5×10^{-6} s
INPUT	4×10^{-6} s
OUTPUT	5×10^{-6} s
JUMP	4×10^{-6} s

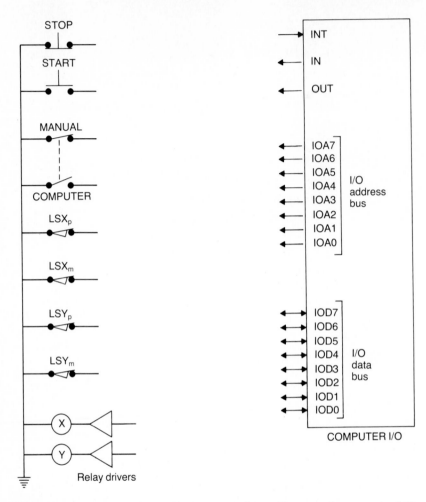

Figure 9.23 Hardware for Problem 9.11. (Refer to Chapter 6 for computer I/O definitions.)

9.12 Using the switching table given in Table 9.10, obtain Boolean equations for the two solenoids and the lamp.

9.13 Using the state-transition diagram in Fig. 9.24, obtain Boolean equations for the states, the two solenoids, and the timer.

9.14 The equations for the two-cylinder system shown in Table 9.9(b) assume that the control logic is always in one of the six defined states. This is not the case for a programmable logic controller that initializes all variables to zero (false) when the system is turned on.

a. How can the equations be modified to include initialization?

b. How can initialization be achieved manually without modifications?

Table 9.10

Switching Table for Problem 9.12

SEQUENCE	START	LS1	LS2	LS3	LS4	SOL1	SOL2	LAMP
0. At rest	0	0	1	0	1	0	0	0
1. SOL1 energized	1	0	1	0	1	1	0	1
2. Transition	0	0	0	0	1	1	0	1
3. SOL2 energized	0	1	0	0	1	1	1	1
4. Transition	0	1	0	0	0	1	1	1
5. SOL2 de-energized	0	1	0	1	0	1	0	1
6. Transition	0	1	0	0	0	1	0	1
7. SOL1 de-energized	0	1	0	0	1	0	0	1
8. Transition	0	0	0	0	1	0	0	1
0. At rest	0	0	1	0	1	0	0	0

417

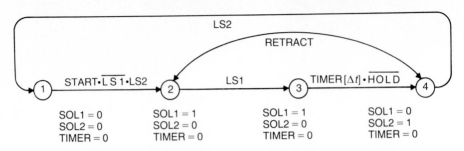

Figure 9.24 State diagram for Problem 9.13.

9.15 Develop a state-transition diagram for the logic shown in the ladder diagram in Fig. 9.19.

9.16 A reaction-time test requires a human subject to jump onto a platform when an amber light is illuminated. An operator sits at the console illustrated in Fig. 9.25, and the subject sits in a cubicle nearby. The sequence of events in the test is as follows:

1. The operator pushes the START button.
2. If the subject is not on the PLATFORM, the AMBER light goes on. Otherwise, the sequence does not proceed.
3. If the subject jumps onto the PLATFORM in less than 1 s, the GREEN light goes on.
4. If the subject does not jump onto the PLATFORM within 1 s, the RED light goes on.

Figure 9.25 Equipment for reaction-time test for Problem 9.16.

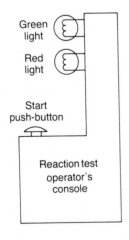

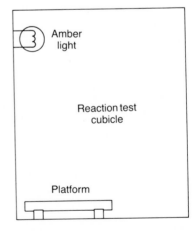

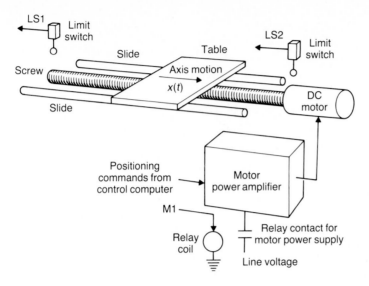

Figure 9.26 Positioning system for Problem 9.17.

5. All lights go off 2 s after the START button is pushed.
6. The process is repeated each time the START button is pushed.

a. Draw a state-transition diagram for the system.
b. Write a Boolean equation for each state in the diagram in part (a).
c. Write a Boolean equation for each output and timer in part (a).

9.17 START and STOP push-buttons are provided in the positioning system shown in Fig. 9.26, along with limit switches LS1 and LS2 that protect the system from overtravel. START is to turn on the motor power supply using a relay M1. Any one of STOP, LS1, LS2, or a signal HALT generated by a position control computer shuts the system off. Design the logic needed to control the relay M1. What type of sequential control hardware should be used to implement this logic?

9.18 A wood-cutting machine is shown schematically in Fig. 9.27. It consists of two pneumatic cylinders that actuate clamps for the workpieces and a third cylinder that moves a saw. The sequence of operation in the system is as follows:

1. Push-button PB1 sets the clamps lightly on the workpiece with 10 psi air pressure. This allows fine adjustment of workpiece position by the operator.
2. A second push-button, PB2, sets the clamps hard on the workpiece with 80 psi air pressure and advances the saw. The speed of the cut can be adjusted by restricting flow into the saw cylinder as shown in Fig. 9.27.
3. When the second push-button is released, the saw returns quickly to the retracted position. This is done by application of full pressure to the saw cylinder.

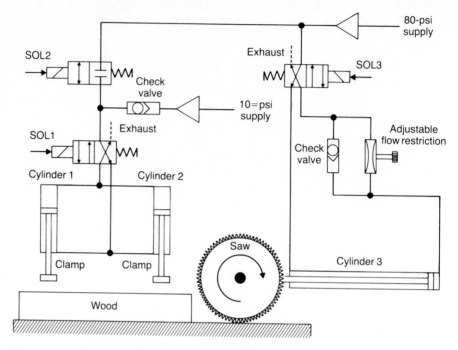

Figure 9.27 Pneumatically operated wood-cutting machine for Problem 9.18.

4. Releasing push-button PB1 releases the clamps, which are held open with 10 psi air pressure.

It is desired to automate this sequence so that a single person can operate two machines at once. Design the logic and hardware required to control the system.

Chapter 10
Process Modeling

INTRODUCTION

Characterizing the dynamic behavior of the process to be controlled can be a challenging part of the control system design problem. Usually a continuous model is developed from the physical properties of the system. The techniques presented in Chapter 2 are then used to obtain a discrete model. This deterministic approach can be very successful when one is working with continuous processes that behave in a linear manner and when the coefficients of the resulting differential and algebraic equations are accurately known.

In practice, however, the results obtained by physical modeling are often disappointing. The technique tends to result in models of high order, making further use of the models difficult. Coefficient values for the models are sometimes not well known and may be difficult to measure. Nonlinearities may be present that are difficult to characterize. In many cases, so many approximations must be made along the way that the final result is only a poor estimate of the dynamics of the actual process. For example, it may be possible to observe the flow of fluid through two tanks in tandem and to say immediately that the system appears to be second-order. But when we inquire what the coefficients of the process model are, the analytical problem is no longer simple. What are the flow characteristics of the valves? Is the flow laminar or turbulent? The final result in some cases may be no closer to the true model than a good guess.

Step-response modeling and stochastic modeling are techniques that can be used as alternatives to the physical modeling approach. In these approaches, known inputs are applied to the system and the outputs are observed. In the case

421

of step-response modeling, a step input is applied to a process and its output is observed. The transient and steady-state characteristics of this response are then analyzed and a model is developed. In the case of stochastic modeling, inputs (often randomly generated) are delivered to a process, and sample sequences of both the process input and the process output are recorded. Coefficients in process models are then determined that result in a model that predicts the outputs observed on the basis of the inputs given.

The objective of this chapter is to review these approaches for process modeling. Physical modeling will be discussed first, along with the problem of linearizing nonlinear models. Step-response modeling will then be discussed. Finally, stochastic modeling using the technique of linear least-squares estimation will be presented. Emphasis is placed on obtaining a useful discrete model for a process so that its behavior can be predicted. This allows control systems to be designed for the process that achieve required levels of closed-loop performance.

10.2

PHYSICAL MODELING

Physical modeling requires a mathematical definition of the dynamic properties of all the physical components that make up a process. Consider the case of a tank from which a liquid flows through an outlet valve. Here we have three basic components: the tank, the valve, and the liquid in the tank. To characterize the process of liquid flowing out of the tank, it is necessary to know the geometry of the tank and flow as a function of liquid level in the tank and valve position. If these can be described by mathematical functions, it is possible to formulate a physical model for the process. Such a model would allow the prediction, for example, of the liquid level in the tank as a function of time and valve position.

Steps in Physical Modeling

The modeling of the physical properties of motors, amplifiers, pumps, valves, linkages, chemical reactions, and the like is a broad field and is outside the scope of this chapter. However, the general technique of physical modeling can be summarized as requiring the following steps. The objective is obtaining a linear process model that can be converted subsequently into discrete form:

1. Partition the process into its component parts.
2. Develop a model for each component.
3. Linearize any nonlinear component models.
4. Combine component models into a process model.
5. Simplify the process model.
6. Verify the process model.

Example 10.1 Liquid-level system Consider the liquid-level system shown in Fig. 10.1. Assume that the tank is cylindrical with an area A. If $q_i(t)$ is the rate of liquid flow into the tank and $q_o(t)$ is the rate of liquid flow out of the tank, then the rate of change in the height of liquid in the tank can be expressed as

$$A \frac{dh(t)}{dt} = q_i(t) - q_o(t)$$

where $h(t)$ is the height of the liquid. Furthermore, valve position is assumed to be constant, and the rate of liquid flow out of the tank is a linear function of liquid height. This can be expressed as

$$q_o(t) = kh(t)$$

where k is a constant. Combining the above equations yields the following differential equation relating height to input flow:

$$A \frac{dh(t)}{dt} + kh(t) = q_i(t)$$

In this example a simple equation is obtained. The process was assumed to behave in a linear manner, and no linearization is required. Referring to Table 2.1, we find that this continuous model can be written in discrete form as

$$h_n = \delta h_{n-1} + \frac{1}{k}(1 - \delta)q_{i_{n-1}}$$

where $\delta = e^{-kT/A}$. □

Linearization of Models

In practice, most physical processes exhibit some degree of nonlinearity and the behavior of the process can be quite different when process variables vary

Figure 10.1 Liquid-level system.

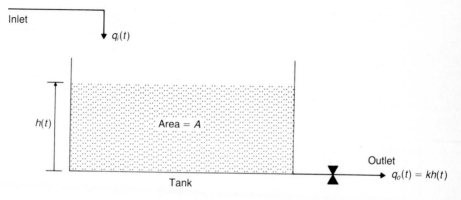

Inlet

$q_i(t)$

$h(t)$

Area $= A$

Outlet

$q_o(t) = kh(t)$

Tank

between extreme values. Fortunately, many processes are relatively linear in normal operation. If it is desired to obtain a linear model that satisfactorily describes a nonlinear process, some consideration of the numerical values and variation of process variables is usually required.

In general, it is possible to pick representative values for the process variables and to linearize the process model about that operating point. For a given function $f(x, y, z, \ldots)$ and a specified operating point $\bar{x}, \bar{y}, \bar{z}, \ldots$, a linearized equation can be obtained by taking the partial derivatives of the function with respect to each variable and evaluating the partial derivatives at that operating point. As long as the variables $x, y, z, \ldots$ remain close to the values specified as the operating point, then

$$f(x, y, z, \ldots) \approx f(\bar{x}, \bar{y}, \bar{z}, \ldots) + (x - \bar{x}) \left.\frac{\partial f}{\partial x}\right|_{\bar{x}, \bar{y}, \bar{z}, \ldots} \tag{10.1}$$

$$+ (y - \bar{y}) \left.\frac{\partial f}{\partial y}\right|_{\bar{x}, \bar{y}, \bar{z}, \ldots} + (z - \bar{z}) \left.\frac{\partial f}{\partial z}\right|_{\bar{x}, \bar{y}, \bar{z}, \ldots} + \cdots$$

This equation is linear in all variables $x, y, z, \ldots$. The magnitude of errors in the linearized model clearly depends on the nature of the nonlinear function, its partial derivatives, the choice of operating point, and the excursion of the process variables away from that operating point.

Example 10.2 Model linearization   Suppose that the tank in Fig. 10.2(a) receives a nearly constant fluid flow input $\bar{q}_i$. Flow out of the tank through the valve is adjustable by valve position $\theta(t)$ and is described by the following nonlinear function of $\theta(t)$ and height $h(t)$:

$$q_o(t) = k\theta(t)\sqrt{h(t)}$$

The differential equation describing the system is

$$A\frac{dh(t)}{dt} + k\theta(t)\sqrt{h(t)} = q_i(t)$$

This is a nonlinear equation, and it is desired to obtain a linearized equation about a particular mean valve position.

If the mean valve position $\bar{\theta}$ is known, the corresponding steady-state height $\bar{h}$ can be found by setting $dh(t)/dt = 0$.

$$k\bar{\theta}\sqrt{\bar{h}} = \bar{q}_i$$

or

$$\bar{h} = \left(\frac{\bar{q}_i}{k\bar{\theta}}\right)^2$$

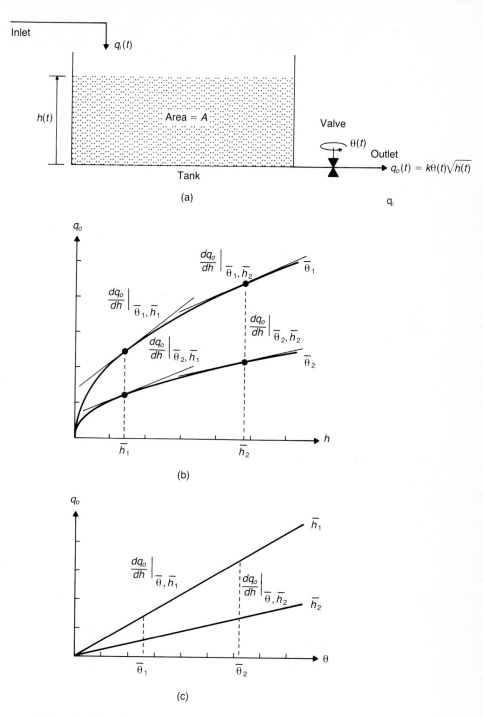

Figure 10.2 (a) Nonlinear liquid-level system; (b) output flow vs. liquid level for two valve positions; (c) output flow vs. valve position for two liquid levels.

Here $\bar{q}_i$, $\bar{\theta}$, and $\bar{h}$ represent the operating point selected for linearizing the process model. Figures 10.2(b) and 10.2(c) show the value of q_o when θ and h are constant, respectively.

Applying Eq. (10.1) results in

$$q_o(t) \approx q_o \bigg|_{\bar{\theta},\bar{h}} + (\theta(t) - \bar{\theta})\frac{\partial q_o}{\partial \theta}\bigg|_{\bar{\theta},\bar{h}} + (h(t) - \bar{h})\frac{\partial q_o}{\partial h}\bigg|_{\bar{\theta},\bar{h}}$$

or

$$q_o(t) \approx -\frac{1}{2}k\bar{\theta}\sqrt{\bar{h}} + k\sqrt{\bar{h}}\,\theta(t) + \frac{k\bar{\theta}}{2\sqrt{\bar{h}}}h(t)$$

The linearized tank model is then

$$A\frac{dh(t)}{dt} + \frac{k\bar{\theta}}{2\sqrt{\bar{h}}}h(t) = q_i(t) + \frac{1}{2}k\bar{\theta}\sqrt{\bar{h}} - k\sqrt{\bar{h}}\,\theta(t)$$

or

$$\tau\frac{dh(t)}{dt} + h(t) = k'(q_i'(t) - \theta'(t))$$

where

$$\tau = \frac{2\bar{q}_i A}{k^2\bar{\theta}^2}$$

$$k' = \frac{2\bar{q}_i^2}{k^2\bar{\theta}^2}$$

and

$$q_i'(t) = \frac{q_i(t)}{\bar{q}_i} - 1$$

$$\theta'(t) = \frac{\theta(t)}{\bar{\theta}} - \frac{3}{2}$$

The discrete form of model is

$$h_n = e^{-T/\tau}h_{n-1} + k'(1 - e^{-T/\tau})(q_{i_{n-1}}' - \theta_{n-1}')$$

where T is the sample period. Note that the effective gain, k', varies with $\bar{q}_i$ and $\bar{\theta}$. If it were necessary to design a control for this system, it would be desirable to know the maximum value of $\bar{q}_i$ and the minimum value of $\bar{\theta}$ so that the maximum expected value of k' would be known. $\square$

Simplification of Process Models

Often, the result of physical modeling of a complex process is a high-order process transfer function. Some of the terms in this process transfer function may correspond to very short time constants that, when compared to other, longer time constants present in the model, are insignificant from a practical point of view. These terms can often be ignored, greatly simplifying the model with little loss of accuracy.

Example 10.3 Model simplification   If a process output is to be controlled with a closed-loop bandwidth of ω_c, then terms in the open-loop transfer function with natural frequencies significantly greater than ω_c or time constants cantly less than $1/\omega$ can often be ignored. Consider the case of a machine drive (motor, amplifier, and position sensor) that is known to behave according to the differential equation

$$\frac{1}{\omega_n{}^2}\frac{d^3\theta(t)}{dt^2} + \frac{2\zeta}{\omega_n}\frac{d^2\theta(t)}{dt^2} + \frac{d\theta(t)}{dt} = kv(t)$$

where $\theta(t)$ is the motor shaft position, ω_n is the natural frequency, ζ is the damping ratio, $v(t)$ is the amplifier command input voltage, and k is the gain of the drive. If the position of the motor is to be controlled, the drive is neither highly underdamped nor highly overdamped, the bandwidth of the closed-loop system is to be ω_c, and if

$$\omega_c \ll \omega_n$$

then a simplified model for the system is

$$\frac{d\theta(t)}{dt} \approx kv(t)$$

For example, if the natural frequency of the drive is 70 Hz, $\zeta = 0.7$, and the desired closed-loop bandwidth is 3 Hz, then

$$3 \text{ Hz} \ll 70 \text{ Hz}$$

There is a good possibility that this simplfied process model will produce as good a result as the original transfer function when it is used to design a closed-loop controller. However, it is important to remember that short time constants and/or high natural frequencies have been dropped from the model, and response to high-frequency inputs is therefore not accurately modeled. Attempts to achieve closed-loop bandwidths of frequencies on the same order of magnitude as the ignored process dynamics can often result in unstable or excessively oscillatory closed-loop systems, a result not predicted with the simplified process model. □

Verification of the Process Model

It is a good idea to verify the accuracy of the process model obtained via the physical modeling approach by comparing the predicted performance with the performance of the actual process. Usually, a comparison can be made between the response to known inputs such as step functions or sinusoids. Any significant deviation of the predicted response from the actual response is an indication that the process model is inadequate. Possible sources of inaccuracy include

1. Inaccurate models for one or more process components
2. Incorrect values of one or more coefficients in the process model
3. Unknown nonlinearities in the process
4. Linearization of nonlinear models

10.3

STEP-RESPONSE MODELING

Step-response modeling has an advantage over physical modeling in that only the experimental, time-domain response of the process to a step input is required to obtain a process model. A step-response test can be carried out relatively quickly and is relatively easy to analyze as long as a high-order dynamic model is not required. The only instrumentation needed is a device to generate the step input and a device capable of recording changes in the process output in response to the step input. An initial inspection of the step response reveals the general form of the process, and the coefficients in the process model can then be determined. Often, simplified models can be obtained by representing higher-order dynamic components as a dead-time (delay) term in the model.

Delay

An example of the step response of a delay process is shown in Fig. 10.3. The discrete process model for this system is

$$c_n = km_{n-d} \tag{10.2}$$

where d is found from the delay D as

$$d = \frac{D}{T} \tag{10.3}$$

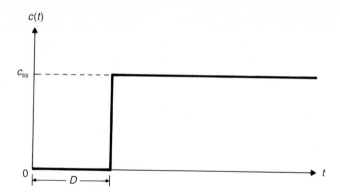

Figure 10.3 Step response of a delay process.

Here d must be an integer, and any fractional part is usually rounded up to the next highest integer, resulting in a more conservative model (more delay). The process gain k can be found from the input step size M as

$$k = \frac{c_{ss}}{M} \tag{10.4}$$

Integration

An example of the step response of an integrating process is shown in Fig. 10.4. The discrete process model for this system is

$$c_n = c_{n-1} + kTm_{n-1} \tag{10.5}$$

Figure 10.4 Step response of an integrating process.

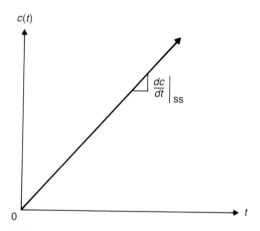

The process gain k can be found from the slope of the response $\left.\dfrac{dc}{dt}\right|_{ss}$, and the size of the step M can be found from the following relationship:

$$k = \frac{\left.\dfrac{dc}{dt}\right|_{ss}}{M} \tag{10.6}$$

Integration with Dead Time

The step response of a process with integration and first-order components is shown in Fig. 10.5. The rate of change in the process output is initially zero and eventually reaches a steady-state value. If a tangent to the response curve is constructed with a slope of $\left.\dfrac{dc}{dt}\right|_{ss}$, the time at which the tangent crosses $c(t) = 0$ can be considered an approximate delay D. An approximate process model can then be obtained by representing the initial portion of the response by the delay D. The discrete model is then

$$c_n = c_{n-1} + kTm_{n-1-d} \tag{10.7}$$

where d is found from the delay D using the relationship

$$d = \frac{D}{T} \tag{10.8}$$

Figure 10.5 Step response of first-order process with integration.

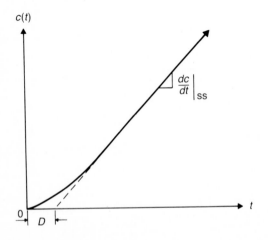

Again d must be an integer and, in practice, d should be rounded up to the next highest integer to produce a more conservative model. The gain k is again obtained from Eq. (10.6).

This approach can also be applied to integrating processes with second- and higher-order components. It is often difficult in practice to obtain enough detail and resolution in the recorded step response to determine the exact dynamics of the process. Hence all dynamic components except the integration are lumped into the dead-time term.

First-Order Systems

Figure 10.6 illustrates the response of a first-order system to a step input. In this case the process output reaches a steady-state value of c_{ss} after a time interval of about 5τ, where τ is the process time constant. The discrete model for the process is

$$c_n = \delta c_{n-1} + k(1 - \delta)m_{n-1} \tag{10.9}$$

where

$$\delta = e^{-T/\tau} \tag{10.10}$$

The time constant τ is estimated by determining the length of time required for the process output to reach 63.2% of its final value. In other words, the time constant τ is found by locating the time when

$$c(\tau) = 0.632c_{ss} \tag{10.11}$$

The gain k is found from the step size M as

$$k = \frac{c_{ss}}{M} \tag{10.12}$$

Figure 10.6 Step response of first-order process.

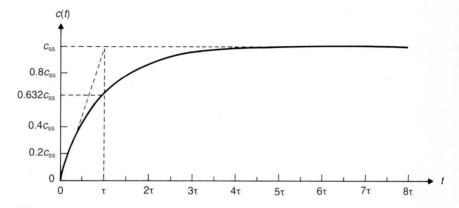

First-Order Systems with Dead Time

Overdamped second-order and higher-order systems can often be approximated by a first-order system with dead time. The approximate discrete model, as shown in Fig. 10.7, is then

$$c_n = \delta c_{n-1} + k(1 - \delta)m_{n-1-d} \tag{10.13}$$

Points on the response curve can again be used to determine model coefficients, and it is convenient to use the point at which the change in the process output reaches 63.2% of its final value. The time at which this point is reached is $t_{0.632}$. Because not only the time constant but also a delay must be determined, another point is required. Here it is convenient to use the point at which 39.3% of the final response is reached. If this time is called $t_{0.393}$, both the time constant τ and the delay D can then be determined from

$$\tau = 2(t_{0.632} - t_{0.393}) \tag{10.14}$$

and

$$D = t_{0.632} - \tau \tag{10.15}$$

It is then possible to determine d, δ, and k for Eq. (10.13) from Eq. (10.8), (10.10), and (10.12), respectively.

Second-Order Systems

The response of a second-order process to a step input can be completely described by the natural frequency ω_n and the damping ratio ζ. Figure 10.8 shows the response of processes with various damping ratios as a function of normalized

Figure 10.7 Step response of first-order process with dead time D.

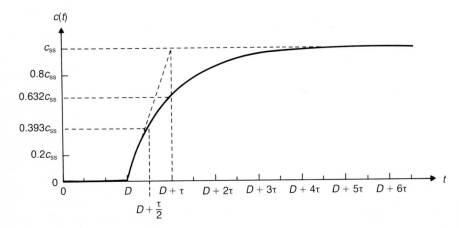

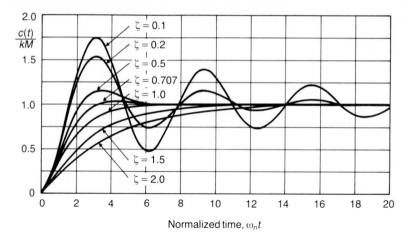

Figure 10.8 Response of a second-order process to a step input.

time $\omega_n t$. The amplitude of the response is also normalized to be independent of step size.

Figure 10.8 can be used to estimate the damping ratio ζ by comparing the form of the observed second-order response to those in the figure. The shape of the response near $t = 0$, the amount of overshoot, and the number of oscillations are important features to consider. The natural frequency ω_n can then be determined by comparing the time scale of the observed response with the normalized time scale in the figure.

Once the damping ratio and the natural frequency have been obtained, the discrete model can be determined. The model is

$$c_n = a_1 c_{n-1} + a_2 c_{n-2} + b_1 m_{n-1} + b_2 m_{n-2} \tag{10.16}$$

where the values of the coefficients a_1, a_2, b_1, and b_2 are calculated by using the expressions given in item 6 of Table 2.1.

A model for a second-order process with a damping ratio that lies outside those given in Fig. 10.8 can also be obtained. Techniques for doing so are described in the following sections. The special case of critical damping is also discussed.

Highly Underdamped Processes

For a highly underdamped process, the log-decrement technique can be used to estimate the damping ratio. If an exponential envelope is drawn tangent to the peaks of the oscillation of the step response shown in Fig. 10.9, it is possible to

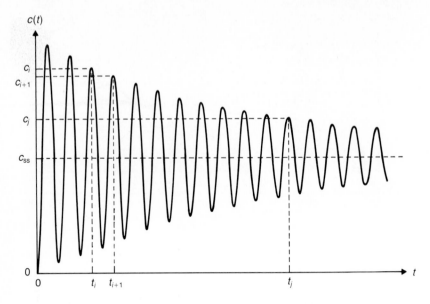

Figure 10.9 Step response of a highly underdamped second-order process.

determine the damping ratio from the height of two peaks, c_i and c_j, by using the
equation

$$\zeta \approx \frac{1}{2\pi(j-i)} \ln\left(\frac{c_i - c_{ss}}{c_j - c_{ss}}\right) \tag{10.17}$$

Equation 10.17 is based on the assumption that

$$\sqrt{1 - \zeta^2} \approx 1 \tag{10.18}$$

For larger damping ratios, a value of ζ can be found from adjacent peaks, using

$$\zeta = \frac{\ln\left(\dfrac{c_i - c_{ss}}{c_{i+1} - c_{ss}}\right)}{\left[4\pi^2 + \ln\left(\dfrac{c_i - c_{ss}}{c_{i+1} - c_{ss}}\right)\right]^{1/2}} \tag{10.19}$$

Once the damping ratio has been determined, the natural frequency can be found
from

$$\omega_n = \frac{2\pi(j-i)}{(t_j - t_i)} \frac{1}{\sqrt{1 - \zeta^2}} \tag{10.20}$$

where t_i and t_j are the time at peaks i and j, respectively.

Highly Overdamped Processes

A highly overdamped process can often be approximated by a first-order process with dead time. These processes possess two time constants, τ_1 and τ_2, where $\tau_1 \gg \tau_2$. The longer time constant τ_1 dominates the response, which is nearly first-order. The effects of the shorter time constant τ_2 can be represented as a delay of

$$d = \frac{\tau^2}{T} \qquad (10.21)$$

where d is an integer. The resulting discrete first-order plus dead-time process model is then

$$c_n = \delta_1 c_{n-1} + k(1 - \delta_1)m_{n-1-d} \qquad (10.22)$$

where

$$\delta_1 = e^{-T/\tau_1} \qquad (10.23)$$

Critically Damped Processes

For the special case of a critically damped process, it can be seen in Fig. 10.8 that for $\zeta = 1$, the response reaches 59.4% of its steady-state value at time $t = 2\tau$, where $\tau = 1/\omega_n$. If the time at which this level is reached is $t_{0.594}$, the time constant τ can be found from

$$\tau = \frac{1}{2}t_{0.594} \qquad (10.24)$$

and hence,

$$\omega_n = 2/t_{0.594} \qquad (10.25)$$

Limitations

The main attractions of the step response method are its simplicity and the fact that it is often possible to represent higher-order processes as lower-order processes with dead time. The step response predicted by models obtained using this method can be compared with the step response of the actual process in order to access the accuracy of the model and to adjust the coefficients of the model, if necessary, to obtain a more accurate fit. However, analysis of higher-order processes by the step-response technique is difficult. The statistical techniques presented in the next section provide a more generalized and more easily implemented way of determining the best discrete process model.

It is important to remember that not all processes are amenable to step-response testing. Processes that incorporate differentiation cannot usually be tested with step inputs. This is because the derivative of a step input has an infinite magnitude. Some physical limit is reached in such a process that causes the observed step response to deviate from that predicted. In general, the magnitude of the step input applied to the process should be carefully considered so as not to cause nonlinear response due to limiting. Often, large step inputs are not found during normal operation of the process.

Example 10.4 Process with limiting element To illustrate the effects of limiting on both observed process step response and system performance under closed-loop control, consider the motor position control system shown in Fig. 10.10(a). If the amplifier has an output voltage limit, the motor velocity cannot exceed a maximum value, regardless of the magnitude of the amplifier input voltage $m(t)$. In this case, the amplifier and motor can be modeled as shown in Fig. 10.10(b), where a limiting function appears between the amplifier gain and the motor.

Suppose that $k_a = 40$ volts/volt, $k_m = 15$ rpm/volt, and the maximum amplifier output voltage is 200 volts. Figure 10.10(c) shows the response of motor position to step inputs of various magnitudes. Note that motor speed does not increase once the magnitude of the step input exceeds 5 volts and the amplifier reaches its output voltage limit of 200 volts. Step-response tests conducted with input magnitude above 5 volts would lead to incorrect conclusions about the gain of the process, which in this case would be underestimated.

The existence of a limiter such as that shown in Fig. 10.10(b) can also lead to deviations in closed-loop performance. Suppose that a proportional control with a sample period of 0.005 s has been designed for the system in Fig. 10.10(a) and that it achieves approximately first-order response to a step command input with a time constant of 0.06 s. As can be seen in the closed-loop step response in Fig. 10.10(d), the expected response is achieved for step inputs of up to 0.25 revolutions. Above this, the amplifier output voltage limit comes into effect, and response is slower in terms of the time required to reach the commanded position.

The effects of limiting become more serious when integration is added to control systems. If the approximate PID control algorithm described by Eq. (2.16) were to be used to control motor position, the response to step commands would be similar to those shown in Fig. 10.10(e). When the voltage limit comes into effect at the higher step-command magnitudes, the controller output integrates to a high value because of the inability of the process to instantaneously follow the step input. This integrated error can be removed only by negative errors accumulated after motor position overshoots the commanded value. The size and duration of the overshoot depend on controller gains and step-command magnitude. This characteristic can be eliminated by removing integration from

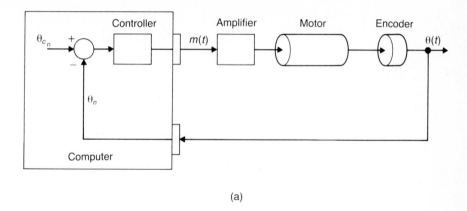

(a)

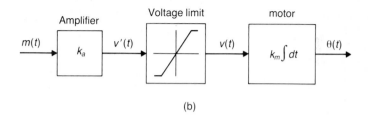

(b)

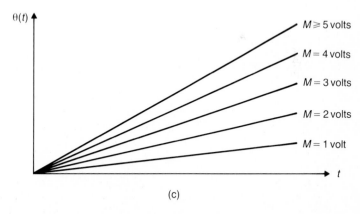

(c)

Figure 10.10 (a) Motor shaft position control system; (b) motor/amplifier process model with amplifier output voltage limit; (c) change in motor position for various step amplifier inputs $m(t) = M$; (d) step response of a closed-loop system with proportional control, illustrating limiting; (e) step response with proportional plus integral control, illustrating "wind up."

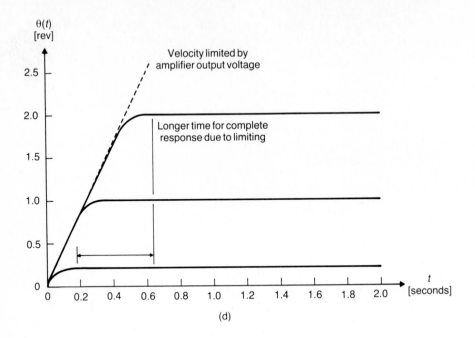

(d)

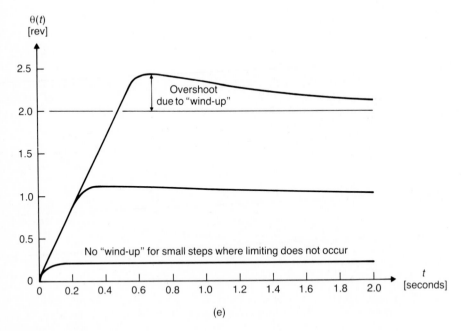

(e)

Figure 10.10 (*cont.*)

the control algorithm, detecting that the limit has been reached and disabling integration of errors in the control, or by ensuring that command inputs do not drive the process into a limiting mode. This last strategy can be readily accomplished by considering the physical limitations of the process and using command generation techniques such as those decribed in Chapter 8 to ensure that the process is not issued commands (such as step inputs) that it is physically incapable of accurately following. □

10.4

STOCHASTIC MODELING

It is possible to estimate the values of coefficients in a discrete model for a process by statistical analysis of samples taken from the system input and output. Although the exact nature of the process may be unknown, the process input m_n can be manipulated and the resulting process output c_n can be measured. Several statistical techniques are available for obtaining estimates of the coefficients in the process model. The technique of linear least-squares estimation is the most straightforward, and our discussion will be limited to this technique. It is beyond the scope of this work to present a complete review of either stochastic modeling or the statistical methods necessary to fully evaluate models obtained using these techniques. The reader is directed to the bibliography at the end of this chapter for further study of these topics.

Least-squares estimation involves finding the coefficients of the model that will "best fit" a set of empirical data. The coefficients that result are called the least-squares estimates. The first step in the estimation problem is deciding on a process model. This step is called **identification**. The second step is finding the model coefficients that "best fit" the data. This step is called **estimation**. The third and final step deals with determining the correctness of the model; it is called **diagnostic checking**.

In the following sections, the special case of fitting a first-order model with dead time to sampled process input and output data is first discussed. Then the technique is generalized to models of higher order by using a matrix approach. Finally, diagnostic checking techniques are described, including analysis of residual errors, autocorrelation, and determination of confidence regions for estimated parameters.

First-Order Process with Dead Time

The difference equation for a first-order system with a delay of d sample intervals can be written as

$$c_n = e^{-T/\tau} c_{n-1} + k(k - e^{-T/\tau}) m_{n-1-d} + \varepsilon_n \tag{10.26}$$

where ε_n is the experimental error caused by various sources of noise in the process. These errors are assumed to be uncorrelated over time and to have a zero mean. Equation (10.26) can be rewritten as

$$c_n = \alpha c_{n-1} + \beta m_{n-1-d} + \varepsilon_n \tag{10.27}$$

where

$$\alpha = e^{-T/\tau} \tag{10.28}$$

and

$$\beta = k(1 - \alpha) \tag{10.29}$$

If the dead time d is assumed to be known, the least-squares estimates of α and β can be found by minimizing the sum of the squared errors. This quantity can be expressed as

$$S(\alpha, \beta) = \sum_{n=1}^{N} \varepsilon_n^2 \tag{10.30}$$

or, from Eq. (10.27),

$$S(\alpha, \beta) = \sum_{n=1}^{N} (c_n - \alpha c_{n-1} - \beta m_{n-1-d})^2 \tag{10.31}$$

The minimum value of the sum-of-squares function can be obtained by setting the two partial derivatives of $S(\alpha, \beta)$ equal to zero. The partial derivatives are

$$\frac{\partial S(\alpha, \beta)}{\partial \alpha} = -2 \sum_{n=1}^{N} (c_n - \alpha c_{n-1} - \beta m_{n-1-d}) c_{n-1} \tag{10.32}$$

$$\frac{\partial S(\alpha, \beta)}{\partial \beta} = -2 \sum_{n=1}^{N} (c_n - \alpha c_{n-1} - \beta m_{n-1-d}) m_{n-1-d} \tag{10.33}$$

where N is the number of data points sampled.

A set of equations called the **normal equations** are found by setting these partial derivatives equal to zero. After simplifcation, the result is

$$\hat{\alpha} \Sigma c_{n-1}^2 + \hat{\beta} \Sigma m_{n-1-d} c_{n-1} = \Sigma c_n c_{n-1} \tag{10.34}$$

$$\hat{\alpha} \Sigma c_{n-1} + m_{n-1-d} + \hat{\beta} \Sigma m_{n-1-d}^2 = \Sigma c_n m_{n-1-d} \tag{10.35}$$

where the limits on the summations are the same as in Eq. (10.30).

The "hats" above the greek letters in the notations $\hat{\alpha}$ and $\hat{\beta}$ denote that these values are estimates of α and β and are not necessarily their true values in the actual process.

Solving for $\hat{\alpha}$ and $\hat{\beta}$ in Eq. (10.34) and (10.35) gives the results

$$\hat{\alpha} = \frac{\Sigma c_n c_{n-1} \Sigma m_{n-1-d}^2 - \Sigma c_{n-1} m_{n-1-d} \Sigma c_n m_{n-1-d}}{\Delta} \qquad (10.36)$$

and

$$\hat{\beta} = \frac{\Sigma c_{n-1}^2 \Sigma c_n m_{n-1-d} - \Sigma c_n c_{n-1} \Sigma c_{n-1} m_{n-1-d}}{\Delta} \qquad (10.37)$$

where

$$\Delta = \Sigma c_{n-1}^2 \Sigma m_{n-1-d}^2 - (\Sigma c_{n-1} m_{n-1-d})^2 \qquad (10.38)$$

A slight problem can arise in that, if values of c and m are recorded starting at $n = 0$ and $d > 0$, additional samples are required prior to the values taken in the data. There are two ways to solve this problem. The first is to assume that the starting values are zero. The second is to begin the summations at $n = d + 1$ rather than $n = 1$. This decreases the amount of data actually used in computing the estimates but is not normally a problem. Estimates of τ and k are obtained from Eq. (10.28) and (10.29) as

$$\hat{\tau} = \frac{-T}{\ln(\hat{\alpha})} \qquad (10.39)$$

and

$$\hat{k} = \frac{\hat{\beta}}{1 - \hat{\alpha}} \qquad (10.40)$$

The delay d was assumed to be known when $\hat{\alpha}$ and $\hat{\beta}$ were obtained in Eq. (10.36) and (10.37). In general, this may not be the case, and d is an additional parameter that must be determined. The following procedure can then be used to determine a model for the process from sampled data.

1. Find $\hat{\alpha}$, $\hat{\beta}$, and $\Sigma \varepsilon_n^2$ for $d = 0$.
2. Using $d + 1$, find new values of $\hat{\alpha}$, $\hat{\beta}$ and $\Sigma \varepsilon_n^2$.
3. If the value of $\Sigma \varepsilon_n^2$ for $d + 1$ is less than the value for d, then $d + 1$ is a better value.
4. Repeat steps 2 and 3 by increasing d until the value of $\Sigma \varepsilon_n^2$ for $d + 1$ is greater than for d.
5. The values of $\hat{\alpha}$ and $\hat{\beta}$ for d can then be assumed to provide the best fit. Use Eq. (10.39) and (10.40) to determine $\hat{\tau}$ and $\hat{k}$.

Example 10.5 First-order process with no delay In order to test a certain chemical process, a program such as Program 10.1 is used to output a pulse to the

```
program Collect_data;
   var
      mdata, cdata : array[0..100] of integer;    {manipulation and feedback arrays}
      n, i : integer;
      clock : boolean;
begin
   readln(n);                                      {number of manipulations is n+1}
   for i := 0 to n do
      readln(mdata[i]);                            {list of manipulations for process}
   for i := 0 to n do
      begin
         repeat                                    {wait for clock=0}
            LDC(clock, 1)                          {clock is on logic input channel 1}
         until not clock;
         repeat                                    {wait for 0 to 1 clock transition}
            LDC(clock, 1)
         until clock;
         DAC(mdata[i], 1);                         {manipulate using D/A channel 1}
         ADC(cdata[i], 1)                          {sample feedback using A/D channel 1}
      end
end.
```

Program 10.1 Process manipulation and data collection for model fitting.

process and record the process response. Given the data shown in Table 10.1, it is desired to find the time constant and gain of the process if the model is of first order and the data were collected at a sample period of 0.5 s. Assume that there is no process delay ($d = 0$).

The least-squares estimates of α and β follow from Eq. (10.36) and (10.37) and the data in Table 10.1. The results are

$$\hat{\alpha} = 0.718$$

and

$$\hat{\beta} = 0.565$$

From Eqs. (10.39) and (10.40) it follows that

$$\hat{\tau} = 1.5$$

Table 10.1

Data from First-Order Process with $T = 0.5$ s

n	0	1	2	3	4	5	6	7	8
m	1.0	1.0	1.0	1.0	0.0	0.0	0.0	0.0	0.0
c	0.0	0.57	0.97	1.26	1.47	1.06	0.76	0.54	0.39

and

$$\hat{k} = 2.0$$

where $\hat{\tau}$ and $\hat{k}$ are estimates of the actual process parameters τ and k. □

Estimating Higher-Order Process Parameters

If the discrete model of a process is put in the form

$$c_n = \alpha_1 c_{n-1} + \alpha_2 c_{n-2} + \cdots + \alpha_r c_{n-r}$$
$$+ \beta_1 m_{n-1-d} + \beta_2 m_{n-2-d} + \cdots + \beta_s m_{n-s-d} + \varepsilon_n \tag{10.41}$$

it is relatively easy to obtain the least-squares estimates of the parameters α_i, $i = 1, 2, 3, \ldots, r$ and β_j, $j = 1, 2, 3, \ldots, s$.

Each observation c_n for $n = 1, 2, 3, \ldots, N$ results in one equation of the form of Eq. (10.41). This set of N equations can be expressed in matrix form as

$$\mathbf{c} = \mathbf{Xb} + \boldsymbol{\varepsilon} \tag{10.42}$$

where

$$\mathbf{c} = \begin{bmatrix} c_1 \\ c_2 \\ c_3 \\ \vdots \\ c_N \end{bmatrix}, \quad \boldsymbol{\varepsilon} = \begin{bmatrix} \varepsilon_1 \\ \varepsilon_2 \\ \varepsilon_3 \\ \vdots \\ \varepsilon_N \end{bmatrix}, \quad \mathbf{b} = \begin{bmatrix} \alpha_1 \\ \alpha_2 \\ \alpha_3 \\ \vdots \\ \alpha_r \\ \beta_1 \\ \beta_2 \\ \beta_3 \\ \vdots \\ \beta_s \end{bmatrix}$$

and

$$\mathbf{X} = \begin{bmatrix} c_0 & c_{-1} & \cdots & c_{1-r} & m_{-d} & m_{-1-d} & \cdots & m_{1-s-d} \\ c_1 & c_0 & \cdots & c_{2-r} & m_{1-d} & m_{-d} & \cdots & m_{2-s-d} \\ \vdots & \vdots & & \vdots & \vdots & \vdots & & \vdots \\ c_{N-1} & c_{N-2} & \cdots & c_{N-r} & m_{N-1-d} & m_{N-2-d} & \cdots & m_{N-s-d} \end{bmatrix}$$

Data $c_{1-r}, c_{2-r}, \ldots, c_{N-1}, c_N$ and $m_{1-s-d}, m_{2-s-d}, \ldots, m_{N-2-d}, m_{N-1-d}$ must be acquired to fill the elements of the arrays and vectors in Eq. (10.42).

The residual errors are

$$\boldsymbol{\varepsilon} = \mathbf{c} - \mathbf{Xb} \tag{10.43}$$

and the sum of the errors squared is

$$S(\mathbf{b}) = \boldsymbol{\varepsilon}^T \boldsymbol{\varepsilon} = (\mathbf{c} - \mathbf{Xb})^T(\mathbf{c} - \mathbf{Xb}) \tag{10.44}$$

The set of normal equations is found by setting

$$\frac{\partial S(\hat{\mathbf{b}})}{\partial \hat{b}_k} = 0 \tag{10.45}$$

for each element $\hat{b}_k$, $k = 1, 2, \ldots, r + s$ in $\hat{\mathbf{b}}$, the vector of estimated model parameters.

From the rules of linear algebra,

$$\frac{\partial S(\mathbf{b})}{\partial b_i} = \frac{\partial(\boldsymbol{\varepsilon}^T \boldsymbol{\varepsilon})}{\partial b_i} = 2\boldsymbol{\varepsilon}^T \frac{\partial \boldsymbol{\varepsilon}}{\partial b_i} \tag{10.46}$$

Combining all $r + s$ equations obtained using Eq. (10.46) yields

$$\mathbf{X}^T[\mathbf{c} - \mathbf{X}\hat{\mathbf{b}}] = 0 \tag{10.47}$$

Solving Eq. (10.47) for $\hat{\mathbf{b}}$ gives the result

$$\hat{\mathbf{b}} = [\mathbf{X}^T\mathbf{X}]^{-1}\mathbf{X}^T\mathbf{c} \tag{10.48}$$

It should be noted that inversion of matrix $\mathbf{X}^T\mathbf{X}$ is required in Eq. (10.48). This matrix is square and of dimension $r + s$. Hence, once the matrix $\mathbf{X}^T\mathbf{X}$ has been calculated, the effort required to compute the inverse is related to the number of parameters in the model of Eq. (10.41), not to the number of observations. The variance is found using Eq. (10.44) as

$$\hat{\sigma}^2 = \frac{S(\mathbf{b})}{N - r - s} \tag{10.49}$$

For a particular value of process delay d, the sum of squares $S(\hat{\mathbf{b}}_d)$ can be found by substituting $\hat{\mathbf{b}}_d$ for $\mathbf{b}$ in Eq. (10.44). The minimum variance for all values of d is then found from

$$\hat{\sigma}_{min}^2 = \min[\hat{\sigma}_d^2, d = 0, 1, 2, \ldots] \tag{10.50}$$

The value of d that results in σ_{min}^2 is then the best value of d. In practice, some knowledge of the probable dead time in the process can be used to limit the number of possible values of d to be evaluated in Eq. (10.50).

Example 10.6 Matrix solution of process model   Reconsider Example 10.5 and the data given in Table 10.1. This problem can be set up for matrix solution by letting

$$\mathbf{X}^T = \begin{bmatrix} 0.0 & 0.57 & 0.97 & 1.26 & 1.47 & 1.06 & 0.76 & 0.54 \\ 1.0 & 1.0 & 1.0 & 1.0 & 0.0 & 0.0 & 0.0 & 0.0 \end{bmatrix}$$

and

$$\mathbf{c}^T = [0.57 \quad 0.97 \quad 1.26 \quad 1.47 \quad 1.06 \quad 0.76 \quad 0.54 \quad 0.39]$$

From Eq. (10.48),

$$\hat{\mathbf{b}} = \begin{bmatrix} \hat{\alpha} \\ \hat{\beta} \end{bmatrix} = (\mathbf{X}^T\mathbf{X})^{-1}\mathbf{X}^T\mathbf{c}$$

$$\hat{\mathbf{b}} = \begin{bmatrix} 0.718 \\ 0.565 \end{bmatrix}$$

The discrete model for the system is therefore

$$c_n = 0.718c_{n-1} + 0.565m_{n-1}$$

Estimates of the process gain and the time constant can again be calculated from Eq. (10.39) and (10.40). However, this step is unnecessary if it is the parameters in the discrete transfer function that are of primary interest. Program 10.2 shows how these calculations can be carried out. □

Residual Errors

After the parameters have been estimated by the method of least squares, the **residual errors** can be computed using Eq. (10.43). These residuals should be carefully examined for obvious trends. It is possible that the residuals might appear as shown in Fig. 10.11(b). This plot reveals that the model is probably the wrong model for the process tested. If the estimation is to be valid, the residuals must be a set of random numbers caused only by experimental error. If the model is correct, the plot of the residuals should appear more as shown in Fig. 10.11(a). Any trend such as a gradual rise or a slight curve in the plot of the residuals invalidates the results of the least-squares analysis. A quick (but less reliable) check of randomness is to count the sign changes in the residuals. If they are random, there should be roughly $N/2$ sign changes.

Autocorrelation

Another check of the randomness of the residuals can be made by computing their **autocorrelation function**. Theoretically, if the residuals are purely random, their autocorrelation function is

$$\rho_i = \begin{cases} 1 & \text{for } i = 0 \\ 0 & \text{for } i \neq 0 \end{cases} \tag{10.51}$$

where

$$\rho_i = \frac{r_i}{r_0} \tag{10.52}$$

```
program Fit_model;
  var
    mdata, cdata : array[0..100] of real;      {manipulation and feedback data }
    c, e : array[0..100] of real;
    b, xTc : array[1..2] of real;
    x : array[1..100, 1..2] of real;
    xTx, xTx1 : array[1..2, 1..2] of real;
    det, sum2, sigma2 : real;
    d : integer;                               {process delay}
    n, i : integer;
begin
  readln(n, d);                                {input n and delay for fitting}
  for i := 0 to n do
    readln(mdata[i], cdata[i]);                {input n+1 data points}
  for i := 1 to n - d do
    begin
      x[i, 1] := cdata[i - 1 + d];            {x matrix and c vector for Eq. 10.48}
      x[i, 2] := mdata[i - 1];
      c[i] := cdata[i + d]
    end;
  xTx[1, 1] := 0.0;
  xTx[1, 2] := 0.0;
  xTx[2, 1] := 0.0;
  xTx[2, 2] := 0.0;
  xTc[1] := 0.0;
  xTc[2] := 0.0;
  for i := 1 to n - d do
    begin
      xTx[1, 1] := xTx[1, 1] + x[i, 1] * x[i, 1];    {calculate xTx}
      xTx[1, 2] := xTx[1, 2] + x[i, 1] * x[i, 2];
      xTx[2, 1] := xTx[2, 1] + x[i, 2] * x[i, 1];
      xTx[2, 2] := xTx[2, 2] + x[i, 2] * x[i, 2];
      xTc[1] := xTc[1] + x[i, 1] * c[i];             {calculate xTc}
      xTc[2] := xTc[2] + x[i, 2] * c[i]
    end;
  det := xTx[1, 1] * xTx[2, 2] - xTx[1, 2] * xTx[2, 1];
  writeln(det);
  xTx1[1, 1] := xTx[2, 2] / det;               {invert xTx}
  xTx1[1, 2] := -xTx[1, 2] / det;
  xTx1[2, 1] := -xTx[2, 1] / det;
  xTx1[2, 2] := xTx[1, 1] / det;
  b[1] := xTx1[1, 1] * xTc[1] + xTx1[1, 2] * xTc[2];   {result of Eq. 10.48}
  b[2] := xTx1[2, 1] * xTc[1] + xTx1[2, 2] * xTc[2];
  writeln(b[1], b[2]);                         {fitted model coefficients}
  for i := 1 to n - d do
    e[i] := c[i] - (x[i, 1] * b[1] + x[i, 2] * b[2]);  {residual errors from Eq. 10.50}
  sum2 := 0.0;
  for i := 1 to n - d do
    sum2 := sum2 + e[i] * e[i];               {sum of errors squared}
  writeln(SUM2);
  sigma2 := sum2 / (n - d - 2);                {variance}
  writeln(sigma2)
end.
```

Program 10.2 Fit first-order model to data using Eq. (10.48).

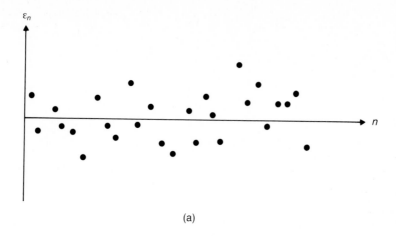

(a)

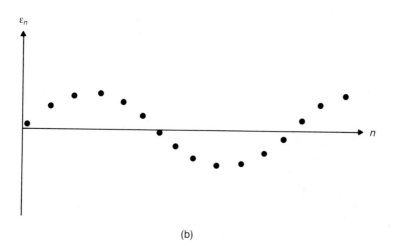

(b)

Figure 10.11 (a) Plot of residuals for (a) correct process model and (b) incorrect process model.

and

$$r_i = \frac{1}{N-i} \sum_{n=1}^{N-i} \varepsilon_n \varepsilon_{n+i} \tag{10.53}$$

Equation (10.51) states that each residual ε_n is not correlated with or related to any other residual ε_{n+i}. The correlation of each residual with itself, $\rho_0 = 1$, is always present. If there were trends in the residuals, each residual would be related to others following it. In other words, if there is a trend, ρ_i is nonzero for $i \neq 0$ and may be large if the trend is significant. Equation (10.51) almost never

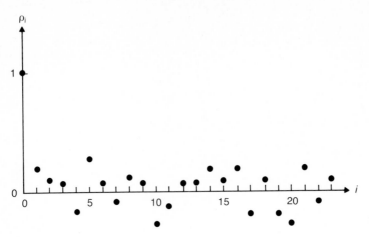

Figure 10.12 Plot of autocorrelation function for correct process model.

holds exactly for data taken from an actual experiment. Instead, the autocorrelation function usually exhibits small numbers for $i \neq 0$, as shown in Fig. 10.12. Statistical tests can be used to determine whether these values are insignificant at given confidence levels.

Normality

Another important check of the residuals is the check for **normality**. The chi-square test can be used to determine whether the residuals are normally distributed. This test is conclusive only when there are a relatively large number of residuals available for the test. If nothing is known about the distribution of the experimental errors, it is usually assumed that the residuals are normally distributed. It is, however, important to be aware of the fact that experimental errors are not always normal. Confidence regions, for example, are based on the assumption of normality, as are other tests discussed in this section.

Confidence Regions for the Parameters

If the residuals are normally distributed, the confidence region of the parameters can be computed. The confidence region of the parameters is a valuable piece of information. When a parameter that is expected to lie between 0 and 1 turns out to be 0.475 with a 95% confidence region of ± 0.005, it is safe to say that it is a good estimate of that parameter. However, when the 95% confidence region around the parameter is ± 2.0, the estimate for the parameter of 0.475 is probably meaningless.

The confidence region for the parameters is bounded by

$$(\mathbf{b} - \hat{\mathbf{b}})^T \mathbf{X}^T \mathbf{X} (\mathbf{b} - \hat{\mathbf{b}}) \leq (r + s)\hat{\sigma}^2 F_\alpha(r + s, N - r - s) \qquad (10.54)$$

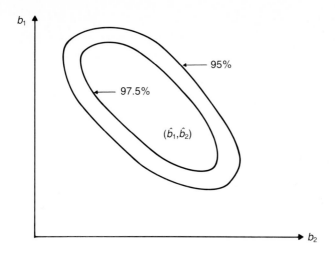

Figure 10.13 95% and 97.5% confidence regions for estimates $\hat{b}_1$ and $\hat{b}_2$.

where $r + s$ is the number of parameters, N is the number of data points, and $F_\alpha(r + s, N - r - s)$ is the F statistic for a $100(1 - \alpha)\%$ confidence region and for $r + s$ and $N - r - s$ degrees of freedom.

In the case of two parameters, the inequality in Eq. (10.54) defines an elliptical region in the parameter space as shown in Fig. 10.13. For three parameters the region is elliptical, but for more than three, the region is hyperelliptical and difficult to represent in pictorial form. For this situation, the simplest way to state the confidence information is by computing the individual confidence limits of the parameters on the basis of an assumption of independence of their estimation.

The individual confidence limits give a rough idea of the significance of the estimates. These individual confidence limits are

$$\hat{b}_i \pm t_\alpha(N - r - s)\sqrt{V_{ii}} \tag{10.55}$$

for $i = 1, 2, \ldots, k$ where $t_\alpha(N - r - s)$ is the Student's t statistic for a $100(1 - 2\alpha)\%$ confidence region and $N - r - s$ degrees of freedom. Appendix D contains a table of t_α for $\alpha = 0.1, 0.05, 0.025, 0.01,$ and 0.005.

V_{ii} is the ith prime diagonal element of the variance–covariance matrix $\mathbf{V}$, which is found from

$$\mathbf{V} = (\mathbf{X}^T\mathbf{X})^{-1}\hat{\sigma}^2 \tag{10.56}$$

Note that the matrix $(\mathbf{X}^T\mathbf{X})^{-1}$ is computed when obtaining the estimates in Eq. (10.48), and it is not necessary to recompute it here. The criterion of Eq. (10.55) must be applied with care. Strictly interpreted, it is valid only under the condition that off-diagonal elements of $\mathbf{V}$ are small compared to the main diagonal elements. This is rather restrictive. In practice, the requirement can often be relaxed, and Eq. 10.55 (along with an examination of matrix $\mathbf{V}$) is usually sufficient for deciding whether an estimation should be accepted or rejected.

Example 10.7 Individual confidence limits   The variance–covariance matrix for the data given in Example 10.5 is found from Eq. (10.57) as

$$\mathbf{V} = \begin{bmatrix} 0.198 & -0.139 \\ -0.139 & 0.347 \end{bmatrix}(0.0000174)$$

The 95% confidence limits based on the assumption of independence from Eq. (10.55) are

$$\alpha = 0.718 \pm 0.005$$

$$\beta = 0.565 \pm 0.006$$

The rather large magnitude of the off-diagonal terms in the variance–covariance matrix indicates that there is a strong interaction between the estimates. This is typical of estimation results for dynamic systems. The estimates are definitely not independent, but the confidence regions are so small that the estimates can be accepted without much reservation. □

Example 10.8 Model for second-order process   The data shown in Table 10.2 were collected from a second-order process. The general second-order model is

$$c_n = \alpha_1 c_{n-1} + \alpha_2 c_{n-2} + \beta_1 m_{n-1} + \beta_2 m_{n-2} + \varepsilon_n$$

Using the definition of X and $\mathbf{c}$, in Eq. (10.42),

$$\mathbf{X}^T = \begin{bmatrix} 0.0 & 0.24 & 0.77 & 1.4 & 2.0 & 2.5 & 3.0 & 3.4 \\ 0.0 & 0.0 & 0.24 & 0.77 & 1.4 & 2.0 & 2.5 & 3.0 \\ 1.0 & 1.0 & 1.0 & 1.0 & 1.0 & 1.0 & 1.0 & 1.0 \\ 0.0 & 1.0 & 1.0 & 1.0 & 1.0 & 1.0 & 1.0 & 1.0 \end{bmatrix}$$

$$\mathbf{c}^T = \begin{bmatrix} 0.24 & 0.77 & 1.4 & 2.0 & 2.5 & 3.0 & 3.4 & 3.7 \end{bmatrix}$$

From Eq. (10.48),

$$\hat{\mathbf{b}} = \begin{bmatrix} \hat{\alpha}_1 \\ \hat{\alpha}_2 \\ \hat{\beta}_1 \\ \hat{\beta}_2 \end{bmatrix} = \begin{bmatrix} 1.35 \\ -0.444 \\ 0.240 \\ 0.213 \end{bmatrix}$$

Table 10.2

Data from Second-Order Process

n	0	1	2	3	4	5	6	7	8	9
m	0.0	1.0	1.0	1.0	1.0	1.0	1.0	1.0	1.0	1.0
c	0.0	0.0	0.24	0.77	1.40	2.00	2.50	3.00	3.40	3.70

The estimate of the residual variance is found from Eq. (10.49) as

$$\hat{\sigma}^2 = 0.0012$$

The variance–covariance matrix from Eq. (10.56) is then

$$
V = \begin{bmatrix}
9.7 & -9.8 & -0.0 & -4.6 \\
-9.8 & 10.0 & 0.0 & 4.4 \\
-0.0 & 0.0 & 1.0 & -1.0 \\
-4.6 & 4.4 & -1.0 & 3.5
\end{bmatrix} (0.0012)
$$

On the basis of the assumption of independence, estimates of the individual 95% confidence region for the parameters are found from Eq. (10.55) as

$$\hat{\alpha}_1: \quad 1.350 \pm 0.22$$
$$\hat{\alpha}_2: \quad -0.444 \pm 0.22$$
$$\hat{\beta}_1: \quad 0.240 \pm 0.069$$
$$\hat{\beta}_2: \quad 0.213 \pm 0.13$$

The difference equation for the process is then

$$c_n = 1.350c_{n-1} - 0.444c_{n-2} + 0.240m_{m-1} - 0.213m_{n-2} \quad \square$$

Example 10.9 Model for first-order process with delay The data in Table 10.3 were taken from a process that has been identified as being of the form

$$c_n = \alpha c_{n-1} + \beta m_{n-1-d} + \varepsilon_n$$

Table 10.3

Data from First-Order Process with Dead Time

n	0	1	2	3	4	5	6	7	8	9	10	11	12
m	1.0	1.0	1.0	1.0	0.0	0.0	0.0	0.0	0.0	0.0	1.0	1.0	1.0
c	0.0	0.0	0.0	0.39	0.63	0.77	0.86	0.52	0.31	0.19	0.11	0.07	0.041

Table 10.4

Estimates of $\hat{\alpha}$ and $\hat{\beta}$ with $d = 0$, 1, 2, and 3

d	$\hat{\alpha}$	$\hat{\beta}$	$\hat{\sigma}^2$
0	0.883	0.104	0.034
1	0.820	0.187	0.027
2	0.604	0.392	0.00001
3	0.458	0.391	0.015

The results shown in Table 10.4 are obtained from Eq. (10.48) and the data given in Table 10.3. Because $\hat{\sigma}^2$ is a minimum for $d = 2$, we can conclude that the least-squares estimates are

$$\hat{\alpha} = 0.604 \quad \text{and} \quad \hat{\beta} = 0.392$$

The estimate of the residual variance is found from Eq. (10.49) to be

$$\hat{\sigma}^2 = 0.0000108$$

The variance–covariance matrix for $d = 2$ from Eq. (10.56) is

$$\mathbf{V} = \begin{bmatrix} 0.667 & -0.298 \\ -0.298 & 0.384 \end{bmatrix} (0.0000108)$$

The 95% confidence limits based on the assumption of independence from Eq. (10.55) are

$$\hat{\alpha} = 0.604 \pm 0.006$$
$$\hat{\beta} = 0.392 \pm 0.005$$

The difference equation describing the process can then be accepted as

$$c_n = 0.604c_{n-1} + 0.392m_{n-3} \quad \square$$

10.5

SUMMARY

This chapter has presented a number of techniques for obtaining discrete models for physical processes. The traditional techniques of physical modeling and step-response testing have been discussed, along with the technique of stochastic modeling. Perhaps the most significant difference between these techniques is in the level at which the analyst interacts with the modeling process. In the case of physical modeling, the analyst is required to know the nature of the inner workings of the process and to predict the process behavior by constructing a process model from these physical properties. The models developed are often complex and can be simplified only by a close examination of model parameters.

In the case of step-response modeling, the output of the process as a function of input is observed in such a way that the results can be interpreted in the form of standard model types. Step-response plots can be used for low-order processes and for processes that resemble low-order processes. In both cases, the subjective judgment of the analyst is required to determine the model parameters.

Stochastic modeling is quite different. Here, if the form of the model can be identified, then estimation techniques can be used to obtain an estimate of the coefficients in the process model. Mathematical tools exist for evaluating the errors in the models and for comparing one form of model with another. The role of the analyst is therefore changed; it is no longer necessary to determine the model coefficients because that is done mathematically. This in no way reduces the responsibility of the analyst for the final accuracy of the model. Because a model can be obtained with little or no knowledge of the actual physical nature of the process, it is all the more important that the analyst carefully examine the models obtained to determine their validity.

The linear least-squares method presented for stochastic modeling is not the only method available, and there are texts on statistical analysis that can be consulted on this subject. The linear least-squares method is, however, easily implemented on a small computer. In fact, it is possible to use a control computer for model estimation. Therefore, this method is a powerful technique for the rapid acquisition of process models without any need for additional testing equipment. The concept of stochastic modeling can be applied in **adaptive control** where the model for a process is continually estimated while the process is being controlled. In this case, the manipulations generated by the controller and the sampled process feedback are used as data for model fitting. The updated process model that is obtained can then be used to automatically recalculate controller gains, using the methods given in Chapter 4 for example. Techniques such as this can be very effective in tracking time-variant processes and in modifying controller characteristics so that closed-loop performance stays within specified limits.

BIBLIOGRAPHY

Astrom, K. J., and B. Wittenmark. *Computer Controlled Systems: Theory and Design.* Englewood Cliffs, NJ: Prentice-Hall, 1984.

Banks, S. *Control Systems Engineering: Modeling and Simulation, Control Theory, and Microprocessor Implementation.* Englewood Cliffs, NJ: Prentice-Hall, 1986.

Box, G. E. P., and G. M. Jenkins. *Time Series Analysis: Forecasting and Control*. San Francisco, CA: Holden-Day, 1970.

Close, C. M., and D. K. Frederick. *Modeling and Analysis of Dynamic Systems*. Boston, MA: Houghton-Mifflin, 1978.

Doebelin, E. O. *Control System Princples and Design*. New York, NY: Wiley, 1985.

Doebelin, E. O. *Systems Modeling and Response*. New York: Wiley, 1980.

Franklin, G. F., and J. D. Powell. *Digital Control of Dynamic Systems*. Reading, MA: Addison-Wesley, 1980.

Kumar, P. R., and P. Varaiya. *Stochastic Systems: Estimation, Identification and Control*. Englewood Cliffs, NJ: Prentice-Hall, 1987.

Ljung, P. *System Identification*. Englewood Cliffs, NJ: Prentice-Hall, 1987.

Ogata, K. *System Dynamics*. Englewood Cliffs, NJ: Prentice-Hall, 1978.

Pandit, S. M., and S. Wu. *Time Series and System Analysis with Applications*. New York, NY: Wiley, 1983.

PROBLEMS

10.1 Develop a process model for the hydraulic positioning system shown in Fig. 10.14.

10.2 Develop a model for the two-tank system shown in Fig. 10.15.

10.3 It is desired to design a controller for the solar water-heating system described in Problem 3.19.

    a. Develop a process model for the system. The manipulation input is to be the collector mass flow rate, and the output is to be tank water temperature. Treat solar radiation and heat loss from the collector and tank as disturbance inputs. Linearize the system about a desired tank water temperature of 60°C.

    b. Use the result of part (a) to design a tank water temperature controller of the form

$$\dot{m}(t) = k_p[T_o(t) - T_w(t)]$$

where k_p is a proportional controller gain. The control is to have a sample period of 0.1 s and is to achieve dead-beat response to step changes in solar radiation in maintaining a tank water temperature of 60°C.

    c. Simulate the system (not linearized) with the control designed in part (b), taking into account limits on water temperature of 0°C to 100°C and limits on mass flow rate of 0 kg/h to 270 kg/h. Evaluate its performance.

    d. What happens when the controller is designed using a process model that is linearized around 20°C (building temperature) rather than 60°C (desired water temperature)?

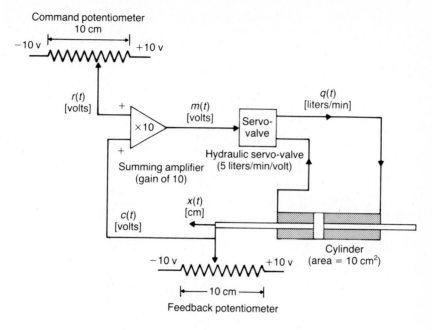

Figure 10.14 Hydraulic positioning system for Problem 10.1.

Figure 10.15 Two-tank system with pump for Problem 10.2.

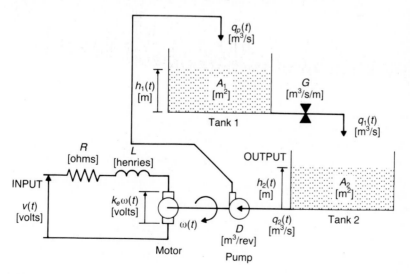

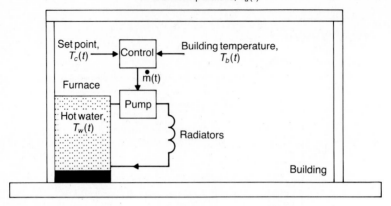

Figure 10.16 Building heated by hot water for Problem 10.4.

10.4   The building shown in Fig. 10.16 is heated by water from a furnace. A pump controls the flow rate of hot water to radiators in the building. The system can be approximately modeled using

$$q_s(t) = q_i(t) - q_o(t)$$

$$q_s(t) = M_b C_b \frac{dT_b(t)}{dt}$$

$$q_i(t) = \dot{m}(t) C_p [T_w(t) - T_b(t)]$$

$$q_o(t) = R_b A_b [T_b(t) - T_a(t)]$$

where $\dot{m}(t)$ = mass flow rate of hot water, in lb_m/h

      $T_b(t)$ = temperature of building, in °F

      $T_w(t)$ = temperature of hot water, in °F

      $T_a(t)$ = ambient temperature (outside building), in °F

      $q_s(t)$ = rate of heat storage in building, in BTU/h

      $q_i(t)$ = rate of heat flow into building, in BTU/h

      $q_o(t)$ = rate of heat flow out of building, in BTU/h

      M_b = mass of air in building, in lb_m

      C_b = heat capacity of air in building, in BTU/lb_m/°F

      C_p = heat capacity of water, in BTU/lb_m/°F

      R_b = resistance of walls and ceiling, in BTU/h/ft²/°F

      A_b = area of walls and ceiling, in ft²

The effective resistance of the walls and ceiling is 0.08 BTU/h/ft²/°F (ignore the floor), and the building is 50 ft long, 30 ft wide, and 20 ft high. It is desired to replace the standard on/off temperature for the building with a controller that maintains the building temperature at 68°F by manipulating the flow rate of hot water to the radiators.

a. Develop a linearized model of the system for a 68°F mean building temperature, a 200°F hot water temperature, and a 35°F ambient temperature operating point.

b. Design a discrete controller that rejects step disturbances in ambient temperature within 15 minutes. The desired building air temperature is $T_d(t)$.

c. Study the response of the system with the control in part (b) for building temperature settings of 55°F, 68°F, and 75°F.

10.5 The inertia of the q axis of the r-θ-z robot shown in Fig. 8.29 changes as the r axis extends. If the mass M_r of the r axis is assumed to be concentrated at the midpoint of the r axis, and the mass M_p of the part is assumed to be concentrated at a point at the end of the r axis, then the inertia of the θ axis is

$$I(t) = I_\theta + M_r(r(t) - 0.5r_{max})^2 + M_p r(t)^2$$

where r_{max} is the maximum extension of the r axis, and I_θ is the fixed inertia associated with the θ axis. Assume that position of the θ axis can be modeled using

$$I(t)\frac{d^2\theta(t)}{dt^2} = k_T v(t)$$

where $v(t)$ is the command voltage delivered to the θ-axis actuator, and k_T is the θ-axis gain.

a. Linearize the above process model about the operating point $r(t) = r_{max}$, $M_p = M_{P_{max}}$, where $M_{P_{max}}$ is the maximum expected part mass.

b. Recalculate the linearized model for operating point $r(t) = r_{max}/2$, $M_p = 0$. How does the process gain at these operating points differ from the gain found in part (a)?

c. How would closed-loop system performance be affected at the operating points in part (b) if the model in part (a) were used to design a controller for the θ axis?

10.6 Approximate proportional plus integral action is used to manipulate the inlet flow valve for liquid-level control in the system shown in Fig. 10.17. Liquid is occasionally removed from the tank by a pump at a constant output flow rate q_o. The area of the tank is A, and the proportional and integral gains are k_p and k_i, respectively. The control is implemented on a computer with sample period T.

a. Develop a linear model for the system.

b. What are the practical limits on the variables $q_i(t)$ and $h(t)$?

c. If the pump is manually turned on for time Δt, what is the response of the system modeled in part (a)? How does it differ from the system with the limits you found in part (b)?

10.7 Determine approximate continuous and discrete process models from the step responses shown in Fig. 10.18. The input to the process in every case is a unit step, and the sample period is $T = 1$ s.

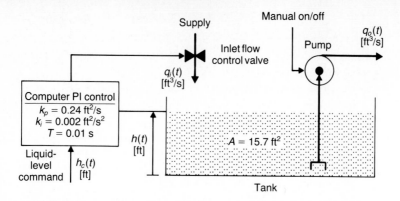

Figure 10.17 Tank with computer PI level control for Problem 10.6.

Figure 10.18 Responses of systems to unit step inputs for Problem 10.7.

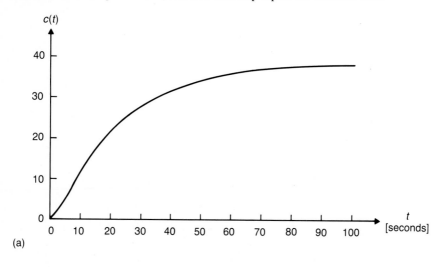

(a)

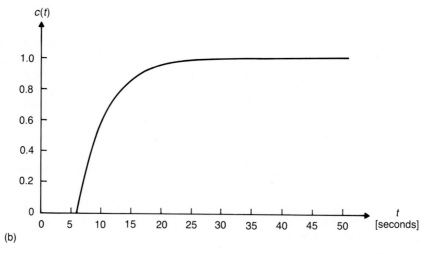

(b)

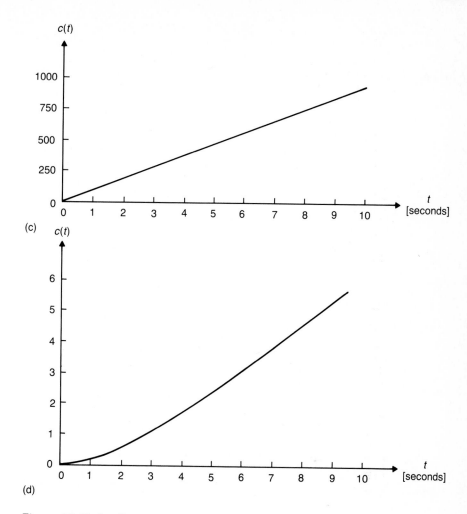

(c)

(d)

Figure 10.18 (*cont.*)

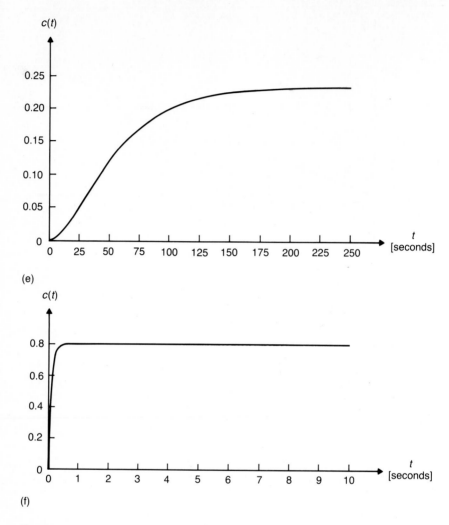

(e)

(f)

Figure 10.18 (*cont.*)

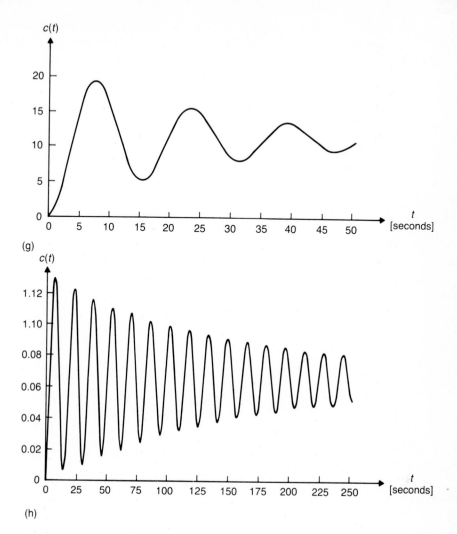

(g)

(h)

Figure 10.18 (*cont.*)

10.8 Determine an approximate discrete model for a positioning system that responds as shown in Fig. 10.19 to a step input of 3 volts. The sample period is 0.005 s.

10.9 Plot the response to a unit step input of an overdamped second-order system with a gain of 10 and two time constants: 0.024 s and 0.0032 s. Obtain a simplified model by representing the system as a first-order system with delay. Compare the unit step responses of these models. Also compare them to the response of the system modeled by dropping the shortest time constant from the second-order model.

Figure 10.19 Response of positioning system to a 3-volt step input for Problem 10.8.

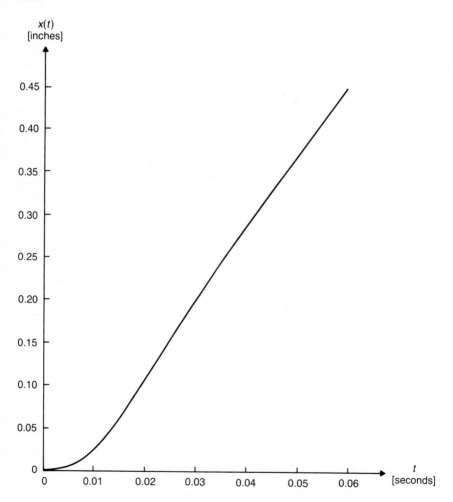

10.10 A motor responds to a step input of 50 volts in an approximately first-order manner and reaches a steady-state speed of 1475 rpm after 3.5 s.

a. Obtain a discrete model for motor position as a function of voltage input.

b. Ignore the time constant in the motor model, and design a discrete proportional position controller that achieves a closed-loop bandwidth of 15 Hz with a sample frequency of 100 Hz.

c. Combine the proportional gain found in part (b) with the model found in part (a) and plot the response of the closed-loop system to a unit step input. What is the approximate damping ratio? Is this response likely to be acceptable?

d. Adjust the gain of the proportional controller using the model in part (a) to obtain an approximate damping ratio of 0.7. Compare the result to the gain you found in part (b). What is the natural frequency of the resulting system in Hz? How does this compare to the bandwidth in part (b)?

e. What limits the performance of this system under proportional closed-loop control?

f. Redesign the controller so that an approximately first-order system with a bandwidth of 15 Hz is achieved.

10.11 A process is believed to have a discrete model of the form

$$c_n = e^{-T/\tau}c_{n-1} + k(1 - e^{-T/\tau})m_{n-1}$$

The following data were obtained from an experiment in which the process is given a unit step and its output is sampled every 2 s:

t:	0	2	4	6
c_n:	0	14.7	18.2	19.6

a. Using the linear least-squares technique, estimate the process parameters k and τ.

b. How good is the estimate made in part (a)?

10.12 The following data were obtained by sampling the input and output of a process at a frequency of 200 Hz:

n	m_n	c_n
0	0.2	0.0
1	0.2	0.13
2	0.0	0.22
3	0.0	0.15

Table 10.5

Data Sampled from a Process with Input $m(t)$ and Output $c(t)$

n	m_n	c_n	n	m_n	c_n
0	3.767	2.315	26	0.187	-8.191
1	-2.837	2.398	27	3.972	-7.412
2	3.750	3.699	28	-4.029	-5.683
3	4.422	2.978	29	1.355	-2.457
4	0.962	3.792	30	-0.900	-2.033
5	-0.152	6.332	31	-3.410	0.263
6	2.309	7.433	32	1.792	0.849
7	1.054	8.086	33	3.527	0.759
8	-1.476	8.782	34	3.762	0.619
9	1.965	8.207	35	-0.162	2.188
10	2.614	7.387	36	-4.859	4.692
11	0.536	6.774	37	3.615	5.595
12	-3.965	5.986	38	-0.687	4.119
13	-3.142	5.669	39	-1.768	4.674
14	0.616	3.207	40	-3.211	3.511
15	-0.460	-0.467	41	-1.636	2.559
16	2.622	-2.631	42	-2.033	-0.277
17	4.416	-3.941	43	-3.696	-3.446
18	-3.838	-4.564	44	-0.586	-5.987
19	-1.595	-2.125	45	4.670	-8.317
20	-4.286	-1.946	46	1.503	-10.580
21	-2.956	-2.594	47	-4.000	-8.647
22	2.909	-4.118	48	-3.219	-7.189
23	-4.384	-5.634	49	-3.553	-5.959
24	-3.177	-6.130	50	3.524	-6.349
25	3.684	-7.652			

Assuming that the process is first-order with no delay, design a discrete controller for it that will result in dead-beat response to step commands.

10.13 The data in the following table were collected by observing a process input m_n and a process outptut c_n. Use the least-squares technique to obtain the process model that best fits these data.

n	m_n	c_n
0	-1.0	1.7
1	0.5	0.3
2	0.7	-3.9
3	0.2	-2.1
4	-0.8	0.4
5	0.1	1.0

10.14 The data given in Table 10.5 were acquired from the input and output of a process sampled at a rate of 0.005 s.

   a. Find the discrete model representing a first-order process with delay that best fits these data.
   b. Plot the residuals from the result of part (a) and check the validity of the model.

10.15 Fit a second-order discrete model with no delay to the data given in Problem 10.14. Compare the residual sum of squares to that obtained in Problem 10.14 via fitting a first-order process with delay. Which is the preferred model?

10.16 The residuals obtained in fitting process models to sampled data are shown in Fig. 10.20. What can be said about the correctness of the models?

10.17 Calculate and plot the 95% confidence region for Example 10.7. Compare the result to the individual 95% confidence limits for the fitted parameters.

10.18 Develop equations similar to Eq. (10.36), (10.37), and (10.38) for a process assumed to be an integration with gain k and delay of d sample periods.

10.19 Develop equations similar to Eq. (10.42) and (10.48) for the general case of a system known to contain an integration.

10.20 Modify Program 10.1 so that manipulations are generated randomly with the limits $m_{min} \le m_n \le m_{max}$.

10.21 Generalize Program 10.2 so that higher-order models can be fitted. Test the program by repeating Example 10.8.

10.22 Modify Program 10.2 so that models with various amounts of delay are fitted until the best model is found.

10.23 Modify Program 10.2 to calculate the autocorrelation function.

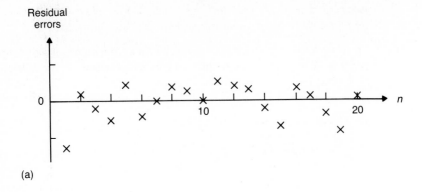

(a)

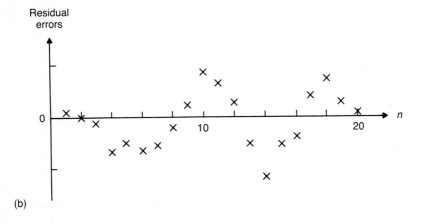

(b)

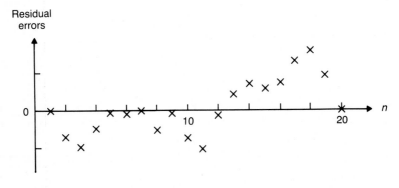

(c)

Figure 10.20 Residual errors from fitting a process model to sampled data for Problem 10.16.

Chapter 11
Analysis Using Transform Methods

11.1

INTRODUCTION

A time-domain approach to modeling, analyzing, and designing discrete control systems has been applied up to this point in the text. Time series have been used extensively, and time-shifting operators have been used to manipulate time series algebraically. This approach makes it possible to obtain transfer functions that represent the input/output response of a discrete system or its components, to determine system responses, and to design discrete controllers.

An alternative approach uses transformations. A **transformation** (or transform) is a technique in calculus in which a function is transformed from dependence on one variable to dependence on another variable. In the case of computer control systems, it is desirable to transform relationships specified in the time domain into a new domain wherein axioms of algebra can be applied rather than axioms of differential or difference equations. The transformations used are the Laplace transformation and the z transformation. The Laplace transformation results in functions of the time variable t being transformed into functions of a new frequency-related variable s. The z transformation is a direct outgrowth of the Laplace transformation and the use of a modulated train of impulses to represent a sampled function mathematically. These frequency-related transformations allow mathematical manipulations that are similar to those we applied in earlier chapters using the B operator. In addition, they allow us to apply the frequency-domain analysis and design techniques of continuous control theory to discrete control systems.

467

11.2

THE LAPLACE TRANSFORMATION

Laplace transformations are used extensively in classical control system analysis and design. The **Laplace transformation** is defined as

$$\mathcal{L}[f(t)] = F(s) = \int_0^\infty f(t)e^{-st}\, dt \tag{11.1}$$

The new variable s is a complex variable defined as $s = \alpha + j\beta$. It is useful here to evaluate the Laplace transforms of several different time functions so that we can use them later in control system analysis and the development of the z transformation.[1]

Laplace Transform of Function Multiplied by a Constant

The Laplace transform of a function multiplied by a constant k is simply the Laplace transform of the function multiplied by the same constant. This can be shown using Eq. (11.1) as follows:

$$\mathcal{L}[kf(t)] = \int_0^\infty kf(t)e^{-st}\, dt = k\int_0^\infty f(t)e^{-st}\, dt \tag{11.2}$$

Hence

$$\mathcal{L}[kf(t)] = k\mathcal{L}[f(t)] \tag{11.3}$$

Laplace Transform of the Sum of Two Functions

The Laplace transform of the sum of the two functions $x(t)$ and $y(t)$, is the sum of their respective Laplace transforms. This can be shown using Eq. (11.1) as follows:

$$\mathcal{L}[(x(t) + y(t)] = \int_0^\infty [x(t) + y(t)]e^{-st}\, dt \tag{11.4}$$

or

$$\mathcal{L}[x(t) + y(t)] = \int_0^\infty x(t)e^{-st}\, dt + \int_0^\infty y(t)e^{-st}\, dt \tag{11.5}$$

---

[1] While the existence and region of validity of the Laplace transform of a given function $f(t)$ [defined by Eq. (11.1)] and the Laplace and z transforms of a given sampled function $f^*(t)$ [defined by Eqs. (11.29) and (11.32)] are a matter of concern mathematically, they need not be a problem in an introductory study of computer control systems for machines and processes.

Hence

$$\mathcal{L}[x(t) + y(t)] = \mathcal{L}[x(t)] + \mathcal{L}[y(t)] \tag{11.6}$$

Equations (11.3) and (11.6) represent the properties of homogeneity and superposition, respectively, and show that the Laplace transformation is a linear operation.

Laplace Transform of a Time-Delayed Function

A time-delayed function has the value

$$\left\{ \begin{array}{ll} 0 & \text{for } t < D \\ f(t - D) & \text{for } t \geq D \end{array} \right\} \tag{11.7}$$

where D is the time delay between $f(t)$ and $f(t - D)$, as shown in Fig. 11.1. The Laplace transform of the delayed function is found by observing that

$$\int_0^\infty f(t - D)e^{-st}\, dt = \int_D^\infty f(t - D)e^{-st}\, dt \tag{11.8}$$

By replacing t with $t + D$ and changing the limits accordingly, we find that

$$\mathcal{L}[f(t - D)] = \int_0^\infty f(t)e^{-s(t + D)}\, dt \tag{11.9}$$

Hence

$$\mathcal{L}[f(t - D)] = e^{-sD}\int_0^\infty f(t)e^{-st}\, dt \tag{11.10}$$

and

$$\mathcal{L}[f(t - D)] = e^{-sD}\mathcal{L}[f(t)] \tag{11.11}$$

Figure 11.1 Delayed function $f(t - D)$.

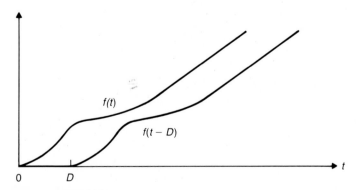

Laplace Transform of Derivatives

The Laplace transform of the derivative of a function can be found by applying the definition in Eq. (11.1).

$$\mathcal{L}\left[\frac{df(t)}{dt}\right] = \int_0^\infty \frac{df(t)}{dt} e^{-st} dt \tag{11.12}$$

Using integration by parts,

$$\mathcal{L}\left[\frac{df(t)}{dt}\right] = f(t)e^{-st}\Big|_0^\infty + s\int_0^\infty f(t)e^{-st} dt \tag{11.13}$$

or

$$\mathcal{L}\left[\frac{df(t)}{dt}\right] = s\mathcal{L}[f(t)] - f(0) \tag{11.14}$$

The $f(0)$ term represents the initial value of the function $f(t)$. If this value is zero, then

$$\mathcal{L}\left[\frac{df(t)}{dt}\right] = s\mathcal{L}[f(t)] \tag{11.15}$$

In other words, differentiating in the time domain is equivalent to multiplying by s in the frequency domain. Equation (11.15) can be generalized to

$$\mathcal{L}\left[\frac{d^i f(t)}{dt^i}\right] = s^i \mathcal{L}[f(t)] \tag{11.16}$$

if $f(0)$, $df(0)/dt, \ldots, d^{i-1}f(0)/dt^{i-1}$ are all zero. This constraint is not usually a problem, because controller design in the frequency domain is normally carried out without regard to initial conditions.

Example 11.1 Laplace transform of an integrating process   It is desired to find Laplace transform of a process defined by

$$\frac{dc(t)}{dt} = km(t)$$

From Eqs. (11.3) and (11.15),

$$sC(s) = kM(s)$$

or

$$C(s) = \frac{k}{s} M(s)$$

where it is assumed that $c(0) = 0$. The more convenient notation

$$C(s) = \mathscr{L}[c(t)]$$

$$M(s) = \mathscr{L}[m(t)]$$

has also been introduced. ☐

Example 11.2 Laplace transform of a third-order process Find the Laplace transform of the process model

$$\frac{1}{\omega_n^2} \frac{d^3 c(t)}{dt^3} + \frac{2\zeta}{\omega_n} \frac{d^2 c(t)}{dt^2} + \frac{dc(t)}{dt} = k\left(\tau \frac{dm(t)}{dt} + m(t) \right)$$

Using Eqs. (11.3), (11.6), and (11.16), we find that

$$\left(\frac{1}{\omega_n^2} s^3 + \frac{2\zeta}{\omega_n} s^2 + s \right) C(s) = k(\tau s + 1) M(s)$$

or

$$\frac{C(s)}{M(s)} = \frac{k(\tau s + 1)}{s\left(\dfrac{1}{\omega_n^2} s^2 + \dfrac{2\zeta}{\omega_n} s + 1 \right)}$$

This equation relates the transform of the output of the process to the transform of its input and is referred to as the process **transfer function**. ☐

11.3

FREQUENCY RESPONSE

The response of a linear continuous process to a sinusoidal input of frequency ω is also sinusoidal with a frequency ω. However, the ratio between input and output amplitude and the amount of phase lag between the input and output is a function of frequency and process dynamics. If the input to the process shown in Fig. 11.2 has the form

$$m(t) = X \sin(\omega t) \tag{11.17}$$

Figure 11.2 Process input and output.

then after all transients have decayed, the output of the process has the form

$$c(t) = Y(\omega) \sin[\omega t + \phi(\omega)] \tag{11.18}$$

where $Y(\omega)$ is the amplitude of the output at input frequency ω resulting from input amplitude X. The amplitude ratio or **magnitude** of the process **frequency response** is $|Y(\omega)/X|$, and $\phi(\omega)$ is the **phase** of the process frequency response. Both are independent of input amplitude for linear systems.

It is useful to plot the magnitude and phase of the process frequency response versus either frequency or the logarithm of frequency. In addition, the magnitude is often plotted in decibels (db) calculated from $20 \log_{10}|Y(\omega)/X|$. These plots of magnitude and phase versus frequency are referred to as **Bode plots**. They are useful both in the analysis of process dynamics and in process controller design because of the physical significance of the magnitude ratio and phase, the ease with which they can be obtained using sinusoidal testing techniques or spectrum analysis, and the existence of powerful controller design techniques based on analysis and modification of frequency response in the Bode form.

Example 11.3 Frequency response of a first-order process   Calculate the frequency response of a process modeled by

$$\tau \frac{dc(t)}{dt} + c(t) = km(t)$$

The solution of this equation for a sinusoidal input

$$m(t) = X \sin(\omega t)$$

is

$$c(t) = Y(\omega) \sin[\omega t + \phi(\omega)]$$

where

$$Y(\omega) = \frac{kX}{(1 + \tau^2\omega^2)^{1/2}}$$

and

$$\phi(\omega) = -\tan^{-1}\left(\frac{\tau\omega}{1}\right)$$

Here $\phi(\omega)$ is the phase of the frequency response, and the magnitude is found from

$$\left|\frac{Y(\omega)}{X}\right| = \frac{k}{(1 + \tau^2\omega^2)^{1/2}} \quad \square$$

Obtaining Frequency Response
from Laplace Transforms

Process frequency response can be determined directly from the process transfer function. This is accomplished by replacing s in the transfer function by $j\omega$. The magnitude and phase of the frequency response can then be calculated from the real and complex parts of the result, which can be written as

$$\frac{C(j\omega)}{M(j\omega)} = \alpha(\omega) + j\beta(\omega) \tag{11.19}$$

The magnitude is then

$$\left|\frac{C(j\omega)}{M(j\omega)}\right| = [\alpha(\omega)^2 + \beta(\omega)^2]^{1/2} \tag{11.20}$$

and the phase angle is[2]

$$\left/\frac{C(j\omega)}{M(j\omega)}\right. = \tan^{-1}\left[\frac{\beta(\omega)}{\alpha(\omega)}\right] \tag{11.21}$$

Example 11.4 Frequency response of an integrating process An integrating process with a gain k has the model

$$\frac{dc(t)}{dt} = km(t)$$

The transfer function of this process is found from the result of Example 11.1 as

$$\frac{C(s)}{M(s)} = \frac{k}{s}$$

The frequency response of the process is found by substituting $j\omega$ for s in the foregoing equation. The result from Eqs. (11.19), (11.20), and (11.21) is

$$\frac{C(j\omega)}{M(j\omega)} = \frac{k}{j\omega} = -j\frac{k}{\omega}$$

$$\left|\frac{C(j\omega)}{M(j\omega)}\right| = \frac{k}{\omega}$$

$$\left/\frac{C(j\omega)}{M(j\omega)}\right. = \frac{-\pi}{2} \text{ rad} = -90°$$

Figure 11.3(a) shows the magnitude of the frequency response of this process plotted versus a linear frequency scale. Figure 11.3(b) shows the magnitude of the

---

[2] Care must be taken when computing phase to ensure that the quadrant of the result is correct and that phase accumulates, if necessary, beyond $+180°$ or $-180°$, as the case may be.

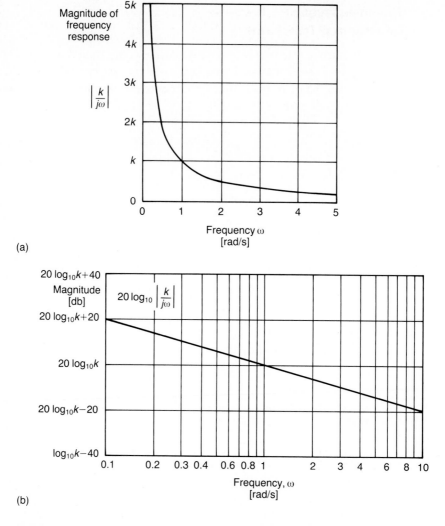

Figure 11.3 Magnitude of integrating process frequency response. (a) Linear magnitude scale, linear frequency scale; (b) logarithmic magnitude scale (in decibels), logarithmic frequency scale.

frequency response plotted in decibels (db) versus a logarithmic frequency scale. □

Approximate Frequency Response

Plots of frequency response for a number of commonly occurring process types are given in Fig. 11.4. The shape of calculated frequency response plots is, of course, related to the mathematical model used to represent process dynamics.

Though Eqs. (11.20) and (11.21) can be used to compute the frequency response of a process directly, it is useful to be able to obtain the approximate process frequency response without extensive computations. A first approximation can be made by assuming that changes in the slope plots of magnitude and phase versus the logarithm of frequency occur at critical frequencies. Examples of this approximation are given in Figs. 11.4(d), (e), and (f). The frequency response of cascaded continuous elements can be obtained by summing the phases and magnitudes (in decibels) of their individual frequency responses. Therefore, the frequency response of many process types not shown in Fig. 11.4 can be obtained by combining those that are shown.

Figure 11.4 Frequency response; (a) gain.

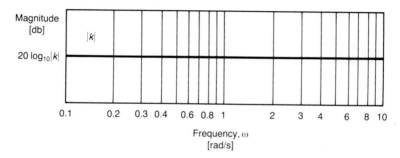

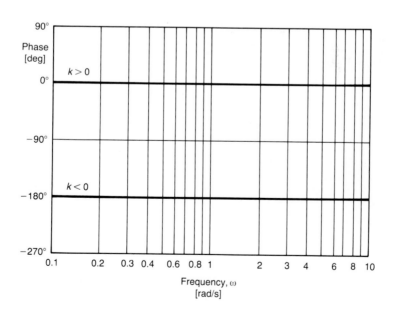

(a)

Figure 11.4 Frequency response; (b) integrators.

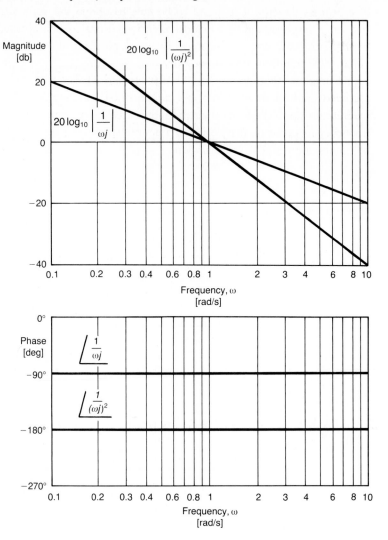

(b)

Figure 11.4 Frequency response; (c) differentiators.

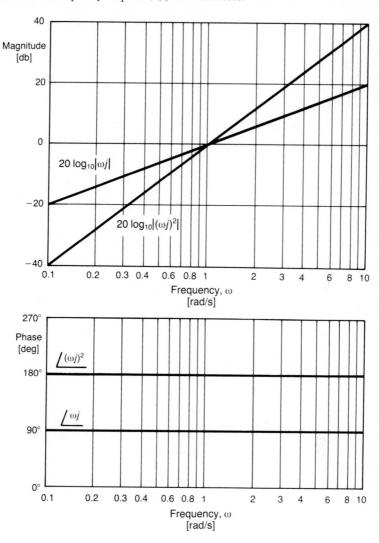

(c)

Figure 11.4 Frequency response; (d) first-order lag.

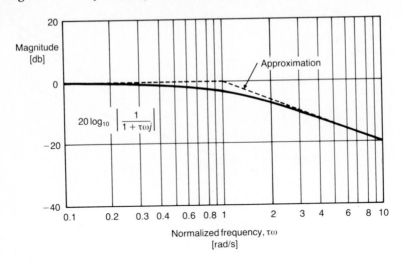

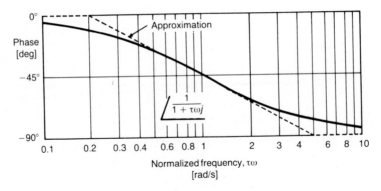

(d)

Figure 11.4 Frequency response; (e) first-order lead.

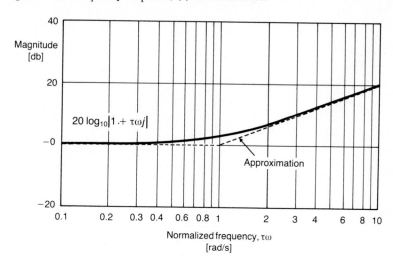

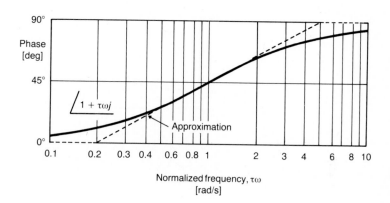

(e)

Figure 11.4 Frequency response; (f) second-order.

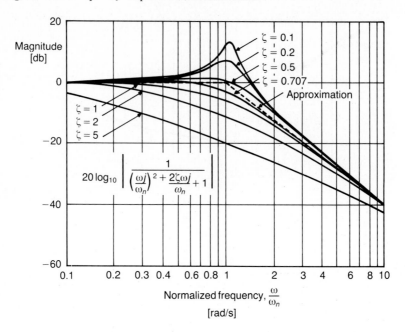

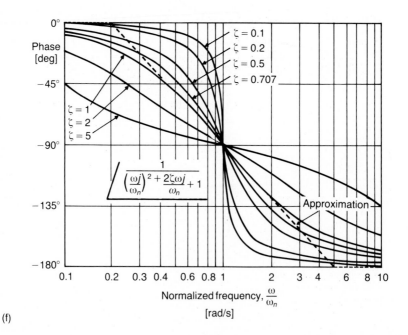

(f)

Example 11.5 Frequency response of an integrating process with a time constant The approximate frequency response of the process modeled by

$$\frac{C(s)}{M(s)} = \frac{k}{s(\tau s + 1)}$$

can be obtained by rewriting the process model as

$$\frac{C(s)}{M(s)} = (k)\left(\frac{1}{s}\right)\left(\frac{1}{\tau s + 1}\right)$$

As shown in Fig. 11.5, the frequency response of these three elements can be combined to give the approximate frequency response of the process. The actual frequency response can be calculated using Eqs. (11.20) and (11.21) as indicated in Program 11.1. □

Figure 11.5 Approximate frequency response of combined system.

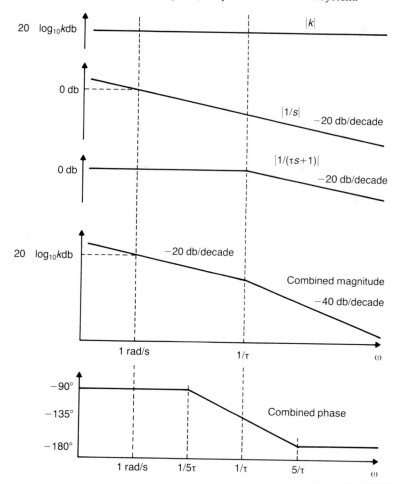

```
program Frequency_response;
  var
    k, tau : real;                                    {gain,time constant [s]}
    w : real;                                         {frequency [ rad/s]}
    alpha, beta : array[1..100] of real;             {real part, complex part}
    mag, phase : array[1..100] of real;              {magnitude, phase}
    i : integer;
begin
  readln(k, tau);                                     {input gain and time constant}
  for i := 1 to 100 do                                {vary w from 0.1 to10 rad/s}
    begin
      w := i / 10.0;
      alpha[i] := -k * tau / (1.0 + sqr(w * tau));    {process is k/(s(tau*s+1))}
      beta[i] := -k / w / (1.0 + sqr(w * tau));
      mag[i] := sqrt(sqr(alpha[i]) + sqr(beta[i]));   {Equation 11.20}
      phase[i] := arctan(beta[i] / alpha[i]);         {Equation 11.21}
      if alpha[i] < 0 then                            {correct arctangent if negative}
        phase[i] := phase[i] - 3.14159;
      mag[i] := 20.0 * ln(mag[i]) / ln(10.0);         {convert magnitude to db}
      phase[i] := phase[i] * 180.0 / 3.14159;         {convert phase to degrees}
      writeln(w, mag[i], phase[i])
    end
end.
```

Program 11.1 Calculate frequency response for Example 11.5.

11.4

THE z TRANSFORMATION

The z **transformation** enables us to represent and analyze discrete systems in a manner analogous to the way we approach continuous systems. In order to define the z transformation, it is necessary to introduce the notion of an ideal sampler. The concepts of the unit impulse and impulse modulation will then be introduced, together with the relationship between the Laplace transformation and the z transformation. Finally, the concept of the zero-order hold will be introduced, allowing the z-transformation technique to be applied to digital control systems that use sample-and-hold devices such as digital-to-analog converters as output for manipulating processes.

Ideal Sampler

An **ideal sampler** is a hypothetical sampling device that has characteristics of *instantaneous* sampling and a time delay T between samples. Given a sampler input $f(t)$ as illustrated in Fig. 11.6, the output of the ideal sampler is a new function $f^*(t)$, which is a train of impulses that have nonzero values only at times

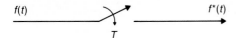

Figure 11.6 Ideal sampler.

$0, T, 2T, 3T, \ldots, nT, \ldots$. To formulate a mathematical representation for $f^*(t)$, it is convenient to define the unit impulse $\delta(t)$. The **unit impulse** is a pulse of zero duration and infinite height. The area under the unit impulse (its "strength") is equal to 1, as shown in Fig. 11.7.

A single unit impulse is shown pictorially in Fig. 11.8(a). The unit impulse can be delayed in time by a factor nT, as shown in Fig. 11.8(b), and multiplied by a constant k, as shown in Fig. 11.8(c). The amplitudes of function $f(t)$ at $t = 0, T, 2T, \ldots$ can be assigned as the areas under the impulses at times $t = 0, T, 2T, \ldots$, respectively. Hence,

$$f^*(t) = f(0)\delta(t) + f(T)\ \delta(t + T) + f(2T)\delta(t - 2T) + \cdots \qquad (11.22)$$

or

$$f^*(t) = \sum_{n=0}^{\infty} f(nT)\ \delta(t - nT) \qquad (11.23)$$

Figure 11.7 Unit impulse (duration $= 0$, amplitude $= \infty$, area $= 1$).

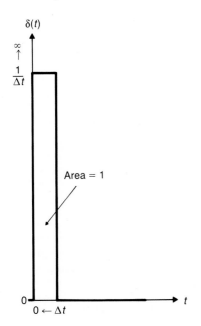

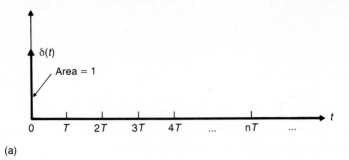

(a)

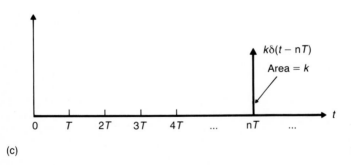

(b)

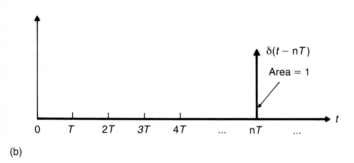

(c)

Figure 11.8 (a) Unit impulse $\delta(t)$; (b) delayed unit impulse $\delta(t - nT)$; and (c) delayed impulse $k\delta(t - nT)$ of area k.

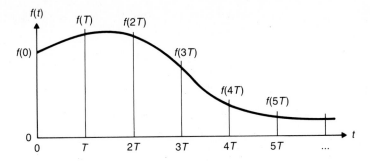

(a)

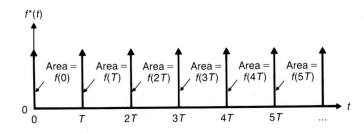

(b)

Figure 11.9 (a) Continuous function $f(t)$ and (b) output of ideal sampler $f^*(t)$.

The result of sampling the function $f(t)$ in Fig. 11.9(a) is illustrated by the impulse train in Fig. 11.9(b).

Laplace Transform of a Unit Impulse

The Laplace transformation can be applied to the unit impulse by using Eq. (11.1). The result is

$$\mathscr{L}[\delta(t)] = \int_0^\infty \delta(t) e^{-st}\, dt \tag{11.24}$$

Because $\delta(t)$ is nonzero only at time $t = 0$ and the area under the impulse is 1,

$$\mathscr{L}[\delta(t)] = \lim_{\Delta t \to 0} \left(\int_0^{\Delta t} \frac{1}{\Delta t} e^{-st}\, dt \right) \tag{11.25}$$

or

$$\mathscr{L}[\delta(t)] = \lim_{\Delta t \to 0} \left(\frac{1 - e^{-s\Delta t}}{s\, \Delta t} \right) \tag{11.26}$$

Using L'Hôpital's rule, we can differentiate the numerator and denominator of Eq. (11.26) with respect to Δt to obtain

$$\mathcal{L}[\delta(t)] = \lim_{\Delta t \to 0} \left(\frac{se^{-s\Delta t}}{s} \right) = 1 \tag{11.27}$$

Laplace Transform of a Modulated Train of Impulses

The terms $f(0)$, $f(T)$, $f(2T)$, ... in Eq. (11.22) are constants. Hence, by applying Eqs. (11.3), (11.6), (11.11), and (11.27), we determine that the Laplace transform of Eq. (11.23) is

$$\mathcal{L}[f^*(t)] = f(0) + f(T)e^{-sT} + f(2T)e^{-2sT} + \cdots \tag{11.28}$$

or

$$\mathcal{L}[f^*(t)] = F^*(s) = \sum_{n=0}^{\infty} f(nT)e^{-nsT} \tag{11.29}$$

Example 11.6 The Laplace transform of a sample unit step   A unit step function $u(t)$ is defined as

$$u(t) = \begin{cases} 0 & \text{for } t < 0 \\ 1 & \text{for } t \geq 0 \end{cases}$$

The Laplace transform of a sampled unit step function $u^*(t)$ is found from Eq. (11.29) as

$$\mathcal{L}[u^*(t)] = \sum_{n=0}^{\infty} u(nT)e^{-nsT} = 1 + e^{-sT} + e^{-2sT} + \cdots$$

The closed form of this infinite series is

$$\mathcal{L}[u^*(t)] = \frac{1}{1 - e^{-sT}} \quad \square$$

Definition of the z Transformation

It can be noted in Example 11.6 and in Eqs. (11.28) and (11.29) that the term e^{-sT} regularly appears. Therefore it is convenient to define a new complex variable:

$$z = e^{sT} \tag{11.30}$$

The transformation from the s domain to the z domain is performed by replacing e^{sT} with z. The z transform of Eq. (11.28) is therefore

$$\mathcal{Z}[f^*(t)] = f(0) + f(T)z^{-1} + f(2T)z^{-2} + \cdots \tag{11.31}$$

or

$$\mathcal{Z}[f^*(t)] = F(z) = \sum_{n=0}^{\infty} f(nT)z^{-n} \tag{11.32}$$

Example 11.7 The z transform of a unit step   From the previous example, the Laplace transform of a sampled unit step function is

$$\mathcal{L}[u^*(t)] = \frac{1}{1 - e^{-sT}}$$

Applying the z transform defined by Eq. (11.30) yields

$$\mathcal{Z}[u^*(t)] = \frac{1}{1 - z^{-1}} \quad \square$$

Obtaining z Transforms from Laplace Transforms

The Laplace transform of an exponential function is useful in obtaining the transfer function of dynamic systems in the z domain. Here,

$$f(t) = e^{-at} \tag{11.33}$$

and, from Eq. (11.4)

$$\mathcal{L}[e^{-at}] = \int_0^\infty e^{-at}e^{-st}\, dt = \int_0^\infty e^{-(s+a)t}\, dt \tag{11.34}$$

The result of this integration is

$$\mathcal{L}[e^{-at}] = \frac{1}{s + a} \tag{11.35}$$

More generally, the Laplace transform $F(s)$ of a function $f(t)$ is a ratio of polynomials in the variable s. It is possible to expand $F(s)$ into a sum of factors, each in the form of Eq. (11.35). If

$$F(s) = \frac{N(s)}{D(s)} = \frac{c_i s^i + c_{i-1}s^{i-1} + \cdots + c_1 s + c_o}{s^n + b_{n-1}s^{n-1} + \cdots + b_1 s + b_o} \tag{11.36}$$

then the polynomial $D(s)$ can be factored with roots $-a_1, -a_2, \ldots, -a_n$ to obtain

$$F(s) = \frac{N(s)}{(s + a_1)(s + a_2)\cdots(s + a_n)} \tag{11.37}$$

If $n > i$ and there are no repeated roots, then[3]

$$F(s) = \sum_{k=1}^{n} \frac{C_k}{(s + a_k)} \tag{11.38}$$

where

$$C_k = [(s + a_k)F(s)]\,|_{s = -a_k} \tag{11.39}$$

In general there may be repeated roots. If each root $-a_k$ has a multiplicity m_k, then

$$F(s) = \frac{N(s)}{(s + a_1)^{m_1}(s + a_2)^{m_2} \cdots (s + a_j)^{m_j}} \tag{11.40}$$

where

$$n = \sum_{k=1}^{j} m_k \tag{11.41}$$

In this case, the partial fraction expansion is

$$F(s) = \sum_{k=1}^{j} \sum_{\ell=1}^{m_k} \frac{C_{k\ell}}{(s + a_k)^{\ell}} \tag{11.42}$$

where

$$C_{k\ell} = \frac{1}{(m_k - \ell)!} \frac{d^{m_k - \ell}}{ds^{m_k - \ell}} [(s + a_k)^{m_k} F(s)] \bigg|_{s = -a_k} \tag{11.43}$$

The z transform of a sampled exponential function is found from Eq. (11.32) as

$$\mathscr{Z}[e^{-at}] = \sum_{n=0}^{\infty} e^{-anT} z^{-n} = \sum_{n=0}^{\infty} (e^{-aT} z^{-1})^n \tag{11.44}$$

The closed form of this infinite series is

$$\mathscr{Z}[e^{-at}] = \frac{1}{1 - e^{-aT} z^{-1}} \tag{11.45}$$

---

[3] If $n \le i$, then $D(s)$ can be divided into $N(s)$ to produce the new equation

$$F(s) = G(s) + \frac{N_1(s)}{D(s)}$$

where the order i_1 of $N_1(s)$ is less than n.

Comparing the result of Eq. (11.45) with that of Eq. (11.35), we note that, given a Laplace transform expanded into partial fractions as in Eq. (11.38), the corresponding z transform when there are no repeated roots is

$$F(z) = \sum_{k=1}^{j} \frac{C_k}{(1 - e^{-a_k T} z^{-1})} \tag{11.46}$$

where C_k is as defined in Eq. (11.39). For the case of repeated roots in Eq. (11.42), the corresponding z transform is

$$F(z) = \sum_{k=1}^{j} \sum_{\ell=1}^{m_k} \frac{(-1)^{\ell-1} C_{k\ell}}{(\ell-1)!} \frac{\delta^{\ell-1}}{\delta a_k^{\ell-1}} \left[\frac{1}{(1 - e^{-a_k T} z^{-1})} \right] \tag{11.47}$$

where $c_{k\ell}$ is as defined in Eq. (11.43).

A table of Laplace transforms and z transforms is given in Appendix E. Much time can be saved by referring to this table when you need a Laplace transform or z transform. When using the table, take care to ensure that all cascaded continuous elements are combined before performing the z transformations. This aspect of the z transform will be described further in following subsections—particularly the need to include a "holding" element with the process model to represent the conversion of a sample function (impulse trains) into "staircase" functions (as commonly output by D/A converters).

Example 11.8 The z transform of a double integration process   Determine the z transform of

$$\frac{C(s)}{M(s)} = \frac{1}{s^2}$$

Here Eq. (11.40) has the form

$$\frac{C(s)}{M(s)} = \frac{1}{(s+a)^m}$$

where $m = 2$. Applying Eq. (11.47) yields

$$\frac{C(z)}{M(z)} = \frac{C_{11}}{(1 - e^{-aT} z^{-1})} + \frac{C_{12} T e^{-aT} z^{-1}}{(1 - e^{-aT} z^{-1})^2}$$

where $C_{11} = 0$, $C_{12} = 1$, and $a = 0$. The result is

$$\frac{C(z)}{M(z)} = \frac{T z^{-1}}{(1 - z^{-1})^2}$$

This result could also be obtained from item 5 in Appendix E.   □

Inverse z Transformation

The **inverse z transformation** is the transformation from the z domain back to the time domain. Power-series expansion is one method for obtaining the inverse z transform of a function $F(z)$. The objective is to find the time domain function

$$f^*(t) = \mathcal{Z}^{-1}[F(z)] \tag{11.48}$$

for $n = 0, 1, 2, \ldots$. If the function $F(z)$ has the form

$$F(z) = \frac{N(z)}{D(z)} = \frac{c_0 + c_1 z^{-1} + c_2 z^{-2} + \cdots}{b_0 + b_1 z^{-1} + b_2 z^{-2} + \cdots} \tag{11.49}$$

then long division can be used to divide the numerator $N(z)$ by the denominator $D(z)$ to obtain

$$F(z) = f_0 + f_1 z^{-1} + f_2 z^{-2} + \cdots \tag{11.50}$$

Comparing Eq. (11.50) to Eq. (11.31) reveals that the terms $f_0, f_1, f_2, \ldots$ correspond to the time-domain terms $f(0), f(T), f(2T), \ldots$. The power-series expansion therefore gives values of the function $f(t)$ at the sample instants $nT = 0, T, 2T, \ldots$.

The table of z transforms in Appendix E can also be used to obtain inverse transformations. If an item in the table matches the z transform, the corresponding function $f(nT)$ is the required inverse transformation. It is important to note that the inverse transformation results in $f^*(t)$ rather than $f(t)$ and that $f(nT)$ is not defined between sample instants. Figure 11.10 shows only two of the infinite number of continuous functions that "fit" $f(nT)$ at the sample instants.

Transfer Functions and Block Diagrams

Transfer functions and block diagrams can be obtained and manipulated in the usual manner with z transforms. However, care must be taken when representing

Figure 11.10 Many continuous functions fit $f(nT)$.

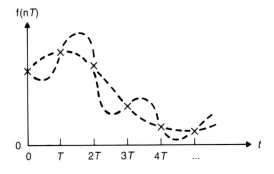

processes with discrete transfer functions when there are cascaded continuous elements in the process. Consider the two cascaded continuous elements $G_1(s)$ and $G_2(s)$ shown in Fig. 11.11(a). These continuous elements must be combined before obtaining the z transform:

$$Y(z) = \mathcal{Z}[G_1(s)G_2(s)]X(z) \tag{11.51}$$

It is not correct to obtain the z transform by using

$$Y(z) = \mathcal{Z}[G_1(s)]\mathcal{Z}[G_2(s)]X(z) \tag{11.52}$$

because this equation represents two continuous elements separated by a sampler, as shown in Fig. 11.11(b). It is important to note that, in general,

$$\mathcal{Z}[G_1(s)G_2(s)] \neq \mathcal{Z}[G_1(s)]\mathcal{Z}[G_2(s)] \tag{11.53}$$

Hold Elements

It is not possible, nor would it be desirable, to output impulses (infinite amplitude, zero duration) from a control computer and input them to a process to be controlled. More commonly, hold elements are used in the system to take the instantaneous manipulation value calculated by a control computer at the beginning of a sample period, transfer that value to the manipulation output, and hold that value constant through the entire sample period. A digital-to-analog (D/A) converter is such a device, and it is used extensively in computer control systems.

This type of hold function is called a zero-order hold. Its transfer function is labeled $G_{ho}(s)$ in Fig. 11.12, and examples of its input and output are illustrated in

Figure 11.11 (a) System with cascaded continuous processes: $Y(z) = \mathcal{Z}[G_1(s)G_2(s)]X(z)$; (b) system represented by $Y(z) = \mathcal{Z}[G_1(s)]\mathcal{Z}[G_2(s)]X(z)$.

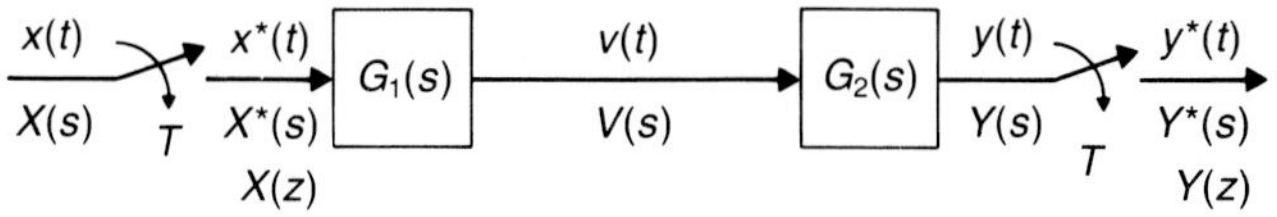

(a)

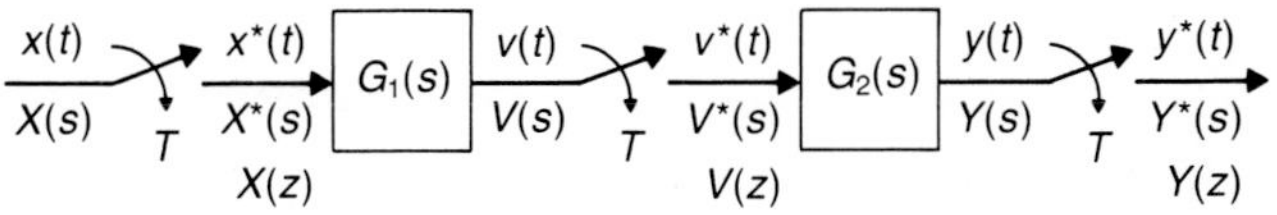

(b)

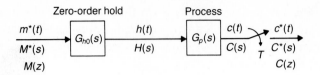

Figure 11.12 Process with zero-order hold input.

Figs. 11.13(a) and 11.13(b), respectively. The output of the zero-order hold $h(t)$ can be constructed from a series of superimposed step functions as follows:

$$h(t) = \sum_{n=0}^{\infty} m(nT)\{u(t - nT) - u[t - (n + 1)T]\} \qquad (11.54)$$

From Eq. (11.1), the Laplace transform of this function is

$$H(s) = \int_{0}^{\infty} m(nT)\left(\sum_{n=0}^{\infty} \{u(t - nT) - u[t - (n + 1)T]\}\right)e^{-st}\,dt \quad (11.55)$$

Moving the summation outside the integral results in

$$H(s) = \sum_{n=0}^{\infty} m(nT)\left\{\int_{0}^{\infty} u(t - nT)e^{-st}\,dt - \int_{0}^{\infty} u[t - (n + 1)T]e^{-st}\,dt\right\}$$

$$(11.56)$$

Figure 11.13 (a) Input to zero-order hold and (b) output of zero-order hold.

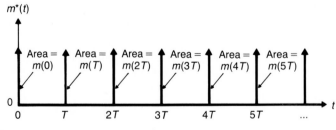

(a)

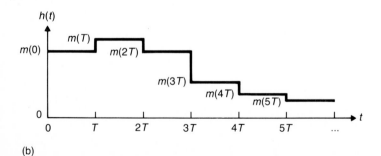

(b)

or

$$H(s) = \sum_{n=0}^{\infty} m(nT)(\mathscr{L}[u(t - nT)] - \mathscr{L}\{u[t - (n + 1)T]\}) \tag{11.57}$$

Taking the Laplace transforms of the delayed unit step functions yields

$$H(s) = \sum_{n=0}^{\infty} m(nT)\left[\frac{e^{-nsT}}{s} - \frac{e^{-(n+1)sT}}{s}\right] \tag{11.58}$$

or

$$H(s) = \frac{(1 - e^{-sT})}{s} \sum_{n=0}^{\infty} m(nT)e^{-nsT} \tag{11.59}$$

Then, from Eq. (11.29),

$$H(s) = \frac{(1 - e^{-sT})}{s} M^*(s) \tag{11.60}$$

Because there is no sampler at the input to process $G_p(s)$ in Fig. 11.12, the discrete transfer function of the process with the zero-order hold is determined using Eq. (11.51) and

$$\frac{C(z)}{M(z)} = \mathscr{Z}\left[\frac{(1 - e^{-sT})}{s} G_p(s)\right] \tag{11.61}$$

Applying Eq. (11.30), we can rewrite (11.61) as

$$\frac{C(z)}{M(z)} = (1 - z^{-1})\mathscr{Z}\left[\frac{G_p(s)}{s}\right] \tag{11.62}$$

Example 11.9 First-order process with zero-order hold   A first-order process is described by the differential equation

$$\tau \frac{dc(t)}{dt} + c(t) = kh(t)$$

where $c(t)$ is the continuous process output and $h(t)$ is the process input supplied by a zero-order hold. Assuming zero initial conditions, the Laplace transform of the process equation is

$$\tau s C(s) + C(s) = kH(s)$$

or

$$C(s) = \frac{\dfrac{k}{\tau}}{\left(s + \dfrac{1}{\tau}\right)} H(s)$$

The corresponding discrete transfer function from Eq. (11.62) is

$$\frac{C(z)}{M(z)} = (1 - z^{-1})\mathfrak{Z}\left[\frac{\dfrac{k}{\tau}}{\left(s + \dfrac{1}{\tau}\right)s}\right]$$

In this case, Eq. (11.46) can be used in obtaining

$$\frac{C(z)}{M(z)} = \frac{k(1 - e^{-T/\tau})z^{-1}}{1 - e^{-T/\tau}z^{-1}}$$

This result could also be obtained by using item 9 in the table of z transforms in Appendix E. □

Example 11.10 Third-order process with zero-order hold Consider the case of a motor/amplifier system that is manipulated by voltage $v(t)$ generated by a D/A converter. The D/A converter can be modeled as a zero-order hold that takes the computer-generated input $m^*(t)$ and then outputs the voltage $v(t)$. The motor velocity is $\Omega(t)$ and its position is $\theta(t)$, as indicated in Fig. 11.14(a).

$$\frac{1}{\omega_n^2}\frac{d^2\Omega(t)}{dt^2} + \frac{2\zeta}{\omega_n}\frac{d\Omega(t)}{dt} + \Omega(t) = k_m v(t)$$

$$\frac{d\theta(t)}{dt} = \Omega(t)$$

The corresponding Laplace transforms are

$$\Omega(s) = \frac{k_m}{\left(\dfrac{s^2}{\omega_n^2} + \dfrac{2\zeta s}{\omega_n} + 1\right)} V(s)$$

$$\theta(s) = \frac{1}{s}\Omega(s)$$

A block diagram for the system is shown in Fig. 11.14(b). The process, including the zero-order hold, consists of three cascaded continuous elements. Equation (11.51) must be used together with Eq. (11.62) in finding the z transform:

$$\theta(z) = \mathfrak{Z}\left[G_{ho}(s)\frac{k_m}{\left(\dfrac{s^2}{\omega_n^2} + \dfrac{2\zeta s}{\omega_n} + 1\right)}\frac{1}{s}\right]M(z)$$

$$\theta(z) = (1 - z^{-1})\mathfrak{Z}\left[\frac{k_m}{s^2\left(\dfrac{s^2}{\omega_n^2} + \dfrac{2\zeta s}{\omega_n} + 1\right)}\right]M(z)$$

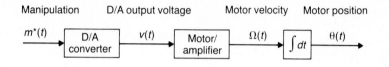

(a)

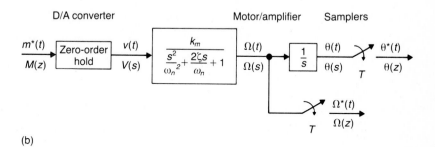

(b)

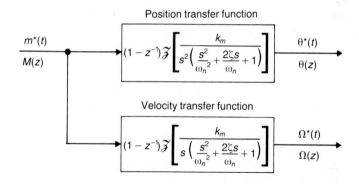

(c)

Figure 11.14 (a) Motor/amplifier system manipulated by output of D/A converter; (b) block diagram for motor/amplifier system; (c) discrete block diagram for motor/amplifier system.

It is not correct to use the product of two z transforms:

$$\Omega(z) = (1 - z^{-1})\mathscr{Z}\left[\frac{k_m}{s\left(\dfrac{s^2}{\omega_n^{\,2}} + \dfrac{2\zeta s}{\omega_n} + 1\right)}\right]M(z)$$

$$\theta(z) \neq \mathscr{Z}\left[\frac{1}{s}\right]\Omega(z)$$

There is no sampler between motor velocity and motor position! □

Stability

The criterion for stability of a system or system component modeled using Laplace transforms is that the real parts of the roots of the characteristic equation obtained from the denominator of its transfer function must be less than zero. The complex variable s can be written as

$$s = \alpha + j\beta \qquad (11.63)$$

If the complex roots of the characteristic equation are $r_{s_1}, r_{s_2}, r_{s_3}, \ldots, r_{s_i}, \ldots$, then the requirement for stability is that for all i,

$$\alpha_i < 0 \qquad (11.64)$$

The corresponding criterion for systems or system components modeled using the z transform can be found by using the definition of z in Eq. (11.30), which can be rewritten as

$$z = e^{(\alpha + j\beta)T} \qquad (11.65)$$

or

$$z = e^{\alpha T} e^{j\beta T} \qquad (11.66)$$

Therefore,

$$|z| = e^{\alpha T} \qquad (11.67)$$

and for $\alpha < 0$,

$$|z| < 1 \qquad (11.68)$$

If the characteristic equation in z has complex roots $r_{z_1}, r_{z_2}, r_{z_3}, \ldots, r_{z_i}, \ldots$, then the requirement for stability is that for all i,

$$|r_{z_i}| < 1 \qquad (11.69)$$

Example 11.11 Stability of integration process with continuous control
The transfer function of the continuous closed-loop control system shown in Fig. 11.15(a) is

$$\frac{C(s)}{R(s)} = \frac{1}{\dfrac{1}{k_c k_p} s + 1}$$

The characteristic equation for the system is

$$\frac{1}{k_c k_p} s + 1 = 0$$

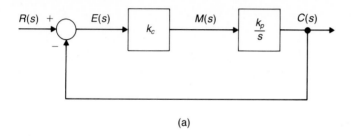

(a)

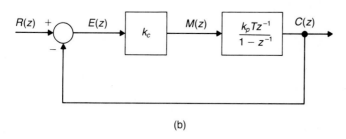

(b)

Figure 11.15 (a) Continuous and (b) discrete closed-loop control system.

which has one root,

$$r_{s_1} = -k_c k_p$$

The requirement for stability from Eq. (11.64) is

$$0 < k_c k_p$$

Regardless of how large the controller gain k_c is, this equation is always satisfied for positive, nonzero k_c and k_p. □

Example 11.12 Stability of integrating process with discrete control   The discrete closed-loop control system shown in Fig. 11.15(b) corresponds closely to the continuous system shown in Fig. 11.15(a). For the discrete system, the closed-loop transfer function is

$$\frac{C(z)}{R(z)} = \frac{k_c k_p T z^{-1}}{1 - (1 - k_c k_p T)z^{-1}}$$

and the characteristic equation is

$$1 - (1 - k_c k_p T)z^{-1} = 0$$

which has one root,

$$r_{z_1} = 1 - k_c k_p T$$

From Eq. 11.69, the criterion for stability is

$$|1 - k_c k_p T| < 1$$

or

$$0 < k_c k_p T < 2$$

Comparing this result to that of the previous example, we note that the presence of sampling and holding elements in the discrete system has placed an upper bound (related to the sample period T) on the gain k_c beyond which the system becomes unstable. □

11.5

FREQUENCY RESPONSE OF THE SAMPLE-AND-HOLD

The sample-and-hold shown in Fig. 11.16(a) consists of a sampler and a zero-order hold. As described in Section 11.3, frequency response can be determined by replacing s with $j\omega$ in the appropriate transfer function. For the zero-order-hold transfer equation in Eq. (11.60), the frequency response is

$$H(j\omega) = \frac{(1 - e^{-j\omega T})}{j\omega} M^*(j\omega) \tag{11.70}$$

This can be rewritten as

$$H(j\omega) = Te^{-j\omega T/2} \left(\frac{\sin\left(\omega \frac{T}{2}\right)}{\omega \frac{T}{2}} \right) M^*(j\omega) \tag{11.71}$$

The frequency response of the sampler can be found by representing the output of the sampler, $m^*(t)$, as the product of $m(t)$ times an impulse train.

$$m^*(t) = m(t) \sum_{n=-\infty}^{\infty} \delta(t - nT) \tag{11.72}$$

This gives the same result as Eq. (11.23) with $m(t) = 0$ for $t < 0$. The impulse train in Eq. (11.72) can be written as a Fourier series:

$$m^*(t) = m(t) \frac{1}{T} \sum_{n=-\infty}^{\infty} e^{jn\omega_s t} \tag{11.73}$$

where the sampling frequency in rad/s is

$$\omega_s = \frac{2\pi}{T} \tag{11.74}$$

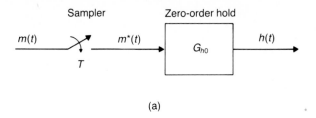

Sampler

Zero-order hold

$m(t)$ $m^*(t)$ G_{h0} $h(t)$

T

(a)

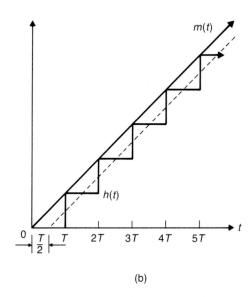

$m(t)$

$h(t)$

$0 \quad \dfrac{T}{2} \quad T \quad 2T \quad 3T \quad 4T \quad 5T$ t

(b)

Figure 11.16 (a) Sample-and-hold; (b) response of sample-and-hold to a ramp input.

Using Eq. (11.1), we find that the Laplace transform of the sampled function is

$$M^*(s) = \int_0^\infty \left[m(t) \frac{1}{T} \sum_{n=-\infty}^{\infty} e^{jn\omega_s t} \right] e^{-st}\, dt \qquad (11.75)$$

This can be rewritten as

$$M^*(s) = \frac{1}{T} \sum_{n=-\infty}^{\infty} \int_0^\infty m(t) e^{-(s-jn\omega_s)t}\, dt \qquad (11.76)$$

or

$$M^*(s) = \frac{1}{T} \sum_{n=-\infty}^{\infty} M(s - jn\omega_s) \qquad (11.77)$$

The frequency response of the sampler is then

$$M^*(j\omega) = \frac{1}{T} \sum_{n=-\infty}^{\infty} M(j\omega - jn\omega_s) \tag{11.78}$$

Combining Eqs. (11.71), (11.74), and (11.78) yields the frequency response of the entire sample-and-hold:

$$H(j\omega) = e^{-jn\omega/\omega_s} \left[\frac{\sin\left(\pi \dfrac{\omega}{\omega_s}\right)}{\pi \dfrac{\omega}{\omega_s}} \right] \sum_{n=-\infty}^{\infty} M(j\omega - jn\omega_s) \tag{11.79}$$

The components of Eq. (11.79) represent very important aspects of sampling that tend to adversely affect closed-loop computer control system performance:

$$\text{Time delay:} \quad e^{-jn(\omega/\omega_s)}$$

$$\text{Amplitude attenuation:} \quad \frac{\sin\left(\pi \dfrac{\omega}{\omega_s}\right)}{\pi \dfrac{\omega}{\omega_s}}$$

$$\text{Possibility of aliasing:} \quad \sum_{n=-\infty}^{\infty} M(j\omega - jn\omega_s)$$

The time delay associated with the sample-and-hold can be illustrated by its response to a ramp input, which is shown in Fig. 11.16(b). The ramp approximated by the "staircase" output of the zero-order hold is delayed by $T/2$ with respect to the ramp input. The time delay that is equivalent to D in Eq. (11.11) is

$$D = \frac{T}{2} = \frac{\pi}{\omega_s} \tag{11.80}$$

The phase of the sample-and-hold frequency response is a result of this time delay and is shown in Fig. 11.17(a). This time-delay characteristic is perhaps the most important detrimental aspect of sampling with regard to control system performance. When considered as a component in a closed-loop system, it tends to limit the bandwidth that can be achieved, and it generally reduces relative system stability.

The magnitude plot of the sample-and-hold frequency response is shown in Fig. 11.17(b). This amplitude attenuation is often relatively unimportant in control systems, because the sample frequency is usually relatively high and

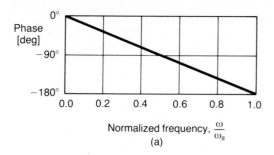

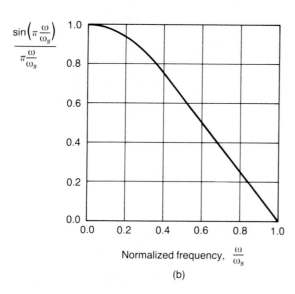

Figure 11.17 (a) Phase and (b) magnitude of
sample-and-hold frequency response.

$\omega/\omega_s \ll 1/2$ for the range of frequencies, ω, of interest. As long as this is the case,
then

$$\frac{\sin\left(\pi\dfrac{\omega}{\omega_s}\right)}{\pi\dfrac{\omega}{\omega_s}} \approx 1 \qquad (11.81)$$

The term $M(j\omega - jn\omega_s)$ in Eq. (11.79) describes the replication of $M(j\omega)$ at
integer multiples of ω_s. When the magnitude of the input to the sample-and-hold,
is as shown in Fig. 11.18(a), the magnitude after sampling is that shown in Fig.
11.18(b) when $\omega_c < \omega_s/2$. There is no problem with aliasing in this case, because
the content of $M^*(j\omega)$ around $\pm\omega_s, \pm 2\omega_s, \pm 3\omega_s, \dots$ generated by sampling
does not overlap the content around $0\omega_s$.

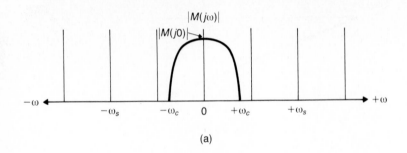

(a)

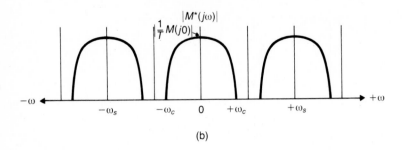

(b)

Figure 11.18 (a) $|M(j\omega)|$ with $\omega_c < \omega_s/2$ and (b) $|M^*(j\omega)|$ with $\omega_c < \omega_s/2$.

Figure 11.19 (a) $|M(j\omega)|$ with $\omega_c > \omega_s/2$ and (b) $|M^*(j\omega)|$ with $\omega_c > \omega_s/2$.

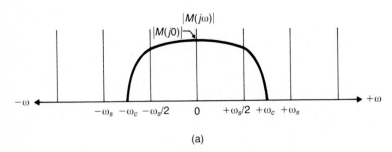

(a)

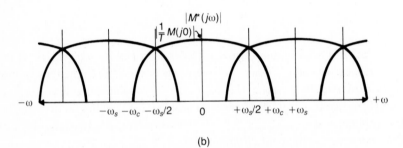

(b)

Figure 11.19(a) shows a case where $\omega_c > \omega_s/2$. It can be observed from Fig. 11.19(b) that the content of $M^*(j\omega)$ generated by the sampler around $\pm\omega_s$ overlaps the content around $0\omega_s$ and that higher frequencies are aliased, appearing at lower frequencies. Aliasing can be eliminated either by increasing the sample frequency ω_s or by using an anti-aliasing filter as discussed in Section 6.5 to reduce ω_c so that $\omega_c < \omega_s/2$.

11.6

SUMMARY

The z transform of a time-domain function is the result of a mathematical transformation and is a new function in the complex variable z. It has been shown that the z transformation is closely related to the Laplace transformation, making it possible to carry out analysis and design in the frequency domain. On the other hand, the B-operator approach developed in the early chapters of this text allows shifting operations to be performed on time series, and hence, analysis and design are performed in the time domain. In spite of this fundamental difference, comparisons of the power series in Eqs. (11.50) and (3.4) and the stability criteria in Eqs. (11.69) and (2.59) indicate that there is a direct correspondence between z^{-1} and B (the stability regions for z and B are switched, however). Additionally, comparison of discrete transfer functions such as the result of Example 11.9 and item 3 in Table 2.1 reveals that the transfer functions are identical (with B replaced by z^{-1}) when a zero-order hold is present at the input of a continuous system. This will be the case for most computer control systems, because zero-order-hold elements such as D/A converters are commonly employed as computer output interfaces. The presence of the zero-order-hold element corresponds to the assumption made in developing B-operator transfer functions that the input to a continuous process is held constant between sample periods.

It has been shown that the presence of sampling and holding elements in systems introduces time delay, amplitude attenuation, and a potential for aliasing into these systems, all of which can adversely affect the performance of closed-loop computer control systems. The time delay characteristics of holding elements are particularly important because of their destabilizing effects on closed-loop computer control systems. Because of stability constraints, gains cannot be set as high in computer control systems as they can be set in comparable continuous control systems. Additional performance limitations —bandwidth and stability in particular—therefore exist in computer control systems. However, the flexibility and versatility of these systems and the fact that many control algorithms and command generation databases simply cannot be implemented with continuous hardware make computer control an attractive alternative.

BIBLIOGRAPHY

Ackermann, J. *Sampled-Data Control Systems: Analysis and Synthesis, Robust System Design.* New York, NY: Springer-Verlag, 1985.

Astrom, K. J., and B. Wittenmark. *Computer Controlled Systems: Theory and Design.* Englewood Cliffs, NJ: Prentice-Hall, 1984.

Doebelin, E. O. *Control Systems: Principles and Design.* New York, NY: Wiley, 1985.

Dorf, R. C. *Modern Control Systems, 4/e.* Reading, MA: Addison-Wesley, 1986.

Franklin, G. F., and J. D. Powell. *Digital Control of Dynamic Systems.* Reading, MA: Addison-Wesley, 1981.

Harrison, H. L., and J. G. Bollinger, *Introduction to Automatic Controls.* New York, NY: Harper and Row, 1969.

Isermann, R. *Digital Control Systems.* New York, NY: Springer-Verlag, 1981.

Jacquot, R. G. *Modern Digital Control Systems.* New York, NY: Dekker, 1981.

Kuo, B. C. *Digital Control Systems.* New York, NY: Holt, Reinhart and Winston, 1980.

Meirovitch, L. *Introduction to Dynamics and Control.* New York, NY: Wiley, 1984.

Ogata, K. *Discrete-Time Control Systems.* Englewood Cliffs, N.J.: Prentice-Hall, 1987.

Phillips, C. L., and H. T. Nagle. *Digital Control System Analysis and Design.* Englewood Cliffs, NJ: Prentice-Hall, 1984.

Van Landingham, H. F. *Introduction to Digital Control Systems.* New York, NY: Macmillan, 1985.

PROBLEMS

11.1 Find the z transforms corresponding to the following functions.

a. $F(s) = \dfrac{1}{s^2 + 1}$

b. $F(s) = \dfrac{1}{s(s + 2)}$

c. $F(s) = \dfrac{10(s + 4)}{s^2}$

d. $F(s) = \dfrac{5}{s(s^2 + 3s + 1)}$

e. $F(s) = \dfrac{0.3}{s(s^2 + s + 1)}$

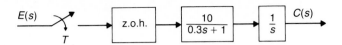

Figure 11.20 Block diagram for Problem 11.3.

11.2 Find the z transforms for the following process models combined with a zero-order hold.

a. $G_p(s) = \dfrac{10}{s}$

b. $G_p(s) = \dfrac{1}{0.5s + 1}$

c. $G_p(s) = \dfrac{0.15}{s^2 + 3s + 1}$

d. $G_p(s) = \dfrac{0.3}{s(s^2 + s + 1)}$

11.3 Find the discrete transfer function for the system shown in Fig. 11.20.

11.4 Find the discrete transfer function of the system shown in Fig. 11.21, and compare the result to that of Problem 11.3. Is this a reasonable block diagram for most machines and processes?

11.5 Using z transforms, draw a discrete block diagram that corresponds to the continuous block diagram shown in Fig. 11.22.

11.6 A process is represented by the block diagram shown in Fig. 11.23. The process is to be manipulated using a D/A converter. Find the discrete open-loop transfer function $C(z)/M(z)$.

11.7 Find the discrete transfer function of the system shown in Fig. 11.24. Is this result a correct discrete open-loop transfer function for the system with the block diagram given in Problem 11.6? Why or why not?

11.8 Consider the system shown in Fig. 11.25, wherein $G_m(s) = 150$, $k_s = 0.2$, and $k_r = 4000$.
   a. Determine the open-loop transfer function $C(z)/E(z)$.
   b. Determine the closed-loop transfer function $C(z)/R(z)$.
   c. Determine the closed-loop transfer function $P(z)/R(z)$.

11.9 Find the z transform of the pulse waveform shown in Fig. 11.26 for a sample period of 0.5 s.

Figure 11.21 Block diagram for Problem 11.4.

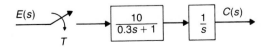

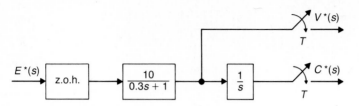

Figure 11.22 Block diagram for Problem 11.5.

Figure 11.23 Block diagram for Problem 11.6.

Figure 11.24 Block diagram for Problem 11.7.

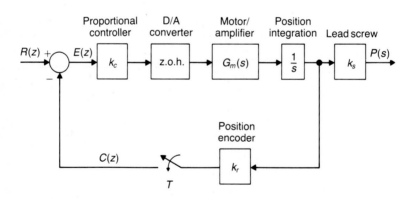

Figure 11.25 Block diagram for Problem 11.8.

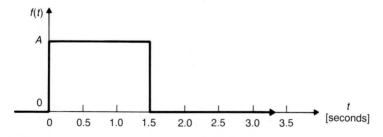

Figure 11.26 Block diagram for Problem 11.9.

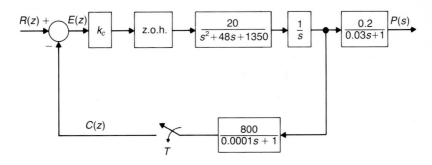

Figure 11.27 Block diagram for Problem 11.14.

11.10 How does the z transform of the waveform in Problem 11.9 change when the duration of the pulse is changed to 1.7 s? What is the significance of this feature of discrete systems?

11.11 A sample rate of 10 Hz is proposed for a closed-loop computer control system for a second-order process with a damping ratio of 0.05 and a natural frequency of 40 rad/s. What potential problems exist with this choice? How can they be avoided?

11.12 Redraw the block diagram in Problem 11.8 so that the continuous open-loop frequency response, $C(j\omega)/E(j\omega)$, can be calculated. Plot the frequency response of the resulting system. (The frequency characteristics of the zero-order hold must be included.)

11.13 Plot the frequency response of the system in Problem 11.12 without the magnitude characteristics of the zero-order hold. Compare the result with that of Problem 11.12. For what frequency range is this approximation valid?

11.14 Because of the complexity of the system shown in Fig. 11.27, it is desired to obtain a simplified model that will make it easier to set the controller gain k_c.

a. Plot the frequency response of each of the system components.

b. If the closed-loop system is not required to respond to frequencies above 1 Hz, draw a new block diagram with simplified blocks wherever possible.

c. Plot and compare the open-loop frequency responses of the original system and the simplified system.

d. Find a controller gain k_c that achieves a closed-loop frequency response bandwidth of 1 Hz.

e. Compare the closed-loop frequency response of the original and simplified systems with the gain you found in part (d).

Chapter 12
Design Using Transform Methods

12.1
INTRODUCTION

Transform techniques can be applied to the design of both continuous systems and discrete systems. Generally, the objective in controller design is to select the controller gain and dynamic characteristics that, when combined with the dynamics of the process, produce desirable closed-loop performance. The effects of sampling and phase lags due to the presence of hold elements must be considered when designing controls for discrete systems. Many engineers prefer to design in the continuous domain, finding it a more convenient and intuitive approach than designing directly in the discrete domain. Once a continuous controller design has been obtained, it can be approximated by a discrete controller. It is also possible to design controllers in the discrete domain without restoring to approximations of continuous controllers. Sample-and-hold effects are implicitly handled when one is designing via discrete methods.

The basic design techniques reviewed in this chapter include root locus, frequency response, integral control, pole/zero cancellation, and direct controller calculation. All of these techniques can be applied in both the continuous domain and the discrete domain. We will discuss how discrete approximations of continuous controllers can be obtained so that controllers designed using continuous techniques can be implemented digitally. Methods for obtaining these controllers include proportional-plus-integral-plus-derivative (PID) controller approximation, the bilinear transformation, hold-equivalence approximation, and pole-zero mapping. The dynamic characteristics of the zero-order holds can be approximated and included in the continuous design so that closed-loop response can be more accurately predicted.

508

Designing controllers in the discrete domain by applying such techniques as root locus, frequency response, and pole-zero cancellation is discussed. In this case, the result is the required discrete controller and there is no need for approximation. The application of the direct design approach, already discussed in Chapter 4, in both the discrete and the continuous domain will be reviewed. The result is the ability to specify the desired closed-loop transfer function from which a controller transfer function can be calculated, given the transfer function of the process to be controlled.

12.2

CONTROLLER DESIGN USING ROOT LOCUS

The gain of a controller is probably its most important characteristic. Regardless of other controller dynamics, gain can be adjusted easily to make controlled systems respond sluggishly or quickly, usually with accompanying changes in relative stability such as how oscillatory the response is and how rapidly transients die out. The **root-locus technique** is a useful graphical method for relating controller gain to resulting closed-loop dynamics. The fundamental idea is that the roots of the characteristic equation of a closed-loop system change as a function of controller gain and can be plotted as a locus of points in the complex plane. By knowing the dynamic characteristics associated with various root locations, we can choose the gain associated with desirable locations if such locations exist in the root locus. Furthermore, the effects of changes in controller formulation can be readily evaluated in terms of the desirability of root locations in the locus obtained. Alternatives can be tried and the best alternative selected. The method can be applied to both continuous and discrete systems in the s plane and the z plane, respectively.

Root Locations in the s Plane

A continuous closed-loop system transfer function has the general form

$$G(s) = \frac{c_i s^i + \cdots + c_2 s^2 + c_1 s + c_0}{b_n s^n + \cdots + b_2 s^2 + b_1 s + b_0}$$

(12.1)

The characteristic equation for the system is obtained from the denominator of the transfer function and can be expressed as

$$(s + a_1)(s + a_2)\cdots(s + a_n) = 0$$

(12.2)

where $-a_1, -a_2, \ldots, -a_n$ are roots of the characteristic equation (the poles of the transfer function). These roots govern the stability of the system and strongly affect transient response.

It is useful to investigate the location of roots of second-order systems because higher-order systems can be factored into real roots and/or pairs of complex conjugate roots. The characteristic equation of a second-order continuous system has the form

$$(s + a_1)(s + a_2) = 0 \qquad (12.3)$$

which has roots

$$s_1 = -a_1 \qquad (12.4)$$

$$s_2 = -a_2 \qquad (12.5)$$

This form is convenient when the roots of the characteristic equation are real (the system is not underdamped). When they are complex conjugate pairs (the system is underdamped), it is more common to express the characteristic equation as

$$s^2 + 2\zeta\omega_n s + \omega_n^2 = 0 \qquad (12.6)$$

where ω_n is the natural frequency and ζ is the damping ratio. Equation (12.6) has roots

$$s_1 = -\omega_n(\zeta + j\sqrt{1 - \zeta^2}) \qquad (12.7)$$

$$s_2 = -\omega_n(\zeta - j\sqrt{1 - \zeta^2}) \qquad (12.8)$$

Figure 12.1 shows the location of roots in the s plane for varying values of ζ and ω_n. The stable region is to the left of the imaginary axis ($\zeta > 0$), with real roots

Figure 12.1 Root locations in the s plane for various values of ζ and ω_n.

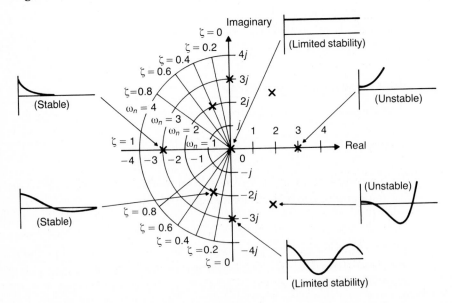

appearing on the negative real axis. Typical transient response is also shown for several root locations, illustrating the decay of a system output from nonzero initial conditions with zero system input.

Root Locations in the z Plane

In the z domain, the general form of the discrete closed-loop transfer function can be written as

$$G(z) = \frac{c_i' z^i + \cdots + c_2' z^2 + c_1' z + c_0'}{b_n' z^n + \cdots + b_2' z^2 + b_1' z + b_0'} \tag{12.9}$$

The corresponding characteristic equation is obtained from the denominator of the transfer function and can be expressed as

$$(z + a_1')(z + a_2') \cdots (z + a_n') = 0 \tag{12.10}$$

which has roots $-a_1', -a_2', \ldots$ and $-a_n'$.

For a second-order system, the characteristic equation is

$$(z + a_1')(z + a_2') = 0 \tag{12.11}$$

which has roots

$$z_1 = -a_1' \tag{12.12}$$

$$z_2 = -a_2' \tag{12.13}$$

These roots can be written in terms of the equivalent roots $-a_1$ and $-a_2$ in the continuous domain, using the definition of the z transform in Eq. (11.30).

$$z_1 = e^{-a_1 T} \tag{12.14}$$

$$z_2 = e^{-a_2 T} \tag{12.15}$$

If a root is real and less than zero, there is no equivalent root in the continuous domain. If the roots are complex conjugate pairs, Eqs. (12.14) and (12.15) can be rewritten as

$$z_1 = e^{-\omega_n(\zeta + j\sqrt{1 - \zeta^2})T} \tag{12.16}$$

$$z_2 = e^{-\omega_n(\zeta - j\sqrt{1 - \zeta^2})T} \tag{12.17}$$

Figure 12.2 shows the location of these roots in the z plane for varying values of ζ and $\omega_n T$. Note in Fig. 12.2 that the stable region is inside the unit circle ($\zeta > 0$). Figure 12.2 also shows the typical transient response for various root locations, illustrating the decay of a system output from nonzero initial conditions with zero system input.

An important feature to note is oscillation at one-half the sample frequency for roots in the left half of the z plane (roots with negative real parts). In a system

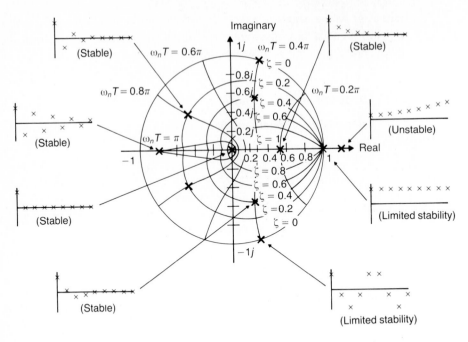

Figure 12.2 Root locations in the z plane for various values of ζ and ω_n.

under computer control, this can represent oscillation forced by the controller. Forced oscillation at one-half the sample frequency is almost always undesirable, so design consideration is normally restricted to the right half-plane. Generally, systems with roots in the left half-plane are a poor approximation of continuous systems.

Setting Controller Gain

The root locus of a closed-loop system with variable controller gain is generated by plotting the roots of its characteristic equation on the complex plane as controller gain is increased from 0 toward infinity. Then, if there are roots in the locus with desirable locations, the gain that results in those roots can be used as the desired controller gain. The damping and natural frequency characteristics shown in Figs. 12.1 and 12.2 are very helpful in identifying these locations. If there are no roots on the locus that possess desirable characteristics, it is necessary to modify the system by adding dynamic elements to the controller. As will be discussed in Section 12.4, the objective is to alter the closed-loop dynamics and the way controller gain manifests itself in the root locus so that favorable root locations will exist.

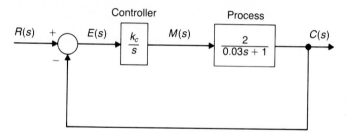

Figure 12.3 First-order process with integral control.

Example 12.1 Design of continuous controller using root locus   Consider the first-order process with integral control shown in Fig. 12.3. The closed-loop, continuous transfer function for this system is

$$G(s) = \frac{k_c 2}{0.03s^2 + s + k_c 2}$$

If a damping ratio of 0.7 is required (reasonably small overshoot for step inputs), the root locus plot shown in Fig. 12.4 indicates that the gain must be somewhat less than 10. Closer examination of the locus reveals that a gain of 8.5 results in a damping ratio of 0.7 and a natural frequency of 24 rad/s. The continuous controller transfer function is therefore

$$G_c(s) = \frac{8.5}{s}$$

Figure 12.4 Root-locus plot for Example 12.1

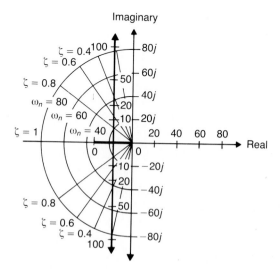

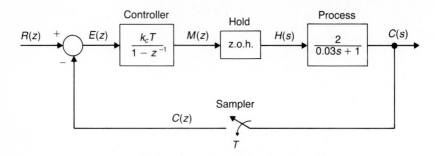

Figure 12.5 First-order process with discrete integral control.

Note that higher natural frequencies cannot be achieved while still maintaining the required level of damping. Hence controller dynamics would have to be modified if higher bandwidths were required. □

Example 12.2 Design of discrete controller using root locus   Figure 12.5 shows the same first-order process used in the previous example, but here we have a sampler, discrete integral controller, and zero-order hold rather than the continuous controller shown in Fig. 12.3. In this case, the discrete process transfer function with the zero-order hold included is

$$G_p(z) = \frac{2(1 - e^{-T/0.03})z^{-1}}{1 - e^{-T/0.03}z^{-1}}$$

The discrete integral controller shown has the form

$$G_c(z) = \frac{k_c T}{1 - z^{-1}}$$

which uses the backward rectangular rule.[1]

The closed-loop system transfer function is

$$G(z) = \frac{k_c T 2(1 - e^{-T/0.03})z^{-1}}{(1 - z^{-1})(1 - e^{-T/0.03}z^{-1}) + k_c T 2(1 - e^{-T/0.03})z^{-1}}$$

or

$$G(z) = \frac{k_c T 2(1 - e^{-T/0.03})z}{z^2 + [k_c T 2(1 - e^{-T/0.03}) - e^{-T/0.03} - 1]z + e^{-T/0.03}}$$

The roots of this transfer function, for controller gain increasing from 0 toward infinity, are plotted in Fig. 12.6 for sample period $T = 0.03$ s. From the plot it can

---

[1] It is useful to retain the factor T in the integral controller transfer function to decouple the gain k_c from the sample period. This also allows comparison of the gain obtained with the result of the continuous case in Example 12.1.

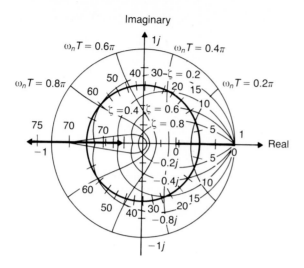

Figure 12.6 Root-locus plot for Example 12.2 with $T = 0.03$ s.

be seen that the gain must be less than 75 ($k_c = 72$ is the limit for stability at this sample period) and should be less than 40 to avoid oscillation at one-half the sample frequency (the design should be restricted to the right half-plane). The gain should be somewhat less than 10 if the object is to achieve a damping ratio of 0.7, as in the previous example. Closer inspection of the root locus results in a desired gain $k_c = 8.2$. The discrete controller transfer function is

$$G_c(z) = \frac{8.2(0.03)}{1 - z^{-1}}$$

and the corresponding control equation is

$$m_n = m_{n-1} + 0.246e_n$$

The gain of 8.2 found in this example is slightly less than that found for the continuous case in Example 12.1 for similar specifications. The resulting closed-loop transfer function is

$$G(z) = \frac{8.2(0.03)2(1 - e^{-T/0.03})z^{-1}}{1 + (k_c T2(1 - e^{-T/0.03}) - e^{-T/0.03} - 1)z^{-1} + e^{-T/0.03}z^{-2}}$$

The roots for $T = 0.03$ s are

$$z_1 = 0.528 + j0.298 \quad \text{and} \quad z_2 = 0.528 - j0.298$$

These can be written in terms of the equivalent continuous roots as

$$z_1 = e^{-(16.7 + j17.1)T} \quad \text{and} \quad z_2 = e^{-(16.7 - j17.1)T}$$

The natural frequency and damping ratio associated with the continuous roots are $\omega_n = 24$ rad/s and $\zeta = 0.7$, respectively.

The sample period of 0.03 s that has been selected can be justified in terms of the closed-loop bandwidth, which can be expected to be approximately 24 rad/s, the natural frequency of the continuous system in Example 12.1. The associated characteristic time is

$$\tau_c = \frac{1}{24 \text{ rad/s}} = 0.04 \text{ s}$$

From the discussion of sample-period selection in Section 4.3, we know that the sample period should be less than or equal to the characteristic time, and in this case, a sample period of 0.03 s has been chosen. If the sample period is decreased, the gain can be increased; it approaches the continuous gain of 8.5. On the other hand, lower gains would be necessary in order to maintain the desired level of damping if the sample period were increased, with an accompanying decrease in closed-loop performance. □

12.3
CONTROLLER DESIGN USING FREQUENCY RESPONSE

The frequency response analysis methods already described in Section 11.3 can serve as a means for designing controllers to obtain a desired closed-loop performance. One approach is to iteratively modify the controller transfer function and plot the resulting closed-loop frequency response to see whether desired characteristics have been obtained. However, it is usually more convenient to manipulate the open-loop frequency response of a system with the objective of achieving an open-loop frequency response from which the nature of the closed-loop frequency response can be predicted. The following discussion focuses on developing a number of "rules of thumb" for controller design using open-loop frequency response plots.

For many machinery and process control systems, it is desirable to have the plot of the magnitude of the closed-loop frequency response be relatively "flat" (neither increasing nor decreasing with frequency) in the range of frequencies over which the system is required to respond to command and disturbance inputs. It should then decrease rapidly at high frequencies. This tends to provide the following desirable characteristics:

□ Uniform closed-loop dynamics without resonance or magnification (higher magnitudes at middle frequencies than at lower frequencies)

□ Significant attenuation at higher frequencies where the system is not required to respond to command and disturbance inputs and should not respond to noise generated in the process, sensors, or controller

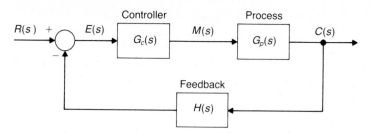

Figure 12.7 Continuous closed-loop control system.

The phase of the closed-loop frequency response ideally should be nearly zero in the range of frequencies wherein the system is required to respond to command and disturbance inputs. There will then be little lag between these inputs and system responses.

The open-loop frequency response for the system shown in Fig. 12.7 is

$$\frac{C(j\omega)}{E(j\omega)} = G_c(j\omega)G_p(j\omega)H(j\omega) \tag{12.18}$$

and the closed-loop frequency response is

$$\frac{C(j\omega)}{R(j\omega)} = \frac{G_c(j\omega)G_p(j\omega)}{1 + G_c(j\omega)G_p(j\omega)H(j\omega)} \tag{12.19}$$

Many processes are characterized by relatively high gain at low frequencies and relatively low gain at high frequencies. The frequency response of an integrating process, as shown in Fig. 11.3, is a good example of these characteristics with infinite gain at $\omega = 0$ and zero gain at $\omega = \infty$. If such a process has a gain k_p, when a controller with unity feedback and gain k_c is added, then the open-loop frequency response of the system is

$$\frac{C(j\omega)}{E(j\omega)} = \frac{k_c k_p}{j\omega} \tag{12.20}$$

The open-loop frequency response magnitude is $k_c k_p/\omega$ for this system, and the open-loop phase is always $-\pi/2$ rad $(-90°)$.

With the feedback loop closed, the frequency response of this system is

$$\frac{C(j\omega)}{R(j\omega)} = \frac{1}{\tau j\omega + 1} \tag{12.21}$$

which represents a first-order closed-loop system with unity gain, a time constant of

$$\tau = \frac{1}{k_c k_p} \tag{12.22}$$

and a bandwidth of

$$\omega_c = k_c k_p \tag{12.23}$$

This first-order closed-loop frequency response is shown in Fig. 11.4(d). The magnitude at $\omega = \omega_c$ is 0.707(-3 db), down from the magnitude of 1 (0 db) at $\omega = 0$. The phase is nearly zero at lower frequencies, but it decreases rapidly after $\omega = \omega_c/5$ to $-\pi/4$ rad ($-45°$) at $\omega = \omega_c$ and to $-\pi/2$ rad ($-90°$) at high frequencies. The transient response of this first-order system to step and ramp inputs will be 95% complete after time 3τ. The steady-state error for the step input is zero, but for a ramp input, the steady-state error is

$$e(\infty) = \dot{r}/k_c k_p \tag{12.24}$$

where $\dot{r}$ is the slope of the ramp input.

Approximately first-order closed-loop frequency response is often a desirable target for control system design. If the combined controller and process transfer functions result in an open-loop frequency response resembling that of an integrator, then nearly first-order frequency response will be obtained when the loop is closed. For this type of system, the following "rules of thumb" can be applied in predicting closed-loop frequency response from open-loop frequency response for continuous systems.

Rule 1

The closed-loop frequency response $C(j\omega)/R(j\omega)$ resembles $1/H(j\omega)$ at low frequencies. This is because the open-loop gain is high at low frequencies and $|G_c(j\omega)G_p(j\omega)H(j\omega)| \gg 1$. If $H(j\omega)$ is simply a gain k_f, then the closed-loop gain at $\omega = 0$ is $1/k_f$. For systems with unity feedback ($H(j\omega) = 1$), the closed-loop gain is 1 (0 db).

Rule 2

The closed-loop frequency response $C(j\omega)/R(j\omega)$ resembles $G_c(j\omega)G_p(j\omega)$ at high frequencies. This is because the open-loop gain is low at high frequencies and $|G_c(j\omega)G_p(j\omega)H(j\omega)| \ll 1$.

Rule 3

The closed-loop bandwidth is approximately the frequency where $|G_c(j\omega)G_p(j\omega)H(j\omega)| = 1$. This frequency is the "lowest-frequency" root of the closed-loop characteristic equation. For the integrating system in Eq. (12.20) through (12.23), the open-loop gain is 1 (0 db) and the closed-loop gain is 0.707 (-3 db) at $\omega = \omega_c$. This is the "lowest-frequency" (and only) root of the closed-loop characteristic equation.

Because of stability considerations and the need to be able to predict how oscillatory a closed-loop system will be, the phase of the open-loop frequency response must also be considered when designing controllers by using frequency response techniques. Gain and phase margins are measures of relative stability that have been developed for this purpose.

Gain Margin

The **gain margin** is measured at the frequency where the phase of the open-loop frequency response becomes $-\pi$ rad ($-180°$). It is the negative (in decibels) of the magnitude of the frequency response at that frequency.

Phase Margin

The **phase margin** is measured at the frequency where the magnitude of the open-loop frequency response becomes 1 (0 db). It is the difference between the phase at that frequency and $-\pi$ rad ($-180°$). It is measured by adding π rad (180°) to the phase at that frequency.

Gain margins and phase margins are always measured using the open-loop frequency response. Both must be positive for the closed-loop system to be stable;[2] the greater their numerical values, the greater the relative stability of the system. Gain and phase margins tend to increase, and systems tend to become less oscillatory, as gain is decreased. The phase margin can be used to predict the damping in the closed-loop system. A rough "rule of thumb" follows.

Rule 4

For underdamped closed-loop systems, the closed-loop damping ratio is roughly the open-loop phase margin (in degrees) divided by 100°. In Fig. 12.8 the closed-loop damping ratio is plotted versus open-loop phase margin for a second-order system consisting of an integration and a time constant.

The gain margin, phase margin, and predicted bandwidth are very useful measures for setting controller gain. The technique is to adjust the controller gain and hence the magnitude of the open-loop frequency response until desirable values of predicted bandwidth, phase margin, and gain margin are obtained. The "rules of thumb" we have outlined often give only approximate results. Because the relationship between closed-loop and open-loop frequency response is usually more complex than indicated by these "rules of thumb," the actual closed-loop frequency response should always be checked experimentally or analytically, once a gain has been selected, to ensure that expected closed-loop

[2] This stability test can be used as long as the system has no poles or zeros with real parts greater than zero in its open-loop transfer function. Most processes are "minimum-phase" systems and are not affected by this limitation,

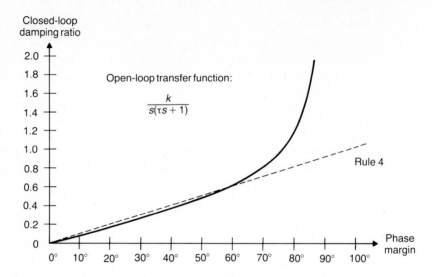

Figure 12.8 Comparison of closed-loop damping ratio with prediction of Rule 4 ("divide by 100°").

characteristics have been obtained. If desired characteristics cannot be achieved simply by manipulating controller gain, it is necessary to add dynamic elements to the controller as discussed in Section 12.4.

Example 12.3 Design of continuous controller using frequency response
Consider the first-order process with integral control shown in Fig. 12.3. The open-loop frequency response of this system is found from

$$G'(j\omega) = \frac{k_c 2}{j\omega(0.03j\omega + 1)}$$

and is shown in Fig. 12.9(a) for the case of a unity-gain controller ($k_c = 1$).

Suppose it is desired to find the value of the controller gain k_c that results in a relatively stable system ($\zeta \approx 0.7$) with reasonably high bandwidth. It can be seen in Fig. 12.9(a) that the system exhibits the desirable open-loop characteristics of high gain at low frequencies and low gain at high frequencies. With unity controller gain, the zero-crossing frequency is about 2 rad/s and the phase margin is about 86°. Rule 4 suggests that it would be appropriate to increase the gain until the phase margin is about 70°. The result is the magnitude plot for a gain $k_c = 7$, also shown in Fig. 12.9(a). The continuous controller transfer function is then

$$G_c(s) = \frac{7}{s}$$

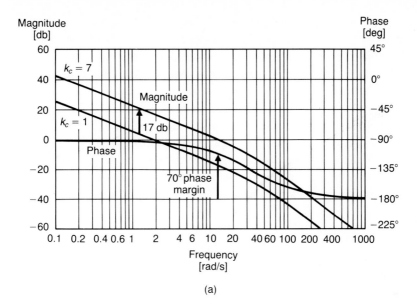

(a)

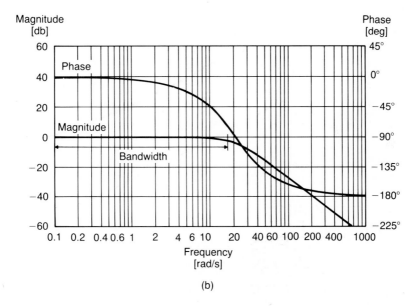

(b)

Figure 12.9 (a) Open-loop frequency response plots for Example 12.3 and (b) closed-loop frequency response plot for Example 12.3 with $k_c = 7$.

Rule 3 would predict a closed-loop bandwidth of about 12 rad/s because that is where the new magnitude plot crosses zero, a point in the area where the phase of the open-loop system begins to deviate from that of an integration. The actual closed-loop frequency response plot is shown in Fig. 12.9(b). It can be seen that the bandwidth is about 17 rad/s, a bit higher than expected. In Example 12.1 we found that a gain of $k_c = 8.5$ results in a damping ratio of 0.7 for this system. The damping ratio with gain $k_c = 7$ chosen here is actually $\zeta = 0.77$. Hence the use of Rule 4 has resulted in a slight underestimation of closed-loop damping. Nevertheless, an acceptable design has been obtained. The bandwidth is reasonably high given the process time constant, and there is no magnification in the closed-loop frequency response, indicating that a relatively stable system has been achieved.

If the controller gain is increased so that the magnitude is 1 (0 db) at a frequency where the phase margin becomes less than $\pi/2$ rad ($90°$) and begins to approach zero, the closed-loop bandwidth of the system increases but its relative stability declines. For example, if the gain of the open-loop system is $k_c = 24$, the phase margin becomes $45°$, and Rule 4 would predict a closed-loop damping ratio of about 0.45. As shown in Fig. 12.10, the closed-loop system that results has significantly higher bandwidth than that shown in Fig. 12.9(b). It has also a maximum magnitude magnification of about 3 db and is clearly underdamped (the actual damping ratio is 0.41). If it were necessary to obtain increased bandwidth without sacrificing damping, dynamic elements would have to be

Figure 12.10 Frequency response plots for Example 12.3 with $k_c = 24$.

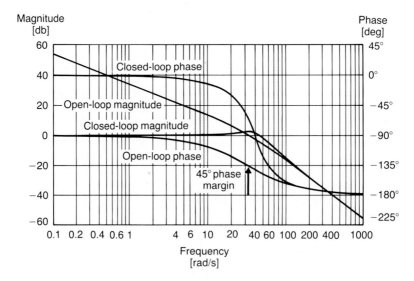

added to the controller to improve the phase margin. This can be done using the pole-zero cancellation technique described in the next section. □

Example 12.4 Design of discrete controller using frequency response
The open-loop transfer function for the discrete system shown in Fig. 12.5 is

$$\frac{C(z)}{E(z)} = \frac{k_c T2(1 - e^{-T/0.03})z^{-1}}{(1 - z^{-1})(1 - e^{-T/0.03}z^{-1})}$$

The open-loop frequency response is found by substituting $z = e^{sT}$ into the foregoing equation and substituting $j\omega$ for s.

$$\frac{C(e^{j\omega T})}{E(e^{j\omega T})} = \frac{k_c T2(1 - e^{-T/0.03})e^{-j\omega T}}{(1 - e^{j\omega T})(1 - e^{T/0.03}e^{-j\omega T})}$$

The result for unity gain ($k_c = 1$) and sample period $T = 0.03$ s is plotted in Fig. 12.11(a). The frequency response for gain $k_c = 6.5$ is also plotted; it results in a phase margin of $70°$ and a gain margin of 21 db. The closed-loop bandwidth can be predicted to be just over 10 rad/s. The corresponding closed-loop frequency response is shown in Fig. 12.11(b). The control equation for this gain and sample period is

$$m_n = m_{n-1} + 0.195e_n$$

It should be noted in both Figs. 12.11(a) and 12.11(b) that the sample period $T = 0.03$ s results in a sample frequency $\omega_s = 209.4$ rad/s. Hence the repetition of the magnitude plots at integer multiples of this frequency. These frequencies are not relevant in controller design because the computer cannot manipulate the process at frequencies greater than one-half the sample frequency. Also note that an anti-aliasing filter is required if there is a possibility of frequencies higher than $\omega_s/2$ being sampled by the system, because the frequency response around integer multiples of the sample frequency is the same as the frequency response at low frequencies. The source of these higher frequencies can be excitation of the process by the "square-edged" nature of the manipulation generated by a D/A converter, noise, or process disturbances that can have arbitrary frequency content.

The plots shown in Figs. 12.11(a) and 12.11(b) change significantly when the sample period is increased. The features on the right side of the magnitude plot move closer to the area of interest for controller design, particularly the rapid decrease in phase. Also, because the frequency response of a discrete system is not a rational function of ω, the asymptotic approximations that often are so convenient for continuous systems cannot be applied. Particular care must be taken in applying the "rules of thumb" presented in this section once frequencies of interest exceed $\omega_s/4$. □

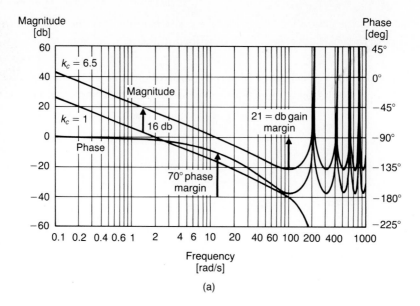

(a)

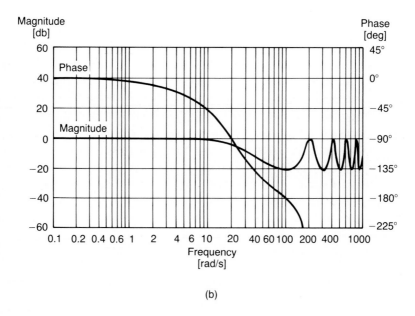

(b)

Figure 12.11 (a) Open-loop frequency response plots for Example 12.4 and (b) closed-loop frequency response plot for Example 12.4 with $k_c = 6.5$.

12.4

COMPENSATION FOR PROCESS DYNAMICS

So far, the only controller design variable we have considered is controller gain, which has been adjusted to achieve desirable closed-loop characteristics. Unfortunately, achievable performance is often limited by process dynamics, and it becomes necessary to introduce compensating dynamics into the controller in order to improve performance further. Integration is often added to controllers when it is desired to reduce or eliminate steady-state errors. Zeros are often introduced into the numerator of the controller to cancel or compensate for poles in the process transfer function that have long time constants, so that the speed of response can be improved without reducing relative stability. Pole-zero cancellation is often used to eliminate undesirable dynamic characteristics (such as large amounts of phase lag at higher frequencies) that can result in oscillatory response or instability as gain is increased.

Addition of Integral Control

The outputs of many closed-loop systems exhibit steady-state errors when changes are made in system inputs. It is often desirable to eliminate this characteristic by using integration in the controller as indicated in Figs. 12.3 and 12.5, taking advantage of infinite gain at frequency $\omega = 0$. It has already been shown in the previous section that integration by itself is a very desirable form of open-loop system. Unfortunately, when it is coupled with other process dynamics, the $-90°$ phase shift associated with integration tends to limit the gain that can be implemented in the control while still satisfying system performance criteria such as bandwidth and relative stability. The following examples illustrate the difference between control systems implemented with and without integral control.

Example 12.5 Continuous control without integration   Suppose that the system shown in Fig. 12.3 is implemented without integral control and has only gain k_c in the controller. The open-loop transfer function is then

$$G'(s) = \frac{k_c 2}{0.03s + 1}$$

As shown in Fig. 12.12, the open-loop frequency response of this system with unity gain ($k_c = 1$) is significantly different from that of the system with integral control shown in Fig. 12.7. The phase never exceeds $-90°$, so the phase margin is never less than $90°$. There is no theoretical gain limit for relative stability, and the closed-loop system is always first-order and never oscillatory.

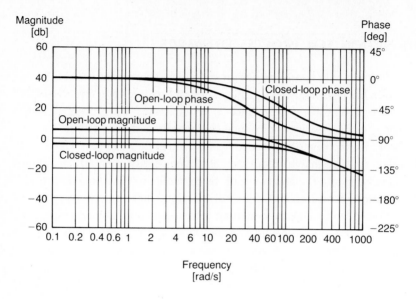

Figure 12.12 Frequency response plots for Example 12.5 with $k_c = 1$.

The closed-loop transfer function is

$$G(s) = \frac{k_c 2}{0.03s + (k_c 2 + 1)}$$

The closed-loop frequency response with unity gain ($k_c = 1$) is also shown in Fig. 12.12. Because the system does not resemble an integration, predicting the bandwidth by using Rule 3 is not particularly accurate.

A root-locus plot for this system (without integration) is shown in Fig. 12.13. Comparison with the root-locus plot for the system with integration in Fig. 12.4 reveals the significant changes in closed-loop dynamics and how the roots of the closed-loop characteristic equation vary with gain. Particularly, there is only one root and it is always real (regardless of gain), whereas in the system with integral control, there was the possibility of complex conjugate roots and resulting oscillatory response.

The time constant of the closed-loop system without integral control is

$$\tau_{cl} = 0.03/(k_c 2 + 1)$$

and the closed-loop bandwidth is

$$\omega_{cl} = (k_c 2 + 1)/0.03$$

Theoretically, τ_{cl} or ω_{cl} can be set to any value desired by making the proper choice of k_c. (In real systems, practical considerations that arise in implementation limit gain and bandwidth, as do unmodeled higher-frequency dynamics.)

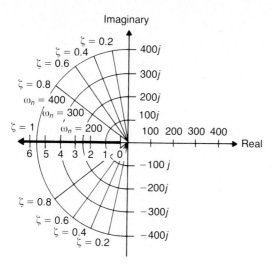

Figure 12.13 Root-locus plot for Example 12.5.

The closed-loop gain at $\omega = 0$ can be found by setting $s = j0$ in the closed-loop transfer function. In this case the gain is

$$|G(j0)| = k_c 2/(k_c 2 + 1)$$

which is the ratio of the output to the input for a constant input. As would be expected, this ratio approaches 1 (no steady-state error) as gain is increased. However, for a finite gain there is a finite steady-state error of the magnitude

$$\frac{|E(j0)|}{|R(j0)|} = \frac{1}{k_c 2 + 1}$$

For the gain $k_c = 1$ shown in Fig. 12.12, the steady-state output is 0.67 times the input (-3.5 db), and the bandwidth is about 100 rad/s. If the gain is increased to $k_c = 10$, the bandwidth increases to about 700 rad/s and the steady-state output increases to 0.95 times the input (-0.4 db).

If it is desired to eliminate this error without resorting to unreasonably high gains, then integration should be retained in the control as indicated in Figure 12.3.[3] Examination of the closed-loop transfer function in Example 12.1 reveals that the closed-loop gain of the system is always 1 (0 db) at $\omega = 0$, and the steady-state error between the output and the input is zero for a constant input.

Example 12.6 Discrete control without integration   When the integral control action is removed from the discrete controller in Fig. 12.5, the open-loop transfer function becomes

$$G'(z) = \frac{k_c 2(1 - e^{-T/0.03})z^{-1}}{1 - e^{-T/0.03}z^{-1}}$$

---

[3] An alternative is to use feedforward control as described in Section 4.4.

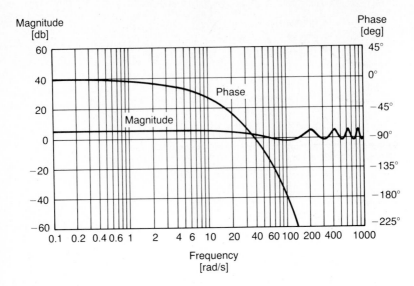

Figure 12.14 Open-loop frequency response plot for Example 12.6 with $T = 0.03$ s and $k_c = 1$.

and the closed-loop transfer function is

$$G(z) = \frac{k_c 2(1 - e^{-T/0.03})z^{-1}}{1 + [k_c 2(1 - e^{-T/0.03}) - e^{-T/0.03}]z^{-1}}$$

A frequency response plot for the open-loop system with a sample period of 0.03 s is shown in Fig. 12.14 for unity controller gain ($k_c = 1$).

It can be observed in Figure 12.14 that phase continually decreases, and, unlike what we saw in the continuous system in the previous example, gain in this case is limited by stability requirements because of the discrete nature of the system. The phase margin is less than 45° with $k_c = 1$, an indication that the system is relatively unstable. This is confirmed by the closed-loop root-locus plot with sample period $T = 0.03$ s shown in Fig. 12.15, where it can be seen that the root for $k_c = 1$ is on the negative real axis and is very near the unit circle. Roots in the left half-plane generally are undesirable, and it is clear from the root-locus plot that a more reasonable gain for this sample period is $k_c = 0.3$ or less. This gain would result in a gain margin of about 12 db, indicating significant improvement in relative stability compared to the 1-db gain margin in Fig. 12.14 with $k_c = 1$.

An alternative is to decrease the sample period so that it is significantly shorter than the process time constant. For example, if the sample period is decreased to $T = 0.005$ s, the open-loop frequency response for gain $k_c = 1$ and the root-locus plot become as shown in Figs. 12.16 and 12.17, respectively. Comparing Fig. 12.16 with Fig. 12.14 reveals that with the decreased sample

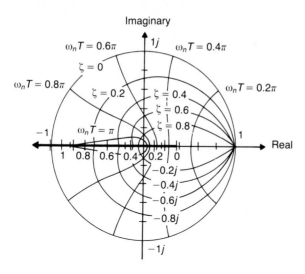

Figure 12.15 Root-locus plot for Example 12.6 with
$T = 0.03$ s.

period, the open-loop frequency response is much more desirable, becoming
more like that of the continuous system shown in Fig. 12.12; however, phase still
decreases more rapidly in the discrete system. The root location for gains around
$k_c = 1$ can be observed to be in a more desirable position in Fig. 12.17 than in
Fig. 12.15, which also indicates the improvement obtained by decreasing the
sample period.

Figure 12.16 Open-loop frequency response plot for Example 12.6 with
$T = 0.005$ s and $k_c = 1$.

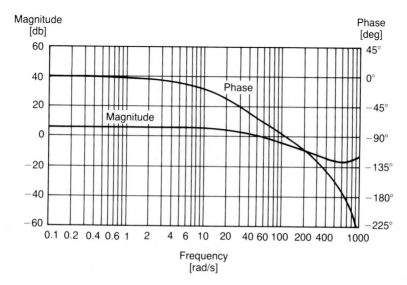

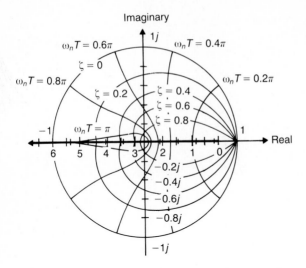

Figure 12.17 Root-locus plot for Example 12.6 with $T = 0.005$ s.

The closed-loop gain is found in the discrete case by setting $z = e^{j0T}(z = 1)$ in the closed-loop transfer function. As was the case with the continuous system in the previous example, the result is

$$|G(e^{j0T})| = k_c 2/(k_c 2 + 1)$$

If a gain of $k_c = 0.2$ and the sample period $T = 0.03$ s are chosen for this system, the controller transfer function is

$$G_c(z) = 0.2$$

and the corresponding control equation is

$$m_n = 0.2e_n$$

The ratio of steady-state output to constant input for this system is 0.29 (-10.9 db). □

Pole-Zero Cancellation

There are usually a number of criteria that must be satisfied in controller design. In order to satisfy conflicting objectives such as minimizing steady-state errors and maximizing bandwidth while maintaining relative system stability, it is necessary to give the designer more degrees of freedom in manipulating the controller to obtain these objectives. Though it is common to use controllers in which more than one gain variable can be adjusted (such as proportional-plus-integral controllers) or to add controller dynamics and observe their effect in frequency response or root-locus plots, it is often easier to introduce specific dynamic elements into the controller to cancel undesirable characteristics in the process. It is then possible to design the rest of the controller around the remaining (and, it is hoped, more favorable) process dynamics. In many cases, combining the cancellation of process dynamics with the addition of gain and integration (if necessary) in the controller results in a closed-loop system that achieves high bandwidth and speed of response without excessive overshoot and the associated long settling times.

Many processes have one or more poles corresponding to relatively long time constants that dominate system response, effectively limiting the bandwidth that can be achieved by adjusting gain. If these poles can be canceled by zeros in the controller transfer function, the limiting effects of these process time constants can be overcome. It is possible, though more difficult, to cancel underdamped process characteristics by canceling pairs of complex poles, and generally to use several zeros to cancel several poles. (Unfortunately, canceled poles still appear in the disturbance transfer function.)

There generally is a practical limit to the exactness of process models, and this in turn limits the success with which process dynamics can be canceled. For example, the location of the poles of a real process may not be exactly known and may vary somewhat with operating point and time. It is often difficult to get good estimates of process damping and natural frequencies, making the cancellation of second-order characteristics difficult.

Example 12.7 Design of a continuous PI controller   Suppose it is desired to design a controller for the process in Example 12.1 that achieves rapid response with no overshoot and zero steady-state error for constant inputs. Integration in the controller tends to eliminate steady-state errors and can be retained in spite of its detrimental effects on relative stability. If a zero is introduced into the controller that cancels the pole in the process transfer function, then, as shown in Fig. 12.18, the controller transfer function is

$$G_c(s) = \frac{k_c(0.03s + 1)}{s}$$

Figure 12.18 First-order process with continuous PI controller (integral control with pole-zero cancellation).

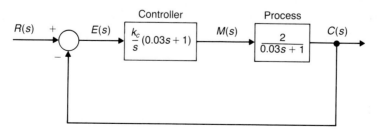

This represents proportional-plus-integral (PI) control because the foregoing equation can be rewritten as

$$G_c(s) = k_p + \frac{k_i}{s}$$

where the integral and proportional gains, k_i and k_p respectively, are

$$k_i = k_c \quad \text{and} \quad k_p = 0.03k_c$$

The resulting open-loop transfer function is

$$G'(s) = \frac{k_c 2}{s}$$

The controller frequency response and resulting open-loop system frequency response are all shown in Fig. 12.19 for unity controller gain ($k_c = 1$). It is important to note that the open-loop frequency response is now identical to that of an integration and hence is very desirable from a control point of view.

The closed-loop transfer function is

$$G(s) = \frac{k_c 2}{s + k_c 2}$$

and the closed-loop system is therefore first-order with a gain of 1. The time constant is

$$\tau_{cl} = 1/k_c 2$$

Figure 12.19 Controller and open-loop frequency response plots for Example 12.7 with $k_c = 1$.

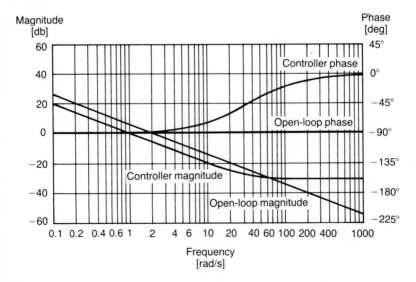

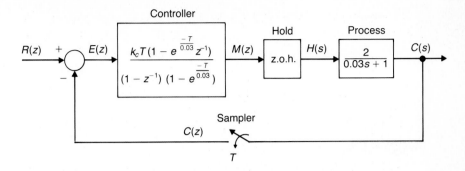

Figure 12.20 First-order process with discrete PI controller (integral control and pole/zero cancellation).

and the closed-loop bandwidth is

$$\omega_{cl} = k_c 2$$

In this case, the use of integral control combined with pole-zero cancellation allows any closed-loop bandwidth to be specified. If controller gain $k_c = 25$ is chosen, the closed-loop bandwidth is 50 rad/s and the closed-loop time constant is 0.02 s. The resulting system is always first-order with no steady-state error for constant inputs.

For the controller gain

$$k_c = 25$$

the corresponding integral and proportional gains, k_i and k_p, are

$$k_i = 25 \quad \text{and} \quad k_p = 0.75 \quad \square$$

Example 12.8 Design of a discrete PI controller   The discrete transfer function of the process shown in Fig. 12.20 is

$$G_p(z) = \frac{2(1 - e^{-T/0.03})z^{-1}}{1 - e^{-T/0.03}z^{-1}}$$

This transfer function contains a pole that can be canceled by adding a zero to the controller[4] as shown in the figure.

$$G_c(z) = \frac{k_c T(1 - e^{-T/0.03}z^{-1})}{(1 - z^{-1})(1 - e^{-T/0.03})}$$

[4] The factor $T/(1 - e^{-T/0.03})$ has been included in the controller transfer function to decouple the gain k_c from the sample period and to allow comparison of the result with that obtained in Example 12.7.

The open-loop system transfer function then is

$$G'(z) = \frac{k_c T2z^{-1}}{1 - z^{-1}}$$

and the closed-loop transfer function is

$$G(z) = \frac{k_c T2z^{-1}}{1 - (1 - k_c T2)z^{-1}}$$

The result is therefore a first-order discrete system.

If a closed-loop bandwidth of about 50 rad/s is desired, as in Example 12.7, a sample frequency of at least 500 rad/s is desirable (an order of magnitude higher than the bandwidth). A sample period of $T = 0.01$ s satisfies this specification. A root-locus plot for the resulting system is shown in Fig. 12.21. It can be observed from the plot that the controller gain should be kept below $k_c = 100$. If a bandwidth of $\omega_{cl} = 50$ rad/s is desired, then

$$\omega_{cl}T = 0.5 = 0.16\pi$$

for $T = 0.01$ s. This is just to the right of the line labeled $\omega_n T = 0.2\pi$ in Fig. 12.21. From the root-locus plot it can be seen that if a gain of about $k_c = 20$ is chosen with this sample period, the closed-loop system should have approximately this bandwidth. For this gain and sample period, the control equation is

$$m_n = m_{n-1} + 0.706e_n - 0.506e_{n-1}$$

Figure 12.21 Root-locus plot for Example 12.8 with $T = 0.01$ s.

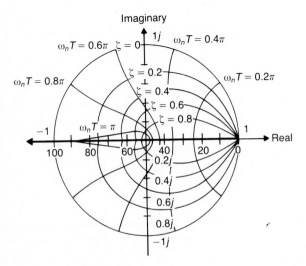

As discussed in Section 2.3, a control equation of this form represents a discrete approximation of continuous PI control where

$$m_n = m_{n-1} + (k_p + k_i T)e_n - k_p e_{n-1}$$

In this case, the corresponding integral and proportional gains, k_i and k_p, are

$$k_i = 20.0 \qquad \text{and} \qquad k_p = 0.506 \quad \square$$

12.5

APPROXIMATION OF ZERO-ORDER HOLD DYNAMICS

If a continuous controller can be designed for a process, it is possible to convert the continuous controller into an equivalent discrete controller that has approximately the same control characteristics. It is necessary, however, to consider the dynamic effects of hold elements in the system when designing the continuous controller from which the equivalent discrete controller will be obtained. One way to incorporate the effect of a hold element into the controller design process is to approximate the discrete system using a continuous system with added sample and hold elements as shown in Fig. 12.22. The time delay associated with the zero-order hold is particularly important because it produces significant phase lags, even at frequencies that are relatively low with respect to the sample frequency, reducing phase margins and relative stability. The amplitude attenuation characteristics of the zero-order hold become significant only at higher frequencies, and often they are not so important. For frequencies less than one-quarter of the sample frequency, the frequency response of the sampler and zero-order-hold elements in the system shown in Fig. 12.22 can be approximated as

$$G_h(j\omega) = \frac{H(j\omega)}{M(j\omega)} \approx e^{-j\omega T/2} \tag{12.25}$$

Figure 12.22 Continuous closed-loop control system with sampler and zero-order hold.

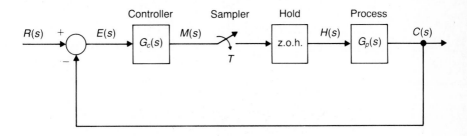

This can be combined with the controller frequency response and the process frequency response during continuous controller design, using

$$G'(j\omega) \approx G_c(j\omega)G_h(j\omega)G_p(j\omega) \tag{12.26}$$

so that the detrimental effects of the hold can be anticipated prior to obtaining an equivalent discrete controller.

Example 12.9 Inclusion of zero-order hold dynamics   Suppose it is desired to design a discrete controller for a process by first designing a continuous controller and later finding an equivalent discrete controller. For the first-order process,

$$G_p(s) = \frac{2}{(0.03s + 1)}$$

When integral control is added and the phase characteristics of the zero-order hold are included, the approximate open-loop frequency response is

$$G'(j\omega) \approx \frac{k_c}{j\omega} e^{-j\omega T/2} \frac{2}{(0.03j\omega + 1)}$$

A plot of this approximate frequency response for sample period $T = 0.03$ s and unity gain ($k_c = 1$) is shown in Fig. 12.23. If closed-loop bandwidth is to be maximized and the damping ratio is to be kept greater than 0.7, it can be seen in

Figure 12.23 Approximate open-loop frequency response plot for Example 12.9 with $T = 0.03$ s and $k_c = 1$.

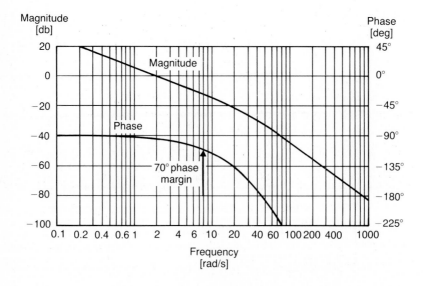

the figure that a gain of $k_c = 4(12\ db)$ results in a phase margin of about $70°$ and a predicted closed-loop bandwidth of about 8 rad/s. The resulting transfer function is

$$G_c(s) = \frac{4}{s}$$

A discrete controller that approximates this integral control action can be obtained using the backward rectangular rule.

$$G_c(z) = \frac{4(0.03)}{1 - z^{-1}}$$

The corresponding discrete control equation is

$$m_n = m_{n-1} + 0.12e_n$$

A number of alternative techniques are presented in the next section for approximating a continuous controller with a discrete controller.

Comparing the frequency response in Fig. 12.23 with that of the continuous system without the hold approximation shown in Fig. 12.9(a) illustrates the difference in the phase plot (the magnitude plot is the same). The phase plot falls off more quickly, reducing the phase margin and achievable bandwidth. This is reflected in the fact that the gain found in this example is $k_c = 4$ compared to the $k_c = 7$ found in Example 12.3 for the same process. □

12.6

DISCRETE APPROXIMATION OF CONTINUOUS CONTROLLERS

Continuous controllers generally cannot be translated directly into the discrete domain by directly applying the z transformation to the continuous controller. The problem is that the input to a continuous controller is not an impulse train, as required by the z transformation, but instead is a continuous function of time. It is therefore necessary to include some kind of extrapolation of the controller input between samples before applying the z transformation to a continuous controller. Thus the discrete controllers obtained are only approximations of continuous controllers, and the equivalency of these approximations tends to degrade as the sample period increases with respect to the characteristic times of the system.

PID Controller Approximation

One approach is to use the discrete approximation of continuous proportional-plus-integral-plus-derivative (PID) controllers developed in Section 2.3. This approximation allows classical PID control actions to be established in a computer control system, with gains calculated from the sample period and the desired proportional, integral, and derivative gains k_p, k_i, and k_d.

Proportional controllers designed in the continuous domain can be implemented directly as discrete controllers. If the continuous control equation is

$$G_c(s) = k_p \qquad (12.27)$$

then the equivalent discrete control equation is

$$G_c(z) = k_p \qquad (12.28)$$

The integral action

$$G_c(s) = \frac{k_i}{s} \qquad (12.29)$$

can be approximated via the backward rectangular rule as

$$G_c(z) = \frac{k_i T}{1 - z^{-1}} \qquad (12.30)$$

The derivative control action

$$G_c(s) = k_d s \qquad (12.31)$$

can be appproximated via the backward difference rule as

$$G_c(z) = \frac{k_d(1 - z^{-1})}{T} \qquad (12.32)$$

Combining Eqs. (12.28), (12.30), and (12.32) results in the PID approximation

$$G_c(z) = \frac{k_0 + k_1 z^{-1} + k_2 z^{-2}}{1 - z^{-1}} \qquad (12.33)$$

where

$$k_0 = k_p + k_i T + k_d/T \qquad (12.34)$$

$$k_1 = -k_p - 2k_d/T \qquad (12.35)$$

$$k_2 = k_d/T \qquad (12.36)$$

Example 12.10 PID approximation   The continuous controller designed in Example 12.7 was of the form

$$G_c(s) = \frac{k_c}{s}(0.03s + 1)$$

This transfer function can be rewritten as

$$G_c(s) = k_c 0.03 + \frac{k_c}{s}$$

which can be recognized as proportional-plus-integral control.

Using Eq. (12.33), we find that the corresponding discrete controller is

$$G_c(z) = \frac{k_c(0.03 + T) - k_c 0.03 z^{-1}}{1 - z^{-1}}$$

From Example 12.7, the closed-loop time bandwidth obtained with the continuous controller is $k_c 2$ rad/s. If a bandwidth of 50 rad/s is desired, then $k_c = 25$. A sample period of $T = 0.01$ s results in a sample frequency that is an order of magnitude higher than the desired closed-loop bandwidth. With this sample period and gain, the corresponding control equation is

$$m_n = m_{n-1} + e_n - 0.75 e_{n-1} \quad \square$$

The Bilinear Transformation

The relationship between the complex variables s and z defined in Eq. (11.30) can be used to establish a direct transformation between the s domain and the z domain. This transformation can be developed by solving Eq. (11.30) for s:

$$s = \frac{1}{T} \ln(z) \tag{12.37}$$

This equation can be expanded into the series

$$s = \frac{2(z-1)}{T(z+1)} + \frac{2(z-1)^3}{3T(z+1)^3} + \frac{2(z-1)^5}{5T(z+1)^5} + \cdots \tag{12.38}$$

Dropping all but the first term of the series results in the following approximation:

$$s \approx \frac{2}{T} \frac{(z-1)}{(z+1)} \tag{12.39}$$

which is referred to as the bilinear transformation or Tustin's rule.

The bilinear transformation maps the left (negative real) half of the s plane (the stable region of the s plane) onto the interior of the unit circle in the z plane (the stable region of the z plane). Similarly, the right half of the s plane (the unstable region of the s plane) is mapped onto the exterior of the unit circle in the z plane (the unstable region).

Some warping in frequency occurs when this transformation is applied, particularly when the sample frequency is low with respect to system dynamics.

The relationship between frequencies in the continuous domain in which controller design is carried out and the z domain in which the transformed controller will be implemented can be investigated by introducing a new frequency variable ω'. The relationship between ω and ω' can be found by substituting $j\omega'$ for s and substituting $e^{j\omega T}$ for z in Eq. (12.39). By noting that

$$e^{j\omega T} = \cos \omega T + j \sin \omega T \tag{12.40}$$

we can write the result as

$$j\omega' = \frac{2}{T} \frac{j \sin \omega t}{\cos \omega T + 1} \tag{12.41}$$

or

$$\omega' = \frac{2}{T} \tan \omega \frac{T}{2} \tag{12.42}$$

When we are given a process transfer function and specifications for closed-loop performance, the procedure is to design a continuous controller $G_c(s)$, using prewarped frequencies ω' if necessary, and then to transform $G_c(s)$ into $G_c(z)$ using Eq. (12.39).

Example 12.11 Approximation using the bilinear transformation   Applying the bilinear transformation to the integral control transfer function

$$G_c(s) = \frac{k_c}{s}$$

results in

$$G_c(z) = \frac{k_c}{\dfrac{2 \, (z - 1)}{T \, (z + 1)}}$$

or

$$G_c(z) = k_c \, \frac{T \, (1 + z^{-1})}{2 \, (1 - z^{-1})}$$

This can be recognized as a trapezoidal integration rule.

For $T = 0.03$ s and $k_c = 4$, the corresponding control equation is

$$m_n = m_{n-1} + 0.06 e_n + 0.06 e_{n-1} \quad \square$$

Example 12.12 Approximation using the bilinear transformation   The continuous controller designed in Example 12.7 has the form

$$G_c(s) = \frac{k_c}{s} (\tau_p s + 1)$$

where τ_p is a process time constant to be canceled. Application of Eq. (12.39) results in

$$G_c(z) = \frac{k_c}{\dfrac{2\,(z-1)}{T\,(z+1)}}\left[\tau_p\,\frac{2}{T}\,\frac{(z-1)}{(z+1)} + 1\right]$$

or

$$G_c(z) = k_c T\,\frac{\dfrac{\tau_p}{T}(1 - z^{-1}) + \dfrac{1}{2}(1 + z^{-1})}{1 - z^{-1}}$$

If it is desired to apply this equation to the controller designed in Example 12.7, where $k_c = 25$ and the process time constant to be canceled is $\tau_p = 0.03$ s, then for a sample period $T = 0.01$ s the foregoing controller transfer function yields the control equation

$$m_n = m_{n-1} + 0.875e_n - 0.625e_{n-1} \quad \square$$

Hold Equivalence

A convenient way to find an equivalent discrete controller, when given an existing continuous controller design, is to insert a zero-order hold before the controller and a sampler after the controller, as shown in Fig. 12.24. The z transform of the resulting controller can then be obtained in a manner identical to that of finding the z transform of a process preceded by a zero-order hold. The discrete controller transfer function can therefore be found by using Eq. (11.62).

$$G_c(z) = (1 - z^{-1})\mathfrak{Z}\left(\frac{G_c(s)}{s}\right) \tag{12.43}$$

Example 12.13 Approximation using hold equivalence Applying Eq. (12.43) to the integral control transfer function

$$G_c(s) = \frac{k_c}{s}$$

Figure 12.24 Continuous controller with added zero-order hold and sampler.

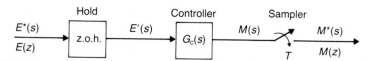

results in the discrete control transfer function

$$G_c(z) = (1 - z^{-1})\mathcal{Z}\left(\frac{k_c}{s^2}\right)$$

Taking the z transform in this equation yields

$$G_c(z) = (1 - z^{-1})\left(\frac{k_c Tz^{-1}}{(1 - z^{-1})^2}\right)$$

or

$$G_c(z) = \frac{k_c Tz^{-1}}{1 - z^{-1}}$$

The corresponding control equation with $T = 0.03$ s and $k_c = 4$ is

$$m_n = m_{n-1} + 0.12e_{n-1}$$

This result corresponds to a forward rectangular integration rule, as opposed to the backward rule used in Eq. (12.30) and Example 12.10. □

Example 12.14 Approximation using hold equivalence Applying Eq. (12.43) to the continuous control transfer function found in Example 12.7, which is

$$G_c(s) = \frac{k_c}{s}(0.03s + 1)$$

results in the discrete controller transfer function

$$G_c(z) = (1 - z^{-1})\mathcal{Z}\left(\frac{k_c(0.03s + 1)}{s^2}\right)$$

which can be rewritten as

$$G_c(z) = (1 - z^{-1})\mathcal{Z}\left(\frac{k_c(0.03)}{s} + \frac{k_c}{s^2}\right)$$

Taking the z transform of this equation results in

$$G_c(z) = (1 - z^{-1})\left(\frac{k_c(0.03)}{1 - z^{-1}} + \frac{k_c Tz^{-1}}{(1 - z^{-1})^2}\right)$$

or

$$G_c(z) = k_c(0.03) + \frac{k_c Tz^{-1}}{1 - z^{-1}}$$

For $T = 0.01$ s and $k_c = 25$, the corresponding control equation is

$$m_n = m_{n-1} + 0.75e_n - 0.5e_{n-1} \quad □$$

Pole-Zero Mapping

In general, a continuous controller transfer function can be written in factored form as

$$G_c(s) = \frac{k_c(s + d_1)(s + d_2)\cdots(s + d_i)}{(s + a_1)(s + a_2)\cdots(s + a_n)} \quad (12.44)$$

where the poles of the controller transfer function are $-a_1, -a_2, \ldots, -a_n$ and the zeros are $-d_1, -d_2, \ldots, -d_i$. Some of the poles may be complex conjugates, as well as some of the zeros. Equation (11.30) can be applied to each pole and each zero in the continuous controller transfer function to obtain an equivalent number of poles and zeros for the discrete controller transfer function. The resulting discrete controller is

$$G_c(z) = \frac{k_c'(1 - e^{-d_1 T}z^{-1})(1 - e^{-d_2 T}z^{-1})\cdots(1 - e^{-d_i T}z^{-1})}{(1 - e^{-a_1 T}z^{-1})(1 - e^{-a_2 T}z^{-1})\cdots(1 - e^{-a_n T}z^{-1})} \quad (12.45)$$

The gain k_c' in Eq. (12.45) is usually set so that the magnitude of the discrete controller frequency response is the same as that of the continuous controller at some critical frequency. If this frequency is $\omega = 0$, then k_c' can be found from

$$k_c' = \frac{k_c(1 - e^{-a_1 T})(1 - e^{-a_2 T})\cdots(1 - e^{-a_n T})}{(1 - e^{-d_1 T})(1 - e^{-d_2 T})\cdots(1 - e^{-d_i T})} \quad (12.46)$$

The procedure can be summarized as follows:

1. Find the poles and the zeros of the continuous controller transfer function.

2. Find the equivalent poles and zeros of the discrete controller transfer function, using Eq. (11.30).

3. Set the magnitude of the discrete controller frequency response to match that of the continuous controller at the desired critical frequency.

Example 12.15 Approximation using pole-zero mapping   The continuous control transfer function designed in Example 12.7 is of the form

$$G_c(s) = \frac{k_c(\tau_p s + 1)}{s}$$

and contains a single pole ($a_1 = 0$) and a single zero ($d_1 = -1/\tau_p$). Applying the pole-zero mapping procedure, Eq. (12.45) with this pole and zero, results in the discrete controller transfer function

$$G_c(z) = \frac{k_c'(1 - e^{-T/\tau_p}z^{-1})}{1 - z^{-1}}$$

In this case, the frequency response of both the continuous controller and the discrete controller have infinite gain at $\omega = 0$, and another low frequency can be chosen at which to match controller gains. Alternatively, the magnitude of the frequency response of the continuous controller is

$$|G_c(j\omega)| = \left|\frac{k_c(\tau_p j\omega + 1)}{j\omega}\right|$$

and

$$|G_c(j\omega)| \approx \frac{k_c}{\omega}$$

for small ω ($\tau_p\omega \ll 1$). Similarly, the magnitude of the frequency response of the discrete controller is

$$|G_c(e^{j\omega T})| = \left|\frac{k_c'(1 - e^{-T/\tau_p}e^{-j\omega T})}{1 - e^{-j\omega T}}\right|$$

and

$$|G_c(e^{j\omega T})| \approx \frac{k_c'(1 - e^{-T/\tau_p})}{\omega T}$$

for small ω ($\tau_p\omega \ll 1$ and $T\omega \ll 1$). Equating these magnitude functions and solving for k_c' yields

$$k_c' = \frac{k_c T}{(1 - e^{-T/\tau_p})}$$

For $\tau_p = 0.03$ s, $T = 0.01$ s, and $k_c = 25$,

$$k_c' = 0.882$$

and the corresponding control equation is

$$m_n = m_{n-1} + 0.882e_n - 0.632e_{n-1} \quad \square$$

Results of Controller Approximation

Because it is difficult to fully anticipate the properties of discrete controllers when designing in the continuous domain, approximations of continuous controllers tend to become less equivalent as control system performance is pushed toward bandwidth and sample period limits. This is illustrated in Figs. 12.25 and 12.26, wherein step responses are plotted for the closed-loop systems designed in Examples 12.7, 12.10, 12.12, and 12.14. The sample period of 0.01 s used in Fig. 12.25 results in discrete system responses that are relatively good approximations of the continuous system response, whereas the sample period of 0.02 s used in Fig. 12.26 gives much poorer results.

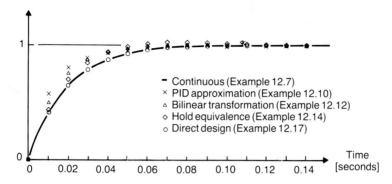

Figure 12.25 Step responses for controllers with $T = 0.01$ s.

In Fig. 12.25 the sample period is 0.01 s and the sample frequency is 628 rad/s, an order of magnitude higher than the 50-rad/s bandwidth for which the continuous closed-loop system was designed in Example 12.7. The step responses are very similar in this case, and any differences can be attributed to the approximation method used to obtain the equivalent discrete controller.

In Fig. 12.26, the sample period is 0.02 s, which is equal to the time constant of the closed-loop continuous system designed in Example 12.7. The sample frequency is 314 rad/s, about 6 times the designed closed-loop bandwidth of 50 rad/s. The differences between the continuous system and its discrete approximations are much more pronounced with this sample period, as are the differences between the individual discrete responses. The effects of sampling and holding were not considered in the continuous design, and these effects clearly are more important at this sample period. The response of the system designed using the PID approximation, for example, reveals undesirable oscillation at one-half the sample frequency. A need for a design change such as reduction in controller gain would have been a logical conclusion had the root locations for this system been checked using a root-locus plot.

Figure 12.26 Step responses for controllers with $T = 0.02$ s.

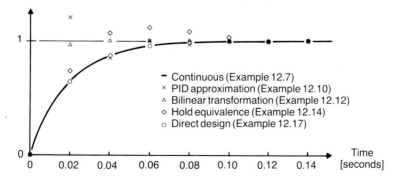

12.7
_____

THE DIRECT DESIGN METHOD

The calculation of controller transfer functions required to obtain desired input/output and input/error relationships was described in Section 4.2. This technique can be used just as easily in the s and z domains. For example, Eq. (4.5) in these cases becomes

$$G_c(s) = \frac{1}{G_p(s)} \frac{G_{dr}(s)}{[1 - G_{dr}(s)]} \qquad (12.47)$$

and

$$G_c(z) = \frac{1}{G_p(z)} \frac{G_{dr}(z)}{[1 - G_{dr}(z)]} \qquad (12.48)$$

where $G_p(s)$ or $G_p(z)$ is the process transfer function, $G_{dr}(s)$ or $G_{dr}(z)$ is the desired command-input/process-output transfer function, and $G_c(s)$ or $G_c(z)$ is the required controller transfer function. Other relationships such as those specified by Eq. (4.7) and (4.10) for error versus input and for output versus disturbance can also be used. Figs. 12.1 and 12.2 can aid in selection of the poles of the desired closed-loop transfer function, and Appendix E can aid in selection of its form.

Example 12.16 Direct continuous controller design   It is desired to calculate the continuous controller that achieves a closed-loop gain of 1 and a closed-loop time constant τ_{dr} for the process shown in Fig. 12.3. In this case, the desired closed-loop transfer function is

$$G_{dr}(s) = \frac{1}{\tau_{dr}s + 1}$$

and, using Eq. (12.47),

$$G_c(s) = \frac{1}{2\tau_{dr}s} (0.03s + 1)$$

The similarity between this result and that of Example 12.7 is due to implicit cancellation of process dynamics in the direct design method. Here, the pole of the process is canceled by the zero of the controller. ☐

Example 12.17 Direct discrete controller design   If it is desired to directly design a discrete controller for the process and zero-order hold shown in Fig. 12.5, the discrete process transfer function is

$$G_p(z) = \frac{2(1 - \delta_p)z^{-1}}{1 - \delta_p z^{-1}}$$

where $\delta_p = e^{-T/0.03}$.

In this case, the desired discrete closed-loop transfer function is

$$G_{dr}(z) = \frac{(1 - \delta_{dr})z^{-1}}{1 - \delta_{dr}z^{-1}}$$

where $\delta_{dr} = e^{-T/\tau_{dr}}$ and τ_{dr} is the desired closed-loop time constant. The required controller transfer function is found from Eq. (12.48) as

$$G_c(z) = \frac{(1 - \delta_{dr})(1 - \delta_p z^{-1})}{2(1 - \delta_p)(1 - z^{-1})}$$

Again, the cancellation of the pole in the process by a zero in the numerator should be noted.

If the desired closed-loop bandwidth is 50 rad/s, the desired closed-loop time constant is $\tau_{dr} = 0.02$ s. With sample period $T = 0.01$ s, the control equation becomes

$$m_n = m_{n-1} + 0.694e_n - 0.497e_{n-1}$$

The response of this system is shown in Fig. 12.25. Response is also shown in Fig. 12.26 for sample period $T = 0.02$ s. In both cases, the response of the discrete system designed by the direct method is, by definition, identical at the sample instants to that of the continuous system designed in Example 12.7. □

12.8

SUMMARY

A number of approaches have been presented in this chapter for designing discrete controllers. Parallel developments have also been presented for continuous controllers because of the close similarity between methods and because of the familiarity many designers have with design in the continuous domain. However, it is generally best to design discrete controllers in the discrete domain whenever possible; the root-locus and direct design methods are often the easiest to use.

From the discussions in this chapter, it should be clear that it is desirable to have a short sample period wherever possible. This tends to make the task of controller design easier and reduces the possibilities of aliasing, instability, and so on. It was pointed out in Section 4.3 that rejection of disturbances can be a more important consideration than bandwidth and speed of response to command inputs in the choice of a sample period. In higher-precision control systems, the effects of disturbances that cannot be predicted and can be detected only by detecting changes in process outputs must be limited by making the sample period short enough so that the process output cannot change significantly between samples. The controller is then able to respond to these disturbances before their effects grow to unacceptable magnitudes.

One must give careful consideration to arithmetic precision and quantization levels when implementing discrete controllers, because the difference between the magnitudes of coefficients in controllers can be just as important as their individual magnitudes. For PI control, for example, the difference in magnitudes between coefficients k_0 and k_1 in Eq. (12.33) represents the integral part of the control action. This difference can be very small if the sample period T is short, yet the value of this difference can have a large effect on performance. Hence it must be accurately represented in control calculations. This phenomenon is similar to the undesirable quantization effects that arise when digital differentiation is used; both tend to become more of a problem when very short sample periods are used.

Implementation issues such as quantization problems and computation times associated with computer hardware, software, and interfaces require that a careful choice of sample periods be made as part of discrete controller design. Often, trade-offs in achievable performance must be made as a result of these computational limitations. It must also be kept in mind that there are physical limits associated with system components. Process and interface components usually have maximum limits on the magnitudes of output variables. Large amounts of energy are often required to manipulate processes at high frequencies, which places a practical limit on achievable bandwidth. In controller design, it is always necessary to consider both the practical and the theoretical aspects of the system.

BIBLIOGRAPHY

Ackermann, J. *Sampled-Data Control Systems: Analysis and Synthesis, Robust System Design.* New York, NY: Springer-Verlag, 1985.

Astrom, K. J., and B. Wittenmark. *Computer Controlled Systems: Theory and Design.* Englewood Cliffs, NJ: Prentice-Hall, 1984.

Doebelin, E. O. *Control Systems: Principles and Design.* New York, NY: Wiley, 1985.

Dorf, R. C. *Modern Control Systems, 4/e.* Reading, MA: Addison-Wesley, 1986.

Franklin, G. F., and J. D. Powell. *Digital Control of Dynamic Systems.* Reading, MA: Addison-Wesley, 1981.

Harrison, H. L., and J. G. Bollinger. *Introduction to Automatic Controls.* New York, NY: Harper and Row, 1969.

Isermann, R. *Digital Control Systems.* New York, NY: Springer-Verlag, 1981.

Jacquot, R. G. *Modern Digital Control Systems.* New York, NY: Dekker, 1981.

Kuo, B. C. *Digital Control Systems.* New York, NY: Holt, Rinehart and Winston, 1980.

Meirovitch, L. *Introduction to Dynamics and Control.* New York, NY: Wiley, 1984.

Ogata, K. *Discrete-Time Control Systems.* Englewood Cliffs, NJ: Prentice-Hall, 1987.

Phillips, C. L., and H. T. Nagle. *Digital Control System Analysis and Design.* Englewood Cliffs, NJ: Prentice-Hall, 1984.

Van Landingham, H. F. *Introduction to Digital Control Systems.* New York, NY: Macmillan, 1985.

PROBLEMS

12.1 Use the discrete root-locus technique to design a discrete proportional controller for the process

$$G_p(s) = \frac{200}{s}$$

which has as its input the output of a D/A converter. The sample period should be 0.5 s, and the closed-loop system should correspond to a first-order continuous system with a time constant of 0.4 s.

12.2 Repeat Problem 12.1, except that dead-beat response (response in one sample period) should be achieved.

12.3 Repeat Problem 12.1 using the discrete frequency response.

12.4 Repeat Problem 12.3 by including sampling and holding effects in the continuous frequency response and designing a continuous proportional controller, the gain of which is then to be used as the gain of the discrete proportional controller.

12.5 Compare the results of Problems 12.3 and 12.4, and identify the source of any differences.

12.6 Show that a first-order continuous process with time constant τ that is manipulated by a D/A converter (zero-order hold) has a discrete process model with a single root on the positive real axis of the z plane.

12.7 Show that systems with roots in the negative real half of the z plane (the left half-plane) experience free oscillation (zero input, nonzero initial conditions) at one-half the sample frequency.

12.8 Replot the root locus for Example 12.8 with $T = 0.03$ s and compare the result with Fig. 12.21.

12.9 Plot the root-locus for Example 12.10 with $T = 0.02$ s. Evaluate the choice of gain $k_c = 25$ for this sample period and choose a new gain if appropriate.

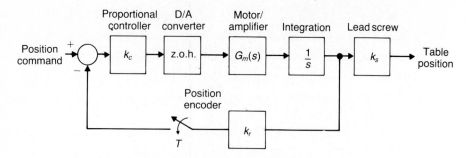

Figure 12.27 Block diagram for Problems 12.10 and 12.11.

12.10 Find the z transform and plot the root locus for the position control system with a sample rate of 200 Hz that is represented by the block diagram shown in Fig. 12.27. The motor/amplifier transfer function is represented simply by a gain:

$$G_m(s) = k_m$$

and $k_m = 4$, $k_s = 0.2$, and $k_r = 4000$. What are the gain limits for stability and forced oscillation at one-half the sample frequency?

12.11 Find the z transform and plot the root locus for the position control system with a sample rate of 200 Hz that is represented by the block diagram shown in Fig. 12.27. The motor/amplifier transfer function is second-order and underdamped:

$$G_m(s) = \frac{k_m}{s^2/\omega_n^2 + 2\zeta s/\omega_n + 1}$$

and $k_m = 4$, $k_s = 0.2$, $k_r = 4000$, $\omega_n = 400$ rad/s, and $\zeta = 0.7$. What are the gain limits for stability and forced oscillation at one-half of the sample frequency?

12.12 Compare the results of Problems 12.10 and 12.11. In particular, comment on whether, for position control purposes, the system in Problem 12.10 is a good approximation of the system in Problem 12.11.

12.13 Use the discrete frequency response to design a discrete proportional controller for the system with the block diagram shown in Fig. 12.27 and the motor/amplifier transfer function given in Problem 12.11. Plot and assess the frequency response of the resulting closed-loop system.

Appendix A
Analysis and Design Using State Variables

A.1
INTRODUCTION

In the state-variable approach to the analysis and design of control systems, sets of first-order differential equations are used to define system dynamics. The state-variable approach is important because it allows a unified representation and analysis of systems of arbitrary order. Systems defined by higher-order differential equations are redefined by a set of first-order differential equations for which a discrete model can be readily obtained. Each first-order equation defines the dynamics of a state variable.

In this chapter, the state-variable approach to system analysis and continuous controller design is first described, followed by a discrete system analysis and controller design strategy for systems modeled using state variables and controlled by a computer. The concept of a state estimator is introduced for cases where not all of the state variables can be measured directly. Matrix methods are used throughout, and the result is a unified representation that can be applied to both single and multiple input/output systems.

A.2
STATE EQUATIONS FOR CONTINUOUS SYSTEMS

System dynamics are represented in the state-variable approach by a set of first-order differential equations. Figure A.1 shows a continuous system or process with manipulation inputs $m_1, m_2, m_3, \ldots, m_p$, disturbance inputs $d_1, d_2, d_3, \ldots, d_v$,

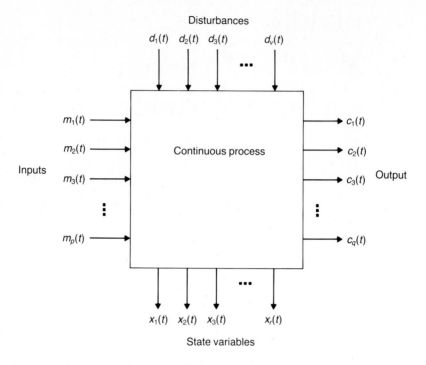

Figure A.1 Continuous process modeled using state variables.

and outputs $c_1, c_2, c_3, \ldots, c_q$. The system has state variables $x_1, x_2, x_3, \ldots, x_r$, and the system dynamics are modeled using

$$\frac{d\mathbf{x}(t)}{dt} = \mathbf{A}\mathbf{x}(t) + \mathbf{B}\mathbf{m}(t) + \mathbf{W}\mathbf{d}(t) \tag{A.1}$$

where

$$\mathbf{x}(t) = [x_1(t)\ x_2(t)\ x_3(t) \cdots x_r(t)]^T \tag{A.2}$$

$$\frac{d\mathbf{x}(t)}{dt} = \left[\frac{dx_1(t)}{dt}\ \frac{dx_2(t)}{dt}\ \frac{dx_3(t)}{dt} \cdots \frac{dx_r(t)}{dt}\right]^T \tag{A.3}$$

$$\mathbf{m}(t) = [m_1(t)\ m_2(t)\ m_3(t) \cdots m_p(t)]^T \tag{A.4}$$

$$\mathbf{d}(t) = [d_1(t)\ d_2(t)\ d_3(t) \cdots d_v(t)]^T \tag{A.5}$$

The system outputs are related to the state variables by

$$\mathbf{c}(t) = \mathbf{H}\mathbf{x}(t) \tag{A.6}$$

where

$$\mathbf{c}(t) = [c_1(t)\ c_2(t)\ c_3(t) \cdots c_q(t)]^T \tag{A.7}$$

Matrix $\mathbf{A}$ is an $r \times r$ matrix, $\mathbf{B}$ is an $r \times p$ matrix, $\mathbf{W}$ is an $r \times v$ matrix, and $\mathbf{H}$ is a $q \times r$ matrix. It will be assumed here that the process is time-invariant and that matrices $\mathbf{A}$, $\mathbf{B}$, $\mathbf{W}$, and $\mathbf{H}$ are constant.

Example A.1 State equation for second-order system   Obtain a set of state equations for a process described by the differential equation

$$\frac{1}{\omega_n^2} \frac{d^2c(t)}{dt^2} + \frac{2\zeta}{\omega_n} \frac{dc(t)}{dt} + c(t) = km(t)$$

In the state-variable approach, it is necessary to represent this second-order differential equation by two first-order differential equations. The two state variables chosen, $x_1(t)$ and $x_2(t)$, are defined as follows:

$$x_1(t) = c(t)$$

$$x_2(t) = \frac{dc(t)}{dt}$$

The two state equations are

$$\frac{dx_1(t)}{dt} = x_2(t)$$

and

$$\frac{dx_2(t)}{dt} = -\omega_n^2 x_1(t) - 2\zeta\omega_n x_2(t) + k\omega_n^2 m(t)$$

or, in the matrix form of Eqs. (A.1) and (A.2),

$$\frac{d}{dt}\begin{bmatrix} x_1(t) \\ x_2(t) \end{bmatrix} = \begin{bmatrix} 0 & 1 \\ -\omega_n^2 & -2\zeta\omega_n \end{bmatrix}\begin{bmatrix} x_1(t) \\ x_2(t) \end{bmatrix} + \begin{bmatrix} 0 \\ k\omega_n^2 \end{bmatrix} m(t)$$

and

$$c(t) = \begin{bmatrix} 1 & 0 \end{bmatrix}\begin{bmatrix} x_1(t) \\ x_2(t) \end{bmatrix}$$

If $k = 1$, $\omega_n = 1.414$ rad/s, and $\zeta = 1.061$, then the continuous state equation for the system is

$$\frac{d}{dt}\begin{bmatrix} x_1(t) \\ x_2(t) \end{bmatrix} = \begin{bmatrix} 0 & 1 \\ -2 & -3 \end{bmatrix}\begin{bmatrix} x_1(t) \\ x_2(t) \end{bmatrix} + \begin{bmatrix} 0 \\ 2 \end{bmatrix} m(t)$$

In this case, the matrices in Eq. (A.1) are

$$\mathbf{A} = \begin{bmatrix} 0 & 1 \\ -2 & -3 \end{bmatrix}$$

$$\mathbf{B} = \begin{bmatrix} 0 \\ 2 \end{bmatrix}$$

and

$$W = [0]$$

Also in Eq. (A.6),

$$H = [1 \quad 0] \ \square$$

Process Transfer Function Matrix

The Laplace transformation can be applied to Eqs. (A.1) and (A.6). The result can be a manipulated to obtain a **process transfer function matrix** that corresponds to the process transfer functions found in Chapters 4 and 11. Assuming zero initial conditions,

$$sX(s) = AX(s) + BM(s) + WD(s) \tag{A.8}$$

$$C(s) = HX(s) \tag{A.9}$$

Rearranging Eq. (A.9) results in

$$X(s) = (sI - A)^{-1}[BM(s) + WD(s)] \tag{A.10}$$

Substituting Eq. (A.10) into Eq. (A.9) and assuming there are no disturbance inputs ($d(t) = [0]$) yields

$$C(s) = G_p(s)M(s) \tag{A.11}$$

where $G_p(s)$ is the process transfer function matrix:

$$G_p(s) = H(sI - A)^{-1}B \tag{A.12}$$

Example A.2 Second-order process transfer function   For the process defined in Example A.1, the continuous state equation is

$$\frac{d}{dt}\begin{bmatrix} x_1(t) \\ x_2(t) \end{bmatrix} = \begin{bmatrix} 0 & 1 \\ -2 & -3 \end{bmatrix}\begin{bmatrix} x_1(t) \\ x_2(t) \end{bmatrix} + \begin{bmatrix} 0 \\ 2 \end{bmatrix}m(t)$$

The process transfer function matrix is found from Eq. (A.12) as

$$G_p(s) = [1 \quad 0]\begin{bmatrix} s & -1 \\ 2 & s+3 \end{bmatrix}^{-1}\begin{bmatrix} 0 \\ 2 \end{bmatrix}$$

or

$$G_p(s) = \begin{bmatrix} \dfrac{2}{s^2 + 3s + 2} \end{bmatrix}$$

The transfer function for this process is therefore

$$\frac{C(s)}{M(s)} = \frac{2}{s^2 + 3s + 2} \ \square$$

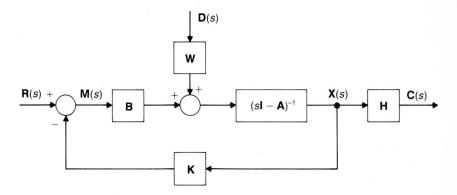

Figure A.2 Continuous closed-loop system with state-variable feedback.

Continuous Closed-Loop Control

A continuous closed-loop control system can be constructed using state-variable feedback as shown in Fig. A.2. Gains are set in the $r \times p$ gain matrix $\mathbf{K}$ to adjust the performance of the system. The control law used is

$$\mathbf{m}(t) = \mathbf{r}(t) - \mathbf{K}\mathbf{x}(t) \tag{A.13}$$

where $\mathbf{r}(t)$ is the command input to the system. Substituting Eq. (A.13) into Eq. (A.1) yields the closed-loop state equation

$$\frac{d\mathbf{x}(t)}{dt} = (\mathbf{A} - \mathbf{B}\mathbf{K})\mathbf{x}(t) + \mathbf{B}\mathbf{r}(t) + \mathbf{W}\mathbf{d}(t) \tag{A.14}$$

Assuming zero initial conditions, taking the Laplace transform of Eq. (A.14), and solving for $\mathbf{X}(s)$ yields

$$\mathbf{X}(s) = (s\mathbf{I} - \mathbf{A} + \mathbf{B}\mathbf{K})^{-1}[\mathbf{B}\mathbf{R}(s) + \mathbf{W}\mathbf{D}(s)] \tag{A.15}$$

Assuming that there are no disturbance inputs and substituting Eq. (A.15) into Eq. (A.9) results in

$$\mathbf{C}(s) = \mathbf{G}(s)\mathbf{R}(s) \tag{A.16}$$

where $\mathbf{G}(s)$ is the closed-loop transfer function matrix

$$\mathbf{G}(s) = \mathbf{H}(s\mathbf{I} - \mathbf{A} + \mathbf{B}\mathbf{K})^{-1}\mathbf{B} \tag{A.17}$$

The characteristic equation for the closed-loop system is

$$det[s\mathbf{I} - \mathbf{A} + \mathbf{B}\mathbf{K}] = 0 \tag{A.18}$$

and has the general form

$$s^r + \alpha_1 s^{r-1} + \alpha_2 s^{r-2} + \cdots + \alpha_{r-1}s^1 + \alpha_r = 0 \tag{A.19}$$

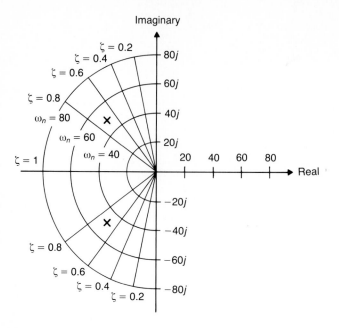

Figure A.3 An s-plane plot for Example A.3, showing root locations for $\omega_n = 50$ rad/s and $\zeta = 0.7$.

Equation (A.19) has roots $s_1, s_2, s_3, \ldots, s_r$, and if either the desired values of these roots or the desired coefficients $\alpha_1, \alpha_2, \alpha_3, \ldots, \alpha_r$ of the characteristic equation are known, then the elements of matrix **K** can be set so that these values are obtained. The closed-loop system can therefore be designed to achieve desired dynamic characteristics.

Example A.3 Closed-loop controller design   Suppose that it is desired to add state-variable feedback to the process defined in Example A.1 to obtain an underdamped closed-loop system with a natural frequency of 50 rad/s and a damping ratio of 0.7. As indicated on the s-plane plot shown in Fig. A.3, the required roots are

$$s_1 = -35 + j35.71$$

$$s_2 = -35 - j35.71$$

and the desired closed-loop characteristic equation is

$$s^2 + 70s + 2500 = 0$$

In this case,

$$\mathbf{K} = [k_1 \quad k_2]$$

From Eq. (A.18) and matrices **A** and **B** defined in Example A.1, the closed-loop characteristic equation is

$$\det\begin{bmatrix} s & -1 \\ 2 + 2k_1 & s + 3 + 2k_2 \end{bmatrix} = 0$$

or

$$s^2 + (3 + 2k_2)s + (2 + 2k_1) = 0$$

By equating like terms in the foregoing equation and the desired characteristic equation, we get

$$2 + 2k_1 = 2500$$

$$3 + 2k_2 = 70$$

The elements of the gain matrix are then

$$k_1 = 1249$$

$$k_2 = 33.5$$

and

$$\mathbf{K} = [1249 \quad 33.5] \quad \square$$

Example A.4 Simulation using state equations   Response of the closed-loop system designed in Example A.3 to command inputs can be studied using the closed-loop state equation in Eq. (A.14). From the result of Example A.3,

$$\frac{d}{dt}\begin{bmatrix} x_1(t) \\ x_2(t) \end{bmatrix} = \left[\begin{bmatrix} 0 & 1 \\ -2 & -3 \end{bmatrix} - \begin{bmatrix} 0 \\ 2 \end{bmatrix}[1249 \quad 33.5] \right]\begin{bmatrix} x_1(t) \\ x_2(t) \end{bmatrix} + \begin{bmatrix} 0 \\ 2 \end{bmatrix}r(t)$$

or

$$\frac{d}{dt}\begin{bmatrix} x_1(t) \\ x_2(t) \end{bmatrix} = \begin{bmatrix} 0 & 1 \\ -2500 & -70 \end{bmatrix}\begin{bmatrix} x_1(t) \\ x_2(t) \end{bmatrix} + \begin{bmatrix} 0 \\ 2 \end{bmatrix}r(t)$$

We assume that the change in the state variables is

$$\Delta \mathbf{x}(t) = \frac{d\mathbf{x}(t)}{dt}\Delta t$$

where Δt is a small interval of time. Then given the state of the system, $\mathbf{x}(t)$, at time t, we can find the state of the system one time interval later, $\mathbf{x}(t + \Delta t)$, from

$$\mathbf{x}(t + \Delta t) \approx \mathbf{x}(t) + \Delta \mathbf{x}(t)$$

In this case, the equations for calculating the next values of the state variables are

$$\Delta x_1(t) = x_2(t)\Delta t$$

$$\Delta x_2(t) = [-2500x_1(t) - 70x_2(t) + 2r(t)]\Delta t$$

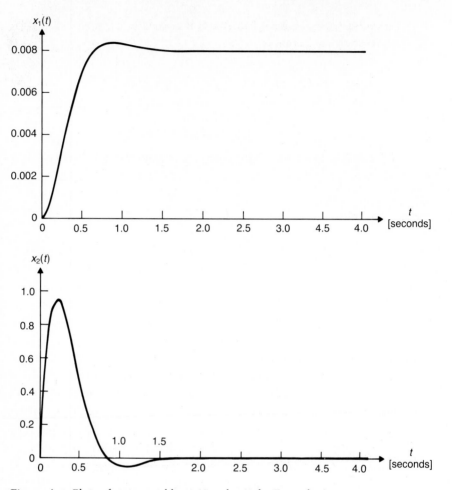

Figure A.4 Plots of state variables $x_1(t)$ and $x_2(t)$ for Example A.4.

and

$$x_1(t + \Delta t) \approx x_1(t) + \Delta x_1(t)$$

$$x_2(t + \Delta t) \approx x_2(t) + \Delta x_2(t)$$

Given the command input function $r(t)$ and initial conditions $\mathbf{x}(0)$ at time $t = 0$, the foregoing equations can be used to calculate the approximate values of the state variables at successive intervals of time. By choosing Δt to be much less than the characteristic times of the system and its inputs, we can obtain a good approximation of the values of the state variables over time. Here the natural frequency of the system was designed in Example A.3 to be 50 rad/s. The characteristic time associated with this natural frequency is

$$t_c = \frac{1}{50} = 0.02 \text{ s}$$

```
program State_Variable_Simulation;
  const
    dt = 0.001;                                        {time interval in seconds}
  var
    x1, x2, dx1, dx2 : real;                           {x1 and x2 are state variables}
    r : real;                                          {r is command input}
    t : real;                                          {time in seconds}
    i : integer;
  begin
    r := 10.0;                                         {step input}
    x1 := 0.0;                                         {zero initial conditions}
    x2 := 0.0;
    for i := 1 to 400 do
      begin
        t := i * dt;
        dx1 := x2 * dt;                                {calculate changes in states}
        dx2 := (-2500.0 * x1 - 70.0 * x2 + 2.0 * r) * dt;   {model from Example A.4}
        x1 := x1 + dx1;                                {update state variables}
        x2 := x2 + dx2;
        writeln(i, t, r, x1, x2)
      end
  end.
```

Program A.1 Simulation of system in Example A.4 using state variables.

and we should choose

$$\Delta t \ll t_c$$

Figure A.4 shows plots of the response of state variables $x_1(t)$ and $x_2(t)$ of the system designed in Example A.3 to the step input $r(t) = 10$ for $t \geq 0$. Note that $x_1(\infty) \neq r(\infty)$ because of the process and feedback gains. A time interval $\Delta t = 0.001$ s was used which satisfies the foregoing requirement, allowing the values of $x_1(t)$ and $x_2(t)$ to be calculated at successive intervals of time using the equations given above in a program such as Program A.1. Simulation of the system in this manner is a valuable tool for evaluating system response. Various input functions can be implemented (step, ramp, sawtooth, sinusoid, and the like, with Δt decreased for rapidly varying inputs), and controller gains can be adjusted if necessary to improve system response. □

A.3

STATE EQUATIONS FOR DISCRETE SYSTEMS

A discrete model for a continuous process can be found by assuming that the manipulation inputs to the process are held constant between sampling instants by sample-and-hold devices as shown in Fig. A.5. This is quite common in

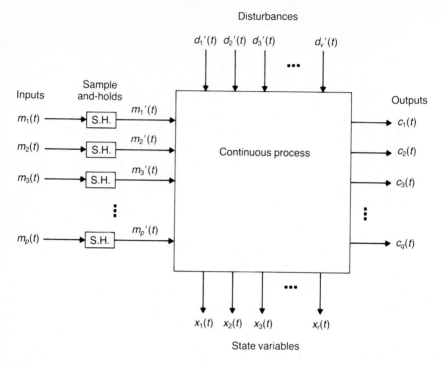

Figure A.5 Continuous process with sample-and-hold inputs for discrete model using state variables.

computer control systems, and it is the case when digital-to-analog (D/A) converters are used by the control computer to manipulate the process. The new process manipulation inputs are $\mathbf{m}'(t) = \mathbf{m}(nT)$ for $nT \leq t < (n + 1)T$, where n is an integer and T is the sample period. If the disturbance inputs are assumed to be constant[1] between sample instants with $\mathbf{d}'(t) = \mathbf{d}(nT)$ for $nT \leq t < [(n + 1)T$, then a discrete form of Eqs. (A.1) and (A.6) can be written as follows:[2]

$$\mathbf{x}[(n + 1)T] = \mathbf{\Phi}(T)\mathbf{x}(nT) + \mathbf{\Theta}(T)\mathbf{m}(nT) + \mathbf{\Psi}(T)\mathbf{d}(nT) \qquad (A.20)$$

and

$$\mathbf{c}(nT) = \mathbf{H}\mathbf{x}(nT) \qquad (A.21)$$

$$\mathbf{\Phi}(T) = e^{\mathbf{A}T} \qquad (A.22)$$

---

[1] It is important to remember that the assumption that the disturbance inputs are constant between sample periods is not completely valid for systems wherein disturbances can appear at any time rather than just at the sample instants. The practical solution to this problem is to use a small enough sample period so that changes in disturbances are relatively small between sample periods.

[2] For proofs of the validity of the developments in this appendix, refer to one of the many excellent texts on modern control theory, some of which are listed in the bibliography.

or from the series expansion

$$\mathbf{\Phi}(T) = \mathbf{I} + \mathbf{A}T + \frac{(\mathbf{A}T)^2}{2!} + \frac{(\mathbf{A}T)^3}{3!} + \cdots \tag{A.23}$$

Matrices $\mathbf{\Theta}(T)$ and $\mathbf{\Psi}(T)$ can be calculated from

$$\mathbf{\Theta}(T) = \mathbf{\Lambda}(T)T\mathbf{B} \tag{A.24}$$

$$\mathbf{\Psi}(T) = \mathbf{\Lambda}(T)T\mathbf{W} \tag{A.25}$$

where

$$\mathbf{\Lambda}(T) = \mathbf{I} + \frac{\mathbf{A}T}{2!} + \frac{(\mathbf{A}T)^2}{3!} + \cdots \tag{A.26}$$

Also,

$$\mathbf{\Phi}(T) = \mathbf{I} + \mathbf{A}T\mathbf{\Lambda}(T) \tag{A.27}$$

Although it is possible to find closed-form expressions for matrices $\mathbf{\Phi}(T)$ and/or $\mathbf{\Lambda}(T)$, these can be evaluated more conveniently using the first few terms of the series in Eqs. (A.23) and (A.26). For example, $\mathbf{\Lambda}(T)$ can computed with k terms by using

$$\mathbf{\Lambda}(T) \approx \mathbf{I} + \frac{\mathbf{A}T}{2}\left\{\mathbf{I} + \frac{\mathbf{A}T}{3}\left[\mathbf{I} + \cdots + \frac{\mathbf{A}T}{k-1}\left(\mathbf{I} + \frac{\mathbf{A}T}{k}\right)\cdots\right]\right\} \tag{A.28}$$

$\mathbf{\Theta}(T)$, $\mathbf{\Psi}(T)$, and $\mathbf{\Phi}(T)$ can then be computed using Eqs. (A.24), (A.25), and (A.27) and used in the state discrete equation in Eq. (A.20).

Example A.5 Discrete state equation   From Example A.1,

$$\mathbf{A} = \begin{bmatrix} 0 & 1 \\ -2 & -3 \end{bmatrix} \qquad \mathbf{B} = \begin{bmatrix} 0 \\ 2 \end{bmatrix} \qquad \mathbf{H} = [1 \quad 0]$$

and the disturbance inputs are zero; that is $\mathbf{d}(t) = [\mathbf{0}]$.

If the sample period to be used is $T = 0.5$ s, then summing 10 terms in the series in Eq. (A.28) yields

$$\mathbf{\Lambda}(T) = \begin{bmatrix} 0.9418 & 0.1548 \\ -0.3096 & 0.4773 \end{bmatrix}$$

and, from Eqs. (A.24) and (A.27),

$$\mathbf{\Phi}(T) = \begin{bmatrix} 0.8452 & 0.2387 \\ -0.4773 & 0.1292 \end{bmatrix} \qquad \mathbf{\Theta}(T) = \begin{bmatrix} 0.1548 \\ 0.4773 \end{bmatrix}$$

Hence the discrete state equation for the system is

$$\begin{bmatrix} x_1((n+1)T) \\ x_2((n+1)T) \end{bmatrix} = \begin{bmatrix} 0.8452 & 0.2387 \\ -0.4773 & 0.1292 \end{bmatrix}\begin{bmatrix} x_1(nT) \\ x_2(nT) \end{bmatrix} + \begin{bmatrix} 0.1548 \\ 0.4773 \end{bmatrix}m(nT)$$

Program A.2 shows how $\Phi(T)$ and $\Theta(T)$ can be calculated using Eqs. (A.24), (A.26), and (A.27). □

Program A.2 Calculation of discrete state-variable model.

```
program Discrete_State_Variable_Model;
  const
    T = 0.5;                              {0.5 s sample period}
    r = 2;                               {2nd-order system}
    terms = 15;                          {number of terms in exp(AT) series}
  type
    matrix = array[1..r, 1..r] of real;
    vector = array[1..r] of real;
  var
    A : matrix;                          {A and B in Eq. A.1}
    B : vector;
    PHI : matrix;                        {PHI and THETA in Eq. A.20}
    THETA : vector;
    LAMDA, LAMDA1 : matrix;              {LAMDA in Eq. A.26}
    i, j, m, n : integer;
  begin
    A[1, 1] := 0.0;                      {A and B from Example A.1}
    A[1, 2] := 1.0;
    A[2, 1] := -2.0;
    A[2, 2] := -3.0;
    B[1] := 0.0;
    B[2] := 2.0;
    for i := 1 to r do                   {calculate LAMDA using Eq. A.26}
      for j := 1 to r do
        LAMDA[i, j] := 0.0;
    for n := 1 to r do
      LAMDA[n, n] := 1.0;                {LAMDA := I}
    for m := terms downto 2 do           {calculate series in factored form}
      begin
        for i := 1 to r do
          for j := 1 to r do
            begin
              LAMDA1[i, j] := 0.0;       {LAMDA := A*(T/m)*LAMDA + I}
              for n := 1 to r do
                LAMDA1[i, j] := LAMDA1[i, j] + A[i, n] * LAMDA[n, j];
            end;
        for i := 1 to r do
          for j := 1 to r do
            LAMDA[i, j] := LAMDA1[i, j] * T / m;
```

```
  for n := 1 to r do
    LAMDA[n, n] := LAMDA[n, n] + 1.0
  end;
    for i := 1 to r do                        {calculate PHI using Eq. A.27}
      for j := 1 to r do
        begin
          PHI[i, j] := 0.0;                    {PHI := A*T*LAMDA + I}
          for n := 1 to r do
          PHI[i, j] := PHI[i, j] + A[i, n] * T * LAMDA[n, j]
        end;
    for n := 1 to r do
      PHI[n, n] := PHI[n, n] + 1.0;
    for i := 1 to r do
      for j := 1 to r do
        writeln(i, j, PHI[i, j]);
    for i := 1 to r do                        {calculate THETA using Eq. A.24}
      begin
        THETA[i] := 0.0;                       {THETA := LAMDA*T*B}
        for n := 1 to r do
        THETA[i] := THETA[i] + LAMDA[i, r] * T * B[n];
        writeln(i, THETA[i])
      end;
  end.
```

Discrete Process Transfer Function Matrix

Applying the z transformation to Eqs. (A.20) and (A.21) yields

$$X(z) = \Phi(T)z^{-1}X(z) + \Theta(T)z^{-1}M(z) + \Psi(T)z^{-1}D(z) \qquad (A.29)$$

and

$$C(z) = HX(z) \qquad (A.30)$$

In a development similar to that already carried out for continuous systems, rearranging Eq. (A.29) results in

$$X(z) = (I - \Phi(T)z^{-1})^{-1}[\Theta(T)z^{-1}M(z) + \Psi(T)z^{-1}D(z)] \qquad (A.31)$$

Assuming that there are no disturbance inputs, substituting Eq. (A.31) into Eq. (A.30) yields

$$C(z) = G_p(z)M(z) \qquad (A.32)$$

where $G_p(z)$ is the **discrete process transfer function matrix**, and

$$G_p(z) = H(I - \Phi(T)z^{-1})^{-1}\Theta(T)z^{-1} \qquad (A.33)$$

Example A.6 Discrete process transfer function Equation 12.33 can be used to obtain the discrete transfer function for the process in Example A.1. From the results of Example A.5, result is

$$G_p(z) = \begin{bmatrix} 1 & 0 \end{bmatrix} \begin{bmatrix} 1 - 0.8452z^{-1} & -0.2387z^{-1} \\ 0.4773z^{-1} & 1 - 0.1292z^{-1} \end{bmatrix}^{-1} \begin{bmatrix} 0.1548z^{-1} \\ 0.4773z^{-1} \end{bmatrix}$$

Then

$$\mathbf{G_p}(z) = [1 \quad 0]$$

$$\times \begin{bmatrix} \dfrac{1 - 0.1292z^{-1}}{1 - 0.9744z^{-1} + 0.2231z^{-2}} & \dfrac{0.2387z^{-1}}{1 - 0.9744z^{-1} + 0.2231z^{-2}} \\[2ex] \dfrac{-0.4773z^{-1}}{1 - 0.9744z^{-1} + 0.2231z^{-2}} & \dfrac{1 - 0.8452z^{-1}}{1 - 0.9744z^{-1} + 0.2231z^{-2}} \end{bmatrix}$$

$$\times \begin{bmatrix} 0.1548z^{-1} \\ 0.4773z^{-1} \end{bmatrix}$$

and

$$\mathbf{G_p}(z) = \begin{bmatrix} \dfrac{0.1548z^{-1} + 0.0939z^{-2}}{1 - 0.9744z^{-1} + 0.2231z^{-2}} \end{bmatrix} \qquad \square$$

Discrete Closed-Loop Control

A closed-loop system can be constructed using state-variable feedback as shown in the block diagram in Fig. A.6. As in the continuous system shown in Figure A.2, the $r \times p$ gain matrix $\mathbf{K}$ allows specification of the roots of the closed-loop characteristic equation. The control law used is

$$\mathbf{m}(nT) = \mathbf{r}(nT) - \mathbf{Kx}(nT) \tag{A.34}$$

where $\mathbf{r}(nT)$ is the command input to the system. Taking the z transform of Eq. (A.34) yields

$$\mathbf{M}(z) = \mathbf{R}(z) - \mathbf{KX}(z) \tag{A.35}$$

Figure A.6 Discrete closed-loop system with state-variable feedback.

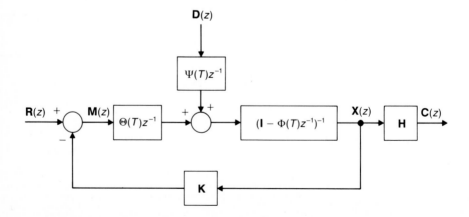

and substituting Eq. (A.35) into Eq. (A.29) results in the following equation describing the closed-loop system:

$$X(z) = (I - \mathbf{\Phi}(T)z^{-1} + \mathbf{\Theta}(T)Kz^{-1})^{-1}[\mathbf{\Theta}(T)z^{-1}R(z) + \mathbf{\Psi}(T)z^{-1}D(z)] \quad (A.36)$$

Closed-Loop Characteristic Equation

The characteristic equation for the discrete closed-loop system described by Eq. (A.36) is

$$det[I - \mathbf{\Phi}(T)z^{-1} + \mathbf{\Theta}(T)Kz^{-1}] = 0 \quad (A.37)$$

As with continuous systems, it is possible to pick the values of the elements of matrix **K** so that the roots of the characteristic equation have desired values. These roots will dictate the response of the system to command and disturbance inputs.

Example A.7 Discrete closed-loop controller design   From Example A.5,

$$\mathbf{\Phi}(T) = \begin{bmatrix} 0.8452 & 0.2387 \\ -0.4773 & 0.1292 \end{bmatrix}$$

and

$$\mathbf{\Theta}(T) = \begin{bmatrix} 0.1547 \\ 0.4774 \end{bmatrix}$$

where $T = 0.5$ s. This system is second-order and

$$\mathbf{K} = [k_1 \quad k_2]$$

From Eq. (A.37) the characteristic equation is

$$det \begin{bmatrix} 1 - 0.8452z^{-1} + 0.1547k_1z^{-1} & -0.2387z^{-1} + 0.1547k_2z^{-1} \\ 0.4773z^{-1} + 0.4774k_1z^{-1} & 1 - 0.1292z^{-1} + 0.4774k_2z^{-1} \end{bmatrix} = 0$$

or

$$1 + (0.1547k_1 + 0.4774k_2 - 0.9744)z^{-1}$$
$$+ (0.0939k_1 - 0.4774k_2 + 0.2231)z^{-2} = 0$$

Suppose that it is desired to obtain a discrete closed-loop system that approximates a continuous system with a damping ratio of 0.7 and a natural frequency of 3 rad/s. As can be seen in the z-plane plot in Fig. A.7, the roots of the desired characteristic equation are approximately

$$z_1 = 0.2 + j0.3$$

$$z_2 = 0.2 - j0.3$$

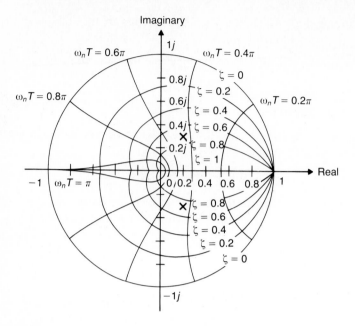

Figure A.7 A z-plane showing root locations for $\omega_n = 3$ rad/s and $\xi = 0.7$ with $T = 0.5$ s.

The corresponding characteristic equation is

$$1 - 0.4z^{-1} + 0.13z^{-2} = 0$$

By equating like terms in the last two equations, we get

$$0.15474k_1 + 0.4774k_2 - 0.9744 = -0.4$$

and

$$0.0939k_1 - 0.4774k_2 - 0.2231 = 0.13$$

Solving these equations for k_1 and k_2 results in

$$k_1 = 1.935 \quad \text{and} \quad k_2 = 0.576$$

The gain matrix is then

$$\mathbf{K} = [1.935 \quad 0.576] \; \square$$

A major problem illustrated in the previous example is that equations had to be manipulated algebraically to obtain the elements in the gain matrix $\mathbf{K}$. An alternative method for finding $\mathbf{K}$ is the application of Ackermann's formula, which can be written as

$$\mathbf{K} = [0 \quad \cdots \quad 0 \quad 0 \quad 1]\mathbf{B}_c(T)^{-1}\mathbf{A}_c(T) \tag{A.38}$$

where

$$\mathbf{B}_c(T) = [\mathbf{\Theta}(T) \quad \mathbf{\Phi}(T)\mathbf{\Theta}(T) \quad \mathbf{\Phi}(T)^2\mathbf{\Theta}(T) \quad \cdots \quad \mathbf{\Phi}(T)^{r-1}\mathbf{\Theta}(T)] \quad (A.39)$$

and

$$\mathbf{A}_c(T) = \mathbf{\Phi}(T)^r + \alpha_1\mathbf{\Phi}(T)^{r-1} + \alpha_2\mathbf{\Phi}(T)^{r-2} + \cdots + \alpha_{r-1}\mathbf{\Phi}(T) + \alpha_r\mathbf{I} \quad (A.40)$$

where $\alpha_1, \alpha_2, \ldots, \alpha_{r-1}$, and α_r are the coefficients of the desired closed-loop characteristic equation

$$1 + \alpha_1 z^{-1} + \cdots + \alpha_{r-1}z^{-r+1} + \alpha_r z^{-r} = 0 \quad (A.41)$$

$\mathbf{B}_c(T)$ is called the controllability matrix, and if $\mathbf{B}_c(T)$ is singular (i.e., det $\mathbf{B}_c(T) = 0$), then $\mathbf{K}$ cannot be calculated in Eq. (A.38). Such a system is called uncontrollable because no control law of the form of Eq. (A.34) exists that achieves the desired characteristic equation in Eq. (A.41).

Example A.8 Control matrix calculation using Ackermann's formula In Example A.7 it was desired to obtain the characteristic equation

$$1 - 0.4z^{-1} + 0.13z^{-2} = 0$$

Therefore in Eq. (12.41),

$$\alpha_1 = -0.4$$

$$\alpha_2 = 0.13$$

and

$$\mathbf{A}_c(T) = \mathbf{\Phi}(T)^2 - 0.4\mathbf{\Phi}(T) + 0.13\mathbf{I}$$

Substituting $\mathbf{\Phi}(T)$ from Example A.5 yields

$$\mathbf{A}_c(T) = \begin{bmatrix} 0.3924 & 0.1371 \\ -0.2742 & -0.0189 \end{bmatrix}$$

From Eq. (A.39)

$$\mathbf{B}_c(T) = \begin{bmatrix} 0.1548 & 0.2448 \\ 0.4773 & -0.0122 \end{bmatrix}$$

The gain matrix $\mathbf{K}$ is then found from Eq. (A.38) as

$$\mathbf{K} = \begin{bmatrix} 0 & 1 \end{bmatrix}\begin{bmatrix} 0.1548 & 0.2448 \\ 0.4773 & -0.0122 \end{bmatrix}^{-1}\begin{bmatrix} 0.3924 & 0.1371 \\ -0.2742 & -0.0189 \end{bmatrix}$$

and the result is

$$\mathbf{K} = [1.935 \quad 0.576]$$

which is the same as that obtained in the previous example. Calculation of $\mathbf{K}$ using Ackermann's formula is illustrated by Program A.3. □

Program A.3 Calculation of gains for state-variable feedback.

```
program State_Variable_Gain_Calculation;
  const
    r = 2;                                    {2nd-order system}
  type
    matrix = array[1..r, 1..r] of real;
    vector = array[1..r] of real;
  var
    PHI : matrix;                             {PHI and THETA in Eq. A.20}
    THETA : vector;
    COEFF : vector;                           {coeffs. of desired char. eqn.}
    ALPHA, ALPHA1, BETA, BETAinv : matrix;    {ALPHA and BETA in Eq. A.38}
    det : real;                               {determinant of BETA}
    K : vector;                               {feedback gains in Eq. A.34}
    i, j, m, n : integer;
  begin
    PHI[1, 1] := 0.8452;                      {PHI and THETA from Ex. A.5}
    PHI[1, 2] := 0.2387;
    PHI[2, 1] := -0.4773;
    PHI[2, 2] := 0.1292;
    THETA[1] := 0.1548;
    THETA[2] := 0.4773;
    for i := 1 to r do                        {desired characteristic eqn.}
      readln(COEFF[i]);
    for i := 1 to r do                        {calc. ALPHA using Eq. A.40}
      for j := 1 to r do
        ALPHA[i, j] := PHI[i, j];             {ALPHA := PHI + COEFF[1]*I}
    for n := 1 to r do
      ALPHA[n, n] := ALPHA[n, n] + COEFF[1];
    for m := 2 to r do
      begin
        for i := 1 to r do
          for j := 1 to r do
            begin
              ALPHA1[i, j] := 0.0;            {ALPHA := PHI*ALPHA + COEFF[m]*I}
              for n := 1 to r do
                ALPHA1[i, j] := ALPHA1[i, j] + PHI[i, n] * ALPHA[n, j]
            end;
        for i := 1 to r do
          for j := 1 to r do
            ALPHA[i, j] := ALPHA1[i, j];
        for n := 1 to r do
          ALPHA[n, n] := ALPHA[n, n] + COEFF[m]
      end;
```

```
for i := 1 to r do                                {calculateBETA using Eq. A.39}
  BETA[i, 1] := THETA[i];                          {first column := THETA}
for j := 2 to r do
  for i := 1 to r do
    begin
      BETA[i, j] := 0.0;                           {next column := previous*PHI}
      for n := 1 to r do
        BETA[i, j] := BETA[i, j] + PHI[i, n] * BETA[n, j - 1]
    end;
  det := BETA[1, 1] * BETA[2, 2] - BETA[2, 1] * BETA[1, 2];   {determinant of BETA}
  BETAinv[1, 1] := BETA[2, 2] / det;               {invert BETA}
  BETAinv[1, 2] := -BETA[1, 2] / det;
  BETAinv[2, 1] := -BETA[2, 1] / det;              {only implemented for 2x2 here}
  BETAinv[2, 2] := BETA[1, 1] / det;
  for j := 1 to r do                               {calculate K using Eq. A.38}
    begin
      K[j] := 0.0;
      for n := 1 to r do                           {K := last row of (BETAinv*ALPHA)}
        K[j] := K[j] + BETAinv[r, n] * ALPHA[n, j];
      writeln(j, K[j])
    end;
end.
```

State Estimator Feedback

State variables are not always available for control feedback because of the expense of using additional sensors or the lack of suitable sensors for measuring them. In this case it is possible to construct a **state estimator** to obtain an estimate $\hat{x}(t)$ of state variables $x(t)$ and to use these estimates in the controller.

The process model in Eqs. (A.29) and (A.30) can be used in the state estimator and

$$\hat{X}(z) = \Phi(T)z^{-1}\hat{X}(z) + \Theta(T)z^{-1}M(z) \qquad (A.42)$$

$$\hat{C}(z) = H\hat{X}(z) \qquad (A.43)$$

Note that manipulations are known at each sample, but not disturbances. Disturbances, then, are not modeled and there may be other errors in the model. The estimate $\hat{x}(t)$ therefore differs from $x(t)$, and a feedback loop must be added to correct the estimator output.

One equation that can be used for the estimator is

$$\hat{X}(z) = \Phi(T)z^{-1}\hat{X}(z) + \Theta(T)z^{-1}M(z) + Lz^{-1}[C(z) - \hat{C}(z)] \qquad (A.44)$$

where L is an $r \times q$ estimator feedback matrix. Substituting Eq. (A.43) into Eq. (A.44) yields

$$\hat{X}(z) = (I - \Phi(T)z^{-1} + LHz^{-1})^{-1}[\Theta(T)z^{-1}M(z) + Lz^{-1}C(z)] \qquad (A.45)$$

The state-estimator approach is illustrated in Fig. A.8.

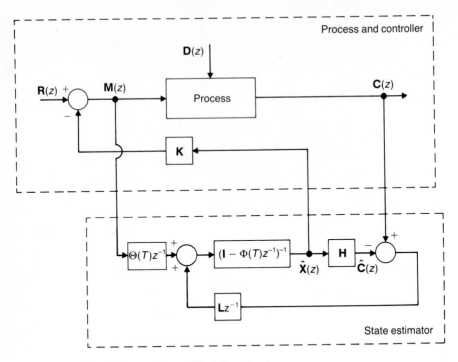

Figure A.8 Closed-loop system with state estimator.

The function of matrix **L** in the estimator is similar to matrix **K** in the controller. The elements in **L** should be selected so that the response of the estimator is significantly faster than that for the process.[3] The characteristic equation of the estimator is

$$det[\mathbf{I} - \mathbf{\Phi}(T)z^{-1} + \mathbf{L}\mathbf{H}z^{-1}] = 0 \qquad (A.46)$$

and the roots of this characteristic equation determine the speed of response. Again, Ackermann's formula can be used, and the elements of **L** can be found from

$$\mathbf{L} = \mathbf{A}_e(T)\mathbf{B}_e(T)^{-1}[0 \quad \cdots \quad 0 \quad 0 \quad 1]^T \qquad (A.47)$$

where

$$\mathbf{B}_e(T) = \begin{bmatrix} \mathbf{H} \\ \mathbf{H}\mathbf{\Phi}(T) \\ \mathbf{H}\mathbf{\Phi}(T)^2 \\ \vdots \\ \mathbf{H}\mathbf{\Phi}(T)^{r-1} \end{bmatrix} \qquad (A.48)$$

[3] State estimators are often not used in high-performance computer control systems because of delays caused by the necessity of computing the estimates, quantization errors, and the detrimental effects of additional estimator dynamics on closed-loop system performance.

and

$$A_e(T) = \mathbf{\Phi}(T)^r + \alpha_{e_1}\mathbf{\Phi}(T)^{r-1} + \cdots + \alpha_{e_{r-1}}\mathbf{\Phi}(T) + \alpha_{e_r}\mathbf{I} \qquad (A.49)$$

where $\alpha_{e_1}, \alpha_{e_2}, \ldots, \alpha_{e_{r-1}},$ and α_{e_r} are the coefficients of the desired estimator characteristic equation

$$1 + \alpha_{e_1}z^{-1} + \cdots + \alpha_{e_{r-1}}z^{r-1} + \alpha_{e_r}z^{-r} = 0 \qquad (A.50)$$

The matrix $\mathbf{B}_e(T)$ in Eq. (A.48) is called the observability matrix. If $\mathbf{B}_e(T)$ is singular, (i.e., det $\mathbf{B}_e(T) = 0$), then $\mathbf{L}$ cannot be calculated in Eq. (A.47) and the system is called unobservable.

Example A.9 State estimator In Example A.7, state-variable feedback was used with a sample period of 0.5 s to achieve a closed-loop damping ratio of 0.7 and a closed-loop natural frequency of 3 rad/s. However, if only the output $c(t)$ can be sensed, then a state estimator can be used to generate the state variables. The response of the state estimator should be significantly faster than the response of the closed-loop system, and a damping ratio of 0.7 and a natural frequency of 30 rad/s can be chosen in establishing the roots of the characteristic equation of the estimator.

An estimator natural frequency of 30 rad/s requires a shorter sample period than 0.5 s. This can be seen both in the location where these roots would fall in Fig. A.7 and in the fact that the characteristic time associated with this natural frequency is $1/30 = 0.033$ s, which is significantly less than 0.5 s. If the sample period is reduced to 0.02 s, then the elements in the discrete state equation must be recalculated for the new sample period. The result is

$$\mathbf{\Phi}(T) = \begin{bmatrix} 0.99961 & 0.01941 \\ -0.03882 & 0.94138 \end{bmatrix}$$

$$\mathbf{\Theta}(T) = \begin{bmatrix} 0.00039 \\ 0.03882 \end{bmatrix}$$

The controller gains also must be recalculated for the new sample period, and the result obtained using Eqs. (12.16) and (12.17) with $\zeta = 0.7$ and $\omega_n = 3$ rad/s is

$$\mathbf{K} = [3.446 \quad 0.609]$$

With a sample period of 0.02 s, a desired estimator damping ratio of 0.7, and a desired estimator natural frequency of 30 rad/s, the desired roots of the characteristic equation of the estimator are obtained using Eqs. (12.16) and (12.17) as

$$z_1 = 0.598 + j0.273 \qquad \text{and} \qquad z_2 = 0.598 - j0.273$$

These roots and those of the new closed-loop system characteristic equation are shown in Fig. A.9.

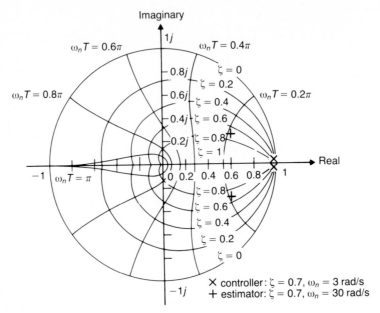

Figure A.9 A z-plane plot showing root locations for Example A.9 with $T = 0.02$ s.

For this system, the state estimator gain matrix in Eq. (A.44) has the form

$$\mathbf{L} = \begin{bmatrix} l_1 \\ l_2 \end{bmatrix}$$

Here l_1 and l_2 can be found either by comparing the result of Eq. (A.46) with like terms in the desired characteristic equation of the estimator,

$$1 - 1.195z^{-1} + 0.43z^{-2} = 0$$

or by using Eqs. (A.47) through (A.50). The result is

$$\mathbf{L} = \begin{bmatrix} 0.746 \\ 9.888 \end{bmatrix} \quad \square$$

Example A.10 Control of motor position   Figure A.10(a) shows a block diagram for control of a DC motor/amplifier system using the state-variable approach. It is assumed that the amplifier has been adjusted so that the open-loop transfer function of the system has a time constant τ and a gain k_m. The figure shows that integrated motor position is being controlled. This can result in zero steady-state error in motor position when ramp (constant velocity) inputs are commanded. The gain k_3 has been added to the command input so that the system will have unity gain. Optional gains k_p and k_d have also been shown. These allow feedforward control of position and velocity.

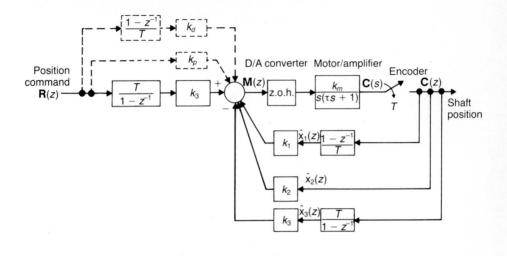

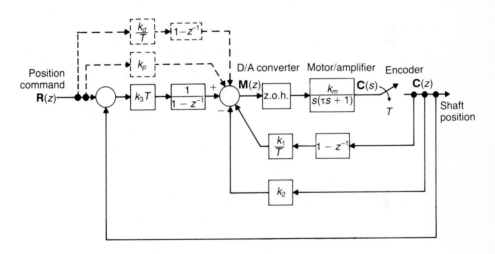

Figure A.10 (a) Block diagram for control of motor shaft position using state-variable feedback; (b) reorganized block diagram for control of motor shaft position using state-variable feedback.

Manipulation of the block diagram in Fig. A.10(a) into the form shown in Fig. A.10(b) results in a more familiar and practical controller implementation. As will be explained later, differencing and summation operations are used instead of the state-estimator approach to obtain estimates $\hat{x}_1(nT)$, $\hat{x}_2(nT)$, and $\hat{x}_3(nT)$ of the state variables $x_1(nT)$, $x_2(nT)$, and $x_3(nT)$. These represent motor velocity, motor position, and integrated motor position, respectively, and are obtained from samples of motor position, $c(t)$.

The state equation for the motor/amplifier system is

$$\frac{d}{dt}\begin{bmatrix} x_1(t) \\ x_2(t) \\ x_3(t) \end{bmatrix} = \begin{bmatrix} -\dfrac{1}{\tau} & 0 & 0 \\ 1 & 0 & 0 \\ 0 & 1 & 0 \end{bmatrix}\begin{bmatrix} x_1(t) \\ x_2(t) \\ x_3(t) \end{bmatrix} + \begin{bmatrix} \dfrac{k_m}{\tau} \\ 0 \\ 0 \end{bmatrix} m(t)$$

and the output equation is

$$c(t) = \begin{bmatrix} 0 & 1 & 0 \end{bmatrix}\begin{bmatrix} x_1(t) \\ x_2(t) \\ x_3(t) \end{bmatrix}$$

If the time constant of the motor/amplifier is 0.0028 s, the gain is 63 rad/s/volt, and a sample period of 0.002 s is chosen for the controller, then the discrete state equation can be obtained using Eqs. (A.24) and (A.27). The result is

$$\begin{bmatrix} x_1((n+1)T) \\ x_2((n+1)T) \\ x_3((n+1)T) \end{bmatrix} = \begin{bmatrix} 0.48954 & 0 & 0 \\ 0.00143 & 1 & 0 \\ 0 & 0.002 & 1 \end{bmatrix}\begin{bmatrix} x_1(nT) \\ x_2(nT) \\ x_3(nT) \end{bmatrix} + \begin{bmatrix} 32.15888 \\ 0.03596 \\ 0.00003 \end{bmatrix} m(nT)$$

and the output equation is

$$c(nT) = \begin{bmatrix} 0 & 1 & 0 \end{bmatrix}\begin{bmatrix} x_1(nT) \\ x_2(nT) \\ x_3(nT) \end{bmatrix}$$

This is a third-order system, and closed-loop control with state-variable feedback is established using three gains:

$$\mathbf{K} = \begin{bmatrix} k_1 & k_2 & k_3 \end{bmatrix}$$

If it is desired to create a closed-loop system with a dominant time constant of 0.053 s and an additional second-order characteristic with a natural frequency of 357 rad/s and a damping ratio of 0.7, then, as shown in Fig. A.11, the desired roots of the closed-loop characteristic equation are

$$z_1 = 0.963$$

$$z_2 = 0.529 + j0.296$$

$$z_3 = 0.529 - j0.296$$

and the desired characteristic equation is

$$1 - 2.02193z^{-1} + 1.38778z^{-2} - 0.354402z^{-3} = 0$$

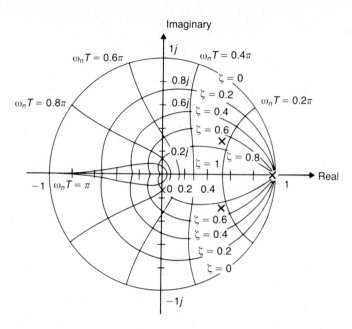

Figure A.11 A z-plane plot showing root locations for Example A.10 with $T = 0.002$ s.

The result of Ackermann's formula in Eqs. (A.38) through (A.41) or the explicit solution for the gains k_1, k_2, and k_3 that produce the above characteristic equation is

$$k_1 = 0.0087$$

$$k_2 = 5.1586$$

$$k_3 = 88.9740$$

Only motor position is sensed in the system and, although the state-estimator approach could be used to produce estimates $\hat{x}_1(nT)$, $\hat{x}_2(nT)$, and $\hat{x}_3(nT)$, it is more practical to use differencing and summation of successive samples cf the output $c(t)$ to obtain

$$\hat{x}_1(nT) = \frac{c(nT) - c[(n-1)T]}{T}$$

$$\hat{x}_2(nT) = c(nT)$$

$$\hat{x}_3(nT) = \sum_{i=1}^{n} c(iT)T$$

or

$$
\begin{bmatrix} \hat{X}_1(z) \\ \hat{X}_2(z) \\ \hat{X}_3(z) \end{bmatrix} = \begin{bmatrix} \dfrac{1 - z^{-1}}{T} \\ 1 \\ \dfrac{T}{1 - z^{-1}} \end{bmatrix} C(z)
$$

These operations are indicated in the block diagrams in Fig. A.10 and, in the form shown in Fig. A.10(b), result in a closed-loop controller that can be readily implemented on a control computer. □

A.4

SUMMARY

The state-variable method is a fundamental concept of modern discrete control theory, and the reader is referred to additional texts on this subject such as those listed in the bibliography of this appendix. Topics of interest include proofs, numerical methods, and additional design techniques. The state-variable method is as well suited to multiple-input/multiple-output systems as it is to single-input/single-output systems and provides a unified approach to the analysis of a wide variety of systems.

It has been shown how continuous and discrete closed-loop control systems with state-variable feedback can be designed by specifying the roots or the coefficients of the characteristic equation to be obtained. For systems for which not all the state variables can be measured, the concept of the state estimator has been introduced. The dynamics of the state estimator are established in much the same way as those of the system with state-variable feedback: by specifying the roots or the coefficients of the desired characteristic equation.

BIBLIOGRAPHY

Ackermann, J. *Sampled-Data Control Systems: Analysis and Synthesis, Robust System Design.* New York, NY: Springer-Verlag, 1985.

Astrom, K. J., and B. Wittenmark. *Computer Controlled Systems: Theory and Design.* Englewood Cliffs, NJ: Prentice-Hall, 1984.

Brogan, W. L. *Modern Control Theory.* New York, NY: Quantum, 1974.

Franklin, G. F., and J. D. Powell. *Digital Control of Dynamic Systems.* Reading, MA: Addison-Wesley, 1981.

Friedland, B. *Control System Design: An Introduction to State-Space Methods,* New York, NY: McGraw-Hill, 1986.

Isermann, R. *Digital Control Systems.* New York, NY: Springer-Verlag, 1981.

Jacquot, R. G. *Modern Digital Control Systems.* New York, NY: Dekker, 1981.

Kuo, B. C. *Digital Control Systems.* New York, NY: Holt, Reinhart and Winston, 1980.

Phelan, R. M. *Automatic Control Systems.* Ithaca, NY: Cornell University Press, 1977.

Phillips, C. L., and H. T. Nagle. *Digital Control System Analysis and Design.* Englewood Cliffs, NJ: Prentice-Hall, 1984.

PROBLEMS

A.1 Develop continuous state equations for the transfer functions given in Problem 11.1. Use $M(s)$ as the input and $X_1(s)$ as the output.

A.2 Develop discrete state equations for the processes given in Problem 11.2. Use $M(z)$ as the input, $X_1(z)$ as the output, and a sample period of 0.1 s.

A.3 Repeat Problem 11.3 using the state-variable approach and a sample period of 0.1 s.

A.4 Develop discrete state-variable models for the systems shown in Fig. 3.20.

A.5 Repeat Problem 11.6 using the state-variable approach and a sample period of 0.02 s.

A.6 Repeat Problem 11.5 using the state-variable approach and a sample period of 0.3 s. The state variables are to be $V(z)$ and $C(z)$.

A.7 Use the state-variable approach to obtain a continuous model for the continuous process shown in the Fig. 11.27. The transform of the process input is to be $H(s)$, the output of the zero-order hold. The transform of the process output is $P(s)$. One of the state variables must be $C(s)$, the transform of the input to the sampler.

A.8 Suppose that $C(s)$ is sampled at a rate of 1000 Hz to obtain $C(z)$ in the process shown in Fig. 11.27. Using the result of Problem A.7, develop a discrete model for the process. The transform of the input is to be $M(z)$, the input to the zero-order hold. The transform of the output is to be $P(z)$.

A.9 Design a state-variable feedback controller for the system in Problem A.8.

A.10 A sample rate of 200 Hz is to be used in a state-variable controller for the process shown in Fig. 11.27. Design a controller that achieves a closed-loop bandwidth of approximately 5 Hz. The process should be simplified in a manner consistent with the foregoing bandwidth and sample period specifications.

A.11 Develop a discrete state-variable model for the system in Problem 2.2. Use the state estimator approach and design a state-variable feedback controller that achieves a closed-loop bandwidth of approximately 10 Hz.

A.12 Develop a discrete state-variable model for the system in Problem 2.2. Design an approximate state-variable feedback controller of the type shown in Figs. A.10(a) and A.10(b) that achieves a closed-loop bandwidth of approximately 10 Hz.

A.13 Design a state-variable controller for the rotary table system described in Example 2.11. The frequency response of the system is to be "flat" to approximately 18 Hz. Assume that current, velocity, and position can be sampled. Choose a sample period that will allow all three variables to be successfully controlled.

A.14 In Example 4.11, feedforward control was used to eliminate following errors in the robot arm shown in Fig. 4.15(a). Redesign this control system so that both its feedforward and its feedback portions are in state-variable form.

A.15 In Example 4.14, an artificial climate is maintained by controlling room temperature and room humidity. Develop a state-variable model that relates these variables to steam-valve position and heater voltage.

A.16 Draw a block diagram for a state-variable controller for the model developed in Problem A.15 that allows room temperature and room humidity to be independently controlled.

Appendix B

Binary Numbers and Arithmetic

The basis for computation in almost all computers is the binary number system. Like all positional number systems, the binary number system consists of a radix (or base) and an ordered alphabet of digits. Generally, a number in a base-R system is represented as

$$N_R = \sum_{i=-\infty}^{\infty} d_i R^i \tag{B.1}$$

where d_i is an alphabet of R characters: $0, 1, 2, \ldots, R - 1$. In the case of the binary system, d_i can be 0 or 1, and R is 2. A number to the base 2, N_2, can therefore be represented in the binary number system as

$$N_2 = \sum_{i=-\infty}^{\infty} d_i 2^i = \cdots + d_2 2^2 + d_1 2^1 + d_0 2^0 + d_{-1} 2^{-1} + d_{-2} 2^{-2} + \cdots \tag{B.2}$$

where d_i is either 0 or 1. The subscript on N denotes the base of the number system. While working in one particular system, we usually drop the radix and express the numbers as an n-tuple of the ordered alphabet.

For example, the binary number 1101.1_2 is the n-tuple that represents the sequence

$$(1)2^3 + (1)2^2 + (0)2^1 + (1)2^0 + (1)2^{-1}$$

In decimal notation, this number is 13.5_{10}. As in the example, a number is usually not represented by an infinite sequence but is truncated by implying the left-most and right-most zeros. In computers, all numbers have to be represented by a finite

579

number of binary digits, or bits, and this forms an important characteristic of a computer. For example, a 16-bit machine is one in which arithmetic is done with numbers represented with 16-bit words. The fixed word length limits the values that the variables have in a program, and certain numbers may have to be scaled to be accommodated in the number of bits available.

For computers with a small word size, multiple-precision arithmetic is used to eliminate the overflow problems caused when numbers become too large and the round-off errors introduced by truncation of numbers. Most computer manufacturers also supply software or hardware that deals with pairs of sequences that represent a rational number in "floating-point" format. One sequence represents the exponent and the other represents the normalized mantissa. The decimal number 13.5_{10} could be represented as $0.135_{10} \times 10^2$ and stored as two sequences: 0.135_{10} (the normalized mantissa) and 2_{10} (the exponent). The binary representation of the number, 1101.1_2, can be rewritten as 0.11011×2^{100} and stored as 0.11011_2 (the normalized mantissa) and 100_2 (the exponent).

In decimal arithmetic, a plus or minus sign can be affixed to denote the sign of the number. However, in a computer the sign has to be denoted by a 0 or a 1. Usually a 0 is used to denote a plus sign and a 1 to denote a minus sign. The sign can be simply appended to create a signed number, but the convention used in most computers is the 2's-complement representation that is explained in detail later.

BINARY-TO-DECIMAL CONVERSION

Conversion from binary form to decimal form is done by using the general representation of a number in a positional number system that is given in Eq. (B.1).

Example B.1 Convert 11011.1001_2 to a decimal number.

Solution

$$11011.1001 = (1)2^4 + (1)2^3 + (0)2^2 + (1)2^1 + (1)2^0 + (1)2^{-1}$$
$$+ (0)2^{-2} + (0)2^{-3} + (1)2^{-4}$$
$$= 16 + 8 + 0 + 2 + 1 + 0.5 + 0 + 0 + 0.0625$$
$$= 27.5625_{10} \quad \square$$

DECIMAL-TO-BINARY CONVERSION

Conversion of an integer from decimal form to binary form is achieved via repeated division by 2. At each step, the remainder is the corresponding bit of the binary number and is not used in further division. The division is continued until the quotient becomes zero.

Example B.2   Convert 59_{10} to a number in the base-2 system.

Solution   First, the integer portion is converted via repeated division by 2.

$$2\underline{\smash{)}59}$$
$$29$$

and a remainder of 1, which becomes the least significant bit of the base-2 number. Repeating the division, we get

$$2\underline{\smash{)}29}$$
$$14$$

and a remainder of 1, which becomes the next bit of the base-2 number. And so on:

	Remainders
$2\underline{\smash{)}14}$	
$2\underline{\smash{)}\ 7}$	0
$2\underline{\smash{)}\ 3}$	1
$2\underline{\smash{)}\ 1}$	1
0	1

Therefore, the result is

$$59_{10} = 111011_2$$

As a check,

$$(1)2^5 + (1)2^4 + (1)2^3 + (0)2^2 + (1)2^1 + (1)2^0 = 59_{10} \quad \square$$

To convert a fraction from decimal form to binary form, we multiply repeatedly by 2. The integer part of the product at every step forms part of the result and is not used in the next multiplication.

Example B.3   Convert 0.8573_{10} to a number in the base-2 system.

Solution   Repeated multiplication by 2 yields

$$
\begin{array}{rr}
 & .8573 \\
\times & 2 \\
\hline
1 \quad & .7146 \\
\times & 2 \\
\hline
1 \quad & .4292 \\
\times & 2 \\
\hline
0 \quad & .8584 \\
\times & 2 \\
\hline
1 \quad & .7168 \\
\times & 2 \\
\hline
1 \quad & .4336 \\
\end{array}
$$

and so on. The result is

$$0.8573_{10} = 0.11011\ldots_2$$

Note that decimal fractions cannot usually be represented by fractions of finite length in other bases. □

OCTAL AND HEXADECIMAL NOTATION

It is useful to point out that although computers perform binary arithmetic, binary numbers are typically lengthy, and the octal (base-8) or hexadecimal (base-16) number system is often used by computer programmers as a means of making their representation more compact. The conversion from binary form to octal form is very simple. Three binary digits are grouped together to form one octal digit, which can have the value of 0 through 7. For the case in Example B.1,

$$111\ 011_2 = 73_8$$

Conversion from binary to hexadecimal form is also simple. The hexadecimal number system requires 16 hexadecimal digits. The 10 digits normally associated with the base-10 number system (0 through 9) are used along with 6 additional digits (A through F) as the hexadecimal digits. Table B.1 shows the correspondence among binary, octal, decimal, and hexadecimal numbers. Here 13_{10} can be seen to be equivalent to 1101_2, 15_8, and D_{16}. To convert from binary form to hexadecimal form, four binary digits can be grouped together to make up one hexadecimal digit. For the case in Example B.1,

$$11\ 1011_2 = 3B_{16}$$

Table B.1

Correspondence among binary, octal, decimal, and hexadecimal numbers

BINARY (base 2)	OCTAL (base 8)	DECIMAL (base 10)	HEXADECIMAL (base 16)
0	0	0	0
1	1	1	1
10	2	2	2
11	3	3	3
100	4	4	4
101	5	5	5
110	6	6	6
111	7	7	7
1000	10	8	8
1001	11	9	9
1010	12	10	A
1011	13	11	B
1100	14	12	C
1101	15	13	D
1110	16	14	E
1111	17	15	F

Conversion from binary numbers with fractions to and from octal or hexadecimal notation can be done by grouping binary bits in groups of three or four. When making the conversion, care must be taken to place the radix point at a group boundary.

Example B.4 Convert 10.11011011_2 to base 8 and base 16.

$$10.110\ 110\ 110_2 = 2.666_8$$
$$10.1101\ 1011\ 1000_2 = 2.D88_{16} \quad \square$$

Example B.5 Convert 10.537_8 and $26.9A5_{16}$ to base 2.

$$10.537_8 = 1\ 000.101\ 011\ 111_2 \qquad = 1000.101011111_2$$
$$26.9A5_{16} = 10\ 0110.1001\ 1010\ 0101_2 = 100110.100110100101_2 \quad \square$$

It is also possible to convert directly from decimal form to octal form or hexadecimal form by adapting the method previously given for decimal-to-binary conversion. Repeated division or multiplication by 8 or 16 is used, as illustrated in the following example.

Example B.6   Convert 59.6_{10} to a number in the base-8 and base-16 systems.

Solution   By repeated division of the integer part by 8,

$$8\underline{|59}\qquad \text{(remainders forming the octal digits)}$$
$$8\underline{|\ 7}\qquad 3$$
$$0\qquad 7$$

Therefore, the integer part of the solution is 73_8. By repeated multiplication of the fractional part by 8,

$$\begin{array}{rr} & .6 \\ & \times\ 8 \\ \hline 4 & .8 \\ & \times\ 8 \\ \hline 6 & .4 \\ & \times\ 8 \\ \hline 3 & .2 \\ & \times\ 8 \\ \hline 1 & .6 \end{array}$$

and so on. The complete answer is therefore

$$59.6_{10} = 73.4631\ldots_8 \quad \square$$

Example B.7   Convert 59.6_{10} to a number in base 16.

Solution   By repeated division by 16,

$$16\underline{|59}\qquad \text{(remainders forming the hexadecimal digits)}$$
$$16\underline{|\ 3}\qquad 11_{10} = B_{16}$$
$$0\qquad 3_{10} = 3_{16}$$

Therefore, the integer part of the solution is $3B_{16}$.
   By repeated multiplication of the fractional part by 16,

$$\begin{array}{rr} & .6 \\ & \times\ 16 \\ \hline 9 & .6 \\ & \times\ 16 \\ \hline 9 & .6 \end{array}$$

and so on. The complete result is therefore

$$59.6_{10} = 3B.99\ldots_{16} \quad \square$$

ADDITION AND SUBTRACTION

The addition of two digits results in a sum and a carry digit. The sum and carry digits for binary addition are shown in Table B.2. They can be used to perform addition of any two numbers by taking two digits at a time.

Example B.8   Add 10011_2 to 00101_2.

Solution

$$
\begin{array}{rl}
 & (0111) \leftarrow \text{carry bits} \\
A & 10011 \\
B & \underline{00101} \\
A + B & 11000 \ \square
\end{array}
$$

For subtraction, similar tables can be drawn up for the difference bit and the borrow bit (see Table B.3).

Example B.9   Subtract the binary number 01001 from 11010.

Solution

$$
\begin{array}{rl}
A & 11010 \\
 & (0001) \leftarrow \text{borrow bits} \\
B & \underline{01001} \\
A - B & 10001 \ \square
\end{array}
$$

As is obvious from the tables, the operations of addition and subtraction are logically different. This causes problems when negative numbers are encountered. When it is desired to add a negative number to a positive number, for example, one approach is to subtract instead of add. Unfortunately, this approach treats negative numbers differently from positive numbers and complicates the

⌐able B.2

Bii. ıry addition: $A + B$

B ╲ A	0	1
0	00	01
1	01	10 ← Sum bit

Carry bit

Table B.3

Binary subtraction: $A - B$

B \ A	0	1
0	00	01
1	11	00

00 ← Difference bit
Borrow bit

computer hardware. To overcome this problem, a signed-number representation is used in a computer so that the addition of a negative number need not be replaced by subtraction.

2'S-COMPLEMENT ARITHMETIC

One signed-number convention that is often used is the radix complement representation. The radix is the base of the numbering system and, in the case of the binary number system, this becomes 2's-complement representation. In general, for the n-digit sequence D_n, the radix complement is given by

$$RC(D_n) = R^n - D_n$$

where the subtraction is performed in the number system with radix R.

Example B.10 For $n = 4$, (a) determine the 2's complement of 0110_2, (b) determine the 8's complement of 7432_8, and (c) determine the 16's complement of $2AF7_{16}$.

Solution

(a) $2^4 = 10000_2$

$$RC(0110_2) = 10000_2 - 0110_2 = 1010_2$$

Thus the 2's complement of 0110_2 is 1010_2.

(b) $8^4 = 10000_8$

$$RC(7432_8) = 10000_8 - 7432_8 = 0346_8$$

Thus the 8's complement of 7432_8 is 0346_8.

(c) $16^4 = 10000_{16}$

$$RC(2AF7_{16}) = 10000_{16} - 2AF7_{16} = D509_{16}$$

Thus the 16's complement of $2AF7_{16}$ is $D509_{16}$. □

One short cut for obtaining the 2's complement of a binary number is to

1. Replace all 0's by 1's and all 1's by 0's. (This is the logical complement of the number and is sometimes called the 1's complement).
2. Add 1 to the number formed in step 1.

For Example B.9(a),

Step (1) gives		1001
Step (2) gives		1001
		+ 1
		1010

Another short cut is to

1. Locate the right-most 1 bit in the number.
2. Take the logical complement of all bits to the left of the bit located in step 1.

For Example B.9(a),

Step (1) gives   0110
                    ↑
Step (2) gives   1010
                    ↑

The 2's complement of a binary number can be used to represent the negative of the number. Consider the case of 8-bit numbers. The sequence 00001100_2 represents the decimal number $+12_{10}$. $RC(00001100_2) = 11110100_2$ and therefore can be used to represent -12_{10}. Similarly, the sequence 01000000 represents $+64_{10}$, and $RC(01000000_2) = 11000000_2$ therefore represents -64_{10}. In this convention, the most significant bit denotes the sign of the number: 0 for positive and 1 for negative. All numbers between 00000000_2 and 01111111_2 are therefore considered positive and all numbers between 10000000_2 and 11111111_2 are considered negative. If one uses octal equivalents instead of writing the long binary sequences, then all 8-bit numbers from 000_8 to 177_8 are considered positive and all numbers between 200_8 and 377_8 are considered negative. If hexadecimal equivalents are used, then all 8-bit numbers from 00_{16} to $7F_{16}$ are considered positive and all numbers between 80_{16} and FF_{16} are considered negative.

Some examples of 16-bit 2's-complement representations of decimal numbers follow.

Decimal Number	Binary Sequence	Octal Sequence	Hexadecimal Sequence
0	0000000000000000	000000	0000
+17	0000000000010001	000021	0011
+60	0000000000111100	000074	003C
+247	0000000011110111	000367	00F7
+1356	0000010101001100	002514	054C
+32765	0111111111111101	077775	7FFD
−12	1111111111110100	177764	FFF4
−357	1111111010011011	177233	FE9B
−2000	1111100000110000	174060	F830
−32766	1000000000000010	100002	8002

With this representation for positive and negative numbers, the operations of addition and subtraction become easier. It is useful to illustrate binary addition and subtraction using 8-bit numbers, because some computer systems use 8-bit words (often called bytes) as a basic unit of operations. This small word length means that, unless two or more bytes are grouped together to form longer length words, only numbers between $+127_{10}$ and -128_{10} can be represented. Figure B.1 shows the 8-bit, 2's-complement number line. It is apparent that all negative numbers have a 1 in the left-most bit position. This bit is often referred to as the sign bit. This organization also has the important property that each number on the line can be obtained by adding 00000001 to the number to its left or by subtracting 00000001 from the number to its right. This is true even around 0 where the sign bit changes value.

Example B.11   Consider two examples of the addition of two 8-bit positive numbers.

	Decimal	Binary		
(a)	+48	00110000		
	+34	+00100010		
	82	01010010	$= +82_{10}$	(correct)
(b)	+56	00111000		
	+105	+01101001		
	161	10100001	= a negative number	(incorrect)

In (b), the limitations of a fixed word length (8 bits in this case) become obvious. Because 161_{10} cannot be represented in 8 bits using the 2's-complement representation, the answer is incorrect. This situation is called "overflow," which

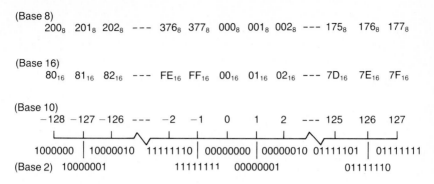

Figure B.1 Eight-bit, 2's-complement number line.

refers to the fact that the result was too big for the word length. The user has to watch for such errors and take corrective action when they occur. Overflow is often detected for the user by the computer hardware. □

Example B.12   A similar problem exists when the sum of two negative numbers is too large a negative number; such is the case in (b) below.

	Decimal	Binary	
(a)	-62	11000010	
	$+(-48)$	$+$ 11010000	
	-110	1 10010010 $= -110_{10}$	(correct)
		↑	
		carry	
		(ignore)	
(b)	-15	11110001	
	$+(-126)$	$+$ 10000010	
	-143	1 01110011 $=$ a positive number	(incorrect)
		↑	
		carry	
		(ignore)	

Eight bit, 2's-complement arithmetic gives the correct results as long as the result is between -128_{10} and $+127_{10}$. □

Example B.13   Addition and subtraction are simplified when 2's-complement arithmetic is used. Consider the subtraction of 124_{10} from 112_{10}.

Decimal	Binary
(+)112	01110000
− (+)124	− 01111100
− 12	1 11110100 $= -12_{10}$

　　　　　　　　　　↑
　　　　　　　　borrow
　　　　　　　（ignore）

Note that the result is negative, as expected, and that the sign bit is included in the operation. □

Example B.14　Consider subtracting -69_{10} from -48_{10}.

Decimal	Binary
(−)48	11010000
− (−)69	− 10111011
+ 21	00010101

Note that the operands are 2's-complement negative numbers and that the positive result is obtained as expected using the subtraction operation. □

　　　In summary, the radix complement representation of signed numbers simplifies addition and subtraction. For addition, overflow errors can occur only when both numbers have the same sign; if the sign of the result is different, the result is too large or too small for the word size and the result is in error. For subtraction, overflow can occur only when the numbers have different signs. Overflow problems can be eliminated by one of the following methods:

1. Scale down all variables so that no numbers exceed the limits imposed by word size.

2. Assume the maximum positive or negative value when overflow occurs.

3. Use floating-point arithmetic.

4. Use two or more binary words to increase the number of bits in the representation of each number. (This is called multiple-precision arithmetic; most computers are capable of performing it in a convenient manner).

MULTIPLE-PRECISION ADDITION AND SUBTRACTION

Computers are usually organized around a fixed word size, and in many minicomputers and microprocessors this size is relatively small (8 or 16 bits). Word size limits the magnitude of numbers that can be stored in one word. If n is the number of bits in a word, the range of numbers that can be stored is

$$\text{Unsigned binary:} \quad 0 \leq N \leq 2^n - 1$$
$$\text{2's-complement binary:} \quad -2^{n-1} \leq N \leq 2^{n-1} - 1$$

If it is necessary to perform arithmetic operations on numbers that are larger than these limits, then it is necessary to use more than one word to represent them. As an example, consider a computer with an 8-bit word length used in an application where it is necessary to handle unsigned numbers between 0 and $10{,}000_{10}$. Fourteen bits are required to represent $10{,}000_{10} (2^{13} < 10{,}000_{10} < 2^{14})$. The largest unsigned number that can be represented by using two words (16 bits) is $2^{16} - 1 = 65535_{10}$. Two words per number is usually called double precision, and it is a good choice for representing numbers in this example because one 8-bit word would be too small ($2^8 - 1 = 255_{10}$) and three 8-bit words would be excessively large ($2^{24} - 1 = 16{,}777{,}215_{10}$).

Addition or subtraction of two double-precision numbers requires operating on four words: two per number. These operations can be performed as follows:

1. Add or subtract the least significant (right-most) words of the two numbers to form the least significant word of the result. Note whether a carry or borrow occurred out of the most significant (left-most) bit of the operation.

2. If a carry (or borrow) occurred, add it to (or subtract it from) the most significant word of the first number.

3. Add the most significant word of the second number to (or subtract it from) the result of step 2 to form the most significant word of the result.

The following examples show the performance of arithmetic operations on double-precision numbers using two 8-bit words.

Example B.15   Double-precision addition.

Decimal	Binary
2025	00000111 11101001
+ 261	+ 00000001 00000101
2286	00001000 11101110

$$
\begin{array}{rl}
\textit{Step 1:} & 11101001 \\
& +00000101 \\
\hline
\text{no carry} \rightarrow \quad 0 & 11101110 \leftarrow \text{least significant} \\
& \text{word of result}
\end{array}
$$

$$
\begin{array}{rl}
\textit{Step 2:} & 00000111 \\
+ & \qquad 0 \leftarrow \text{no carry} \\
\hline
& 00000111 \\
\textit{Step 3:} \ + & 00000001 \\
\hline
& 00001000 \leftarrow \text{most significant} \\
& \text{word of result} \quad \square
\end{array}
$$

Example B.16   Double-precision subtraction.

Decimal	Binary	
2332	00001001	00011100
− 129	− 00000000	10000001
2203	00001000	10011011

$$Step\ 1: \quad \begin{array}{r} 00011100 \\ -\ 10000001 \\ \hline \end{array}$$

borrow → 1 10011011 ← least significant
word of result

$$Step\ 2: \quad \begin{array}{r} 00001001 \\ -\qquad\quad 1 \\ \hline 00001000 \end{array}$$ ← borrow

$Step\ 3:$ − 00000000
$\overline{}$
00001000 ← most significant
word of result □

MULTIPLICATION OF BINARY NUMBERS

In a number system with radix R, multiplication by R^i just means a shift of the radix point i places to the right if i is positive and i places to the left if i is negative. In the decimal system, for example,

$$123.45_{10} \times 10^1 = 1234.5_{10}$$
$$1356.2_{10} \times 10^{-2} = 13.562_{10}$$

Similarly, in the binary system,

$$1101.10_2 \times 2^2 = 110110_2$$
$$1001.01_2 \times 2^{-3} = 1.00101_2$$

This property can be used in formulating a multiplication procedure. Suppose it is desired to multiply a binary number N by the binary number 101.10. $N \times 101.10$ is equivalent to

$$N \times (1)2^2 + N \times (0)2^1 + N \times (1)2^0 + N \times (1)2^{-1} + N \times (0)2^{-2}$$

Each of the foregoing terms is either zero or the binary number N shifted right or left the number of places specified by the exponent. If $N = 11.01$, the terms become

$$
\begin{array}{r}
0.000 \\
1.101 \\
11.010 \\
000.000 \\
\underline{1101.000} \\
10001.111
\end{array}
$$

Example B.17   Multiply 1101 by 1011.

Solution

$$
\begin{array}{r}
1101 \\
\times \quad 1011 \\
\hline
1101 \\
1101 \\
0000 \\
\underline{1101} \\
10001111
\end{array} \quad \square
$$

Note that the number of bits in the product may be more than the number of bits in the numbers that are multiplied, and the user must often program for a double-precision result. Also, the numbers to be multiplied must be positive and, hence, not in 2's-complement notation. If either or both of the numbers to be multiplied are negative, the procedure is to take their absolute values, perform the multiplication, and then negate the result if only one was negative.

DIVISION OF BINARY NUMBERS

Long division of binary numbers is similar to long division of decimal numbers, and an intelligent guess is made regarding the next bit of the dividend at each stage of the division. In some respects, binary division is easier than decimal division because there are only two possibilities to choose from: 0 and 1.

Division can be represented as

$$
\frac{\text{Quotient}}{\text{Divisor}} = \text{Dividend} + \frac{\text{Remainder}}{\text{Divisor}}
$$

After one adjusts the binary points of both quotient and divisor by equal amounts (if necessary) so that the divisor is a whole number, binary division can be carried out as illustrated in the following example. As in the case with binary multiplication, if either or both of the numbers are negative, their absolute values are taken prior to performing the division. The dividend is then negated if only one was negative. The remainder should be negated if the original *quotient* was negative.

Example B.18   Consider the long division of 1011.1 by 11.01.

Solution

$$
\begin{array}{r}
011 \\
1101 \overline{\smash{)}101110} \\
0000 \\
\hline
10111 \\
1101 \\
\hline
10100 \\
1101 \\
\hline
111
\end{array}
$$

The result is then

$$\frac{1011.1}{11.01} = 11 + \frac{1.11}{11.01}$$

Division can be carried out further to obtain a more accurate dividend and a less significant remainder. In this case the repeating result is

$$\frac{1011.1}{11.01} = 11.01010101\ldots \quad \square$$

PROBLEMS

B.1   Convert the following numbers from decimal form to binary form.
   a. 175.0
   b. 3246.0
   c. 44.835
   d. 0.042

B.2   Convert the numbers obtained in Problem B.1 from binary form to octal form and then to hexadecimal form.

B.3   Convert the decimal numbers given in Problem B.1 to numbers in base 3. (Note that the alphabet for the base 3 number system is 0, 1, and 2.)

B.4   Convert the following unsigned numbers from binary form to decimal form.
   a. 110110111
   b. 101000
   c. 0.111
   d. 1011.0111

B.5   Convert the binary numbers in Problem B.4 to hexadecimal form.

B.6   What are the largest positive and negative integers that can be represented using 2's-complement arithmetic and the following word lengths.
   a. 8 bits
   b. 12 bits
   c. 16 bits
   d. 24 bits
   e. 32 bits

B.7   16-bit, 2's-complement arithmetic is used in a control computer. Convert each of the following decimal numbers into the binary form used in this computer.
   a. 127
   b. -346
   c. 2242
   d. -3465
   e. 18543
   f. -32015

B.8   The following binary numbers are found in the memory of a computer that uses 16-bit, 2's-complement, integer arithmetic. What are the corresponding decimal numbers?
   a. 0000011111111000
   b. 0110001101110011
   c. 1111110011110100
   d. 1000000000011101

B.9   Add the binary numbers in Problem B.8 using 16-bit, 2's-complement, integer arithmetic.

B.10 Convert the binary numbers in Problem B.8 into hexadecimal form.

B.11 Convert the binary numbers in Problem B.8 into octal form.

B.12 Give the results of the following 8-bit binary operations.

    a.   00010110
          + 00011101

    b.   00010110
          − 00011101

    c.   00000110
          × 00011101

    d. $00000110 \overline{)00011101}$

    e. Conversion of $3C_{16}$ to binary.

    f. Conversion of $E2_{16}$ to binary.

B.13 Multiply 1001_2 by 101_2 using unsigned binary arithmetic.

B.14 Divide 1101100_2 by 101_2 using unsigned binary arithmetic.

B.15 Multiply -19 by 23 using binary arithmetic.

B.16 Divide -500 by -7 using binary arithmetic.

B.17 The following equations are implemented in software on a control computer:

$$e_n = r_n - c_n$$

$$m_n = m_{n-1} + 2e_n$$

Assume that $m_{n-1} = 00000111_2$, $r_n = 00111111_2$, and $c_n = 00011100_2$ at the nth sample instant. What is the value of m_n after the binary arithmetic represented by the above equations is performed?

B.18 A variable must represent 4000 increments over a range of ± 5? How many bits must be included in the fractional part of a 16-bit binary word used to represent the variable?

B.19 Convert the following decimal fractions into 8-bit, unsigned binary fractions.

    a. 0.2

    b. 1/3

    c. 0.01

    d. 13/16

B.20 Convert the following decimal fractions into 8-bit, signed binary fractions.

    a. 0.25

    b. −0.3125

    c. 0.67

    d. −0.14

B.21 Convert the 8-bit, signed binary fractions obtained in Problem B.20 back to decimal fractions. What is the error between the binary representation and the original decimal representation?

B.22 Repeat Problems B.20 and B.21 using 16-bit, signed binary fractions. Compare the accuracy of the results to the accuracy obtained using 8-bit, signed binary fractions.

B.23 Convert π into a binary number. How many decimal digits of accuracy is your result equivalent to?

B.24 The following hexadecimal numbers represent 16-bit, 2's-complement, signed binary integer. Convert them into decimal form.
  a. FFFC
  b. 73A1
  c. 01D8
  d. 8002

B.25 Two 8-bit binary numbers are multiplied together to obtain a 16-bit result. If these 8-bit numbers are both signed fractions, what operations must be performed on the product to convert it to an 8-bit, signed fraction? How does the accuracy of the result compare with the accuracy of the original numbers?

Appendix C 7-Bit ASCII Character Codes (b_6, b_5, b_4, b_3, b_2, b_1, b_0)

b_3, b_2, b_1, b_0	0	1	2	3	4	5	6	7
0	NUL	DLE	SP	0	@	P	`	p
1	SOH	DC1	!	1	A	Q	a	q
2	STX	DC2	"	2	B	R	b	r
3	ETX	DC3	#	3	C	S	c	s
4	EOT	DC4	$	4	D	T	d	t
5	ENQ	NAK	%	5	E	U	e	u
6	ACK	SYN	&	6	F	V	f	v
7	BEL	ETB	'	7	G	W	g	w
8	BS	CAN	(	8	H	X	h	x
9	HT	EM	)	9	I	Y	i	y
A	LF	SUB	*	:	J	Z	j	z
B	VT	ESC	+	;	K	[	k	{
C	FF	FS	,	<	L	\	l	|
D	CR	GS	-	=	M	]	m	}
E	SO	RS	.	>	N	^	n	~
F	SI	US	/	?	O	_	o	DEL

(b_6, b_5, b_4)

CONTROL CHARACTERS

NUL	Null	DC1	Device control 1
SOH	Start of heading	DC2	Device control 2
STX	Start of text	DC3	Device control 3
ETX	End of text	DC4	Device control 4
EOT	End of transmission	NAK	Negative acknowledge
ENQ	Enquiry	SYN	Synchronous idle
ACK	Acknowledge	ETB	End of transmission block
BEL	Bell, or alarm	CAN	Cancel
BS	Backspace	EM	End of medium
HT	Horizontal tabulation	SUB	Substitute
LF	Line feed	ESC	Escape
VT	Vertical tabulation	FS	File separator
FF	Form feed	GS	Group separator
CR	Carriage return	RS	Record separator
SO	Shift out	US	Unit separator
SI	Shift in	SP	Space
DLE	Data link escape	DEL	Delete

Appendix D
Student's *t* Distribution

			α		
n	0.100	0.050	0.025	0.010	0.005
1	3.078	6.314	12.706	31.821	63.657
2	1.886	2.920	4.303	6.965	9.925
3	1.638	2.353	3.182	4.541	5.841
4	1.533	2.132	2.776	3.747	4.604
5	1.476	2.015	2.571	3.365	4.032
6	1.440	1.943	2.447	3.143	3.707
7	1.415	1.895	2.365	2.998	3.499
8	1.397	1.860	2.306	2.896	2.355
9	1.383	1.833	2.262	2.821	3.250
10	1.372	1.812	2.228	2.764	3.169
11	1.363	1.796	2.201	2.718	3.106
12	1.356	1.782	2.179	2.681	3.055
13	1.350	1.771	2.160	2.650	3.012
14	1.345	1.761	2.145	2.624	2.977
15	1.341	1.753	2.131	2.602	2.947
16	1.337	1.746	2.120	2.583	2.921
17	1.333	1.740	2.110	2.567	2.898
18	1.330	1.734	2.101	2.552	2.878
19	1.328	1.729	2.093	2.539	2.861
20	1.325	1.725	2.086	2.528	2.845
21	1.323	1.721	2.080	2.518	2.831
22	1.321	1.717	2.074	2.508	2.819
23	1.319	1.714	2.069	2.500	2.807
24	1.318	1.711	2.064	2.492	2.797
25	1.316	1.708	2.060	2.485	2.787
26	1.315	1.706	2.056	2.479	2.779
27	1.314	1.703	2.052	2.473	2.771
28	1.313	1.701	2.048	2.467	2.763
29	1.311	1.699	2.045	2.462	2.756
30	1.310	1.697	2.042	2.457	2.750
40	1.303	1.684	2.021	2.423	2.704
60	1.296	1.671	2.000	2.390	2.660
120	1.289	1.658	1.980	2.358	2.617

Appendix E **Table of z Transforms**

	$f(nT)$	$F(s)$
1.	$\begin{cases} f(nT - aT), & n \ge a \\ 0, & n < a \end{cases}$	$e^{-aTs}F(s)$
2.	$x(nT) + y(nT)$	$X(s) + Y(s)$
3.	$kf(nT)$	$kF(s)$
4.	1	$\dfrac{1}{s}$
5.	nT	$\dfrac{1}{s^2}$
6.	$\dfrac{1}{2}(nT)^2$	$\dfrac{1}{s^3}$
7.	e^{-anT}	$\dfrac{1}{s + a}$
8.	nTe^{-anT}	$\dfrac{1}{(s + a)^2}$
9.	$1 - e^{-anT}$	$\dfrac{a}{s(s + a)}$
10.	$\dfrac{1}{a}(anT - 1 - e^{-anT})$	$\dfrac{a}{s^2(s + a)}$
11.	$(1 - anT)e^{-anT}$	$\dfrac{s}{(s + a)^2}$
12.	$1 - e^{-anT}(1 + anT)$	$\dfrac{a^2}{s(s + a)^2}$
13.	$(e^{-anT} - e^{-bnT})$	$\dfrac{b - a}{(s + a)(s + b)}$
14.	$be^{-bnT} - ae^{-anT}$	$\dfrac{(b - a)s}{(s + a)(s + b)}$
15.	$\sin anT$	$\dfrac{a}{s^2 + a^2}$
16.	$\cos anT$	$\dfrac{s}{s^2 + a^2}$
17.	$e^{-anT}\cos bnT$	$\dfrac{s + a}{(s + a)^2 + b^2}$
18.	$e^{-anT}\sin bnT$	$\dfrac{b}{(s + a)^2 + b^2}$
19.	$1 - e^{-anT}\left(\cos bnT + \dfrac{a}{b}\sin bnT\right)$	$\dfrac{a^2 + b^2}{s[(s + a)^2 + b^2]}$

$F(z)$

$z^{-a}F(z)$

$X(z) + Y(z)$

$kF(z)$

$$\dfrac{1}{1 - z^{-1}}$$

$$\dfrac{Tz^{-1}}{(1 - z^{-1})^2}$$

$$\dfrac{T^2(1 + z^{-1})z^{-1}}{2(1 - z^{-1})^3}$$

$$\dfrac{1}{1 - e^{-aT}z^{-1}}$$

$$\dfrac{Te^{-aT}z^{-1}}{(1 - e^{-aT}z^{-1})^2}$$

$$\dfrac{(1 - e^{-aT})z^{-1}}{(1 - z^{-1})(1 - e^{-aT}z^{-1})}$$

$$\dfrac{[(aT - 1 + e^{-aT}) + (1 - e^{-aT} - aTe^{-aT})z^{-1}]z^{-1}}{a(1 - z^{-1})^2(1 - e^{-aT}z^{-1})}$$

$$\dfrac{[1 - e^{-aT}(1 + aT)z^{-1}]z^{-1}}{(1 - e^{-aT}z^{-1})^2}$$

$$\dfrac{[(1 - e^{-aT} - aTe^{-aT}) + (e^{-2aT} - e^{-aT} + aTe^{-aT})z^{-1}]z^{-1}}{(1 - z^{-1})(1 - e^{-nT}z^{-1})^2}$$

$$\dfrac{(e^{-aT} - e^{-bT})z^{-1}}{(1 - e^{-aT}z^{-1})(1 - e^{-bT}z^{-1})}$$

$$\dfrac{[(b - a) - (be^{-aT} - ae^{-bT})z^{-1}]z^{-1}}{(1 - e^{-aT}z^{-1})(1 - e^{-bT}z^{-1})}$$

$$\dfrac{(\sin aT)z^{-1}}{1 - (2 \cos aT)z^{-1} + z^{-2}}$$

$$\dfrac{(1 - \cos aT)z^{-1}}{1 - (2 \cos aT)z^{-1} + z^{-2}}$$

$$\dfrac{(1 - e^{-aT} \cos bT)z^{-1}}{1 - 2e^{-aT}(\cos bT)z^{-1} + e^{-2aT}z^{-2}}$$

$$\dfrac{(e^{-aT} \sin bT)z^{-1}}{1 - 2e^{-aT}(\cos bT)z^{-1} + e^{-2aT}z^{-2}}$$

$$\dfrac{(A + Bz^{-1})z^{-1}}{(1 - z^{-1})(1 - 2e^{-aT}(\cos bT)z^{-1} + e^{-2aT}z^{-2})}$$

$$A = 1 - e^{-aT} \cos bT - \frac{a}{b}e^{-aT} \sin bT; \quad B = e^{-2aT} + \frac{a}{b}e^{-aT} \sin bT - e^{-aT} \cos bT$$

Index